T0245250

CAMBRIDGE LIBRARY COLLECTION

Books of enduring scholarly value

Earth Sciences

In the nineteenth century, geology emerged as a distinct academic discipline. It pointed the way towards the theory of evolution, as scientists including Gideon Mantell, Adam Sedgwick, Charles Lyell and Roderick Murchison began to use the evidence of minerals, rock formations and fossils to demonstrate that the earth was older by millions of years than the conventional, Bible-based wisdom had supposed. They argued convincingly that the climate, flora and fauna of the distant past could be deduced from geological evidence. Volcanic activity, the formation of mountains, and the action of glaciers and rivers, tides and ocean currents also became better understood. This series includes landmark publications by pioneers of the modern earth sciences, who advanced the scientific understanding of our planet and the processes by which it is constantly re-shaped.

The Geology of the South East of England

Gideon Mantell (1790–1852) was an English physician and geologist best known for pioneering the scientific study of dinosaurs. After an apprenticeship to a local surgeon in Sussex, Mantell became a member of the Royal College of Surgeons in 1811. He developed an interest in fossils, and in 1822 his discovery of fossil teeth which he later identified as belonging to an iguana-like creature he named *Iguanadon* spurred research into ancient fossils. This volume, first published in 1833, contains the results of Mantell's research into the geology and fossil remains of south-eastern England, especially Sussex and Kent. Mantell describes each rock stratum of the region in detail, and includes within each section descriptions of fossil remains found in the formation, arranged by species. Copiously illustrated, this volume remains one of Mantell's best known works and contains fascinating detail concerning the development of geology and palaeontology.

Cambridge University Press has long been a pioneer in the reissuing of out-of-print titles from its own backlist, producing digital reprints of books that are still sought after by scholars and students but could not be reprinted economically using traditional technology. The Cambridge Library Collection extends this activity to a wider range of books which are still of importance to researchers and professionals, either for the source material they contain, or as landmarks in the history of their academic discipline.

Drawing from the world-renowned collections in the Cambridge University Library, and guided by the advice of experts in each subject area, Cambridge University Press is using state-of-the-art scanning machines in its own Printing House to capture the content of each book selected for inclusion. The files are processed to give a consistently clear, crisp image, and the books finished to the high quality standard for which the Press is recognised around the world. The latest print-on-demand technology ensures that the books will remain available indefinitely, and that orders for single or multiple copies can quickly be supplied.

The Cambridge Library Collection will bring back to life books of enduring scholarly value (including out-of-copyright works originally issued by other publishers) across a wide range of disciplines in the humanities and social sciences and in science and technology.

The Geology of the South East of England

Gideon Algernon Mantell

CAMBRIDGE UNIVERSITY PRESS

Cambridge, New York, Melbourne, Madrid, Cape Town, Singapore,
São Paolo, Delhi, Dubai, Tokyo, Mexico City

Published in the United States of America by Cambridge University Press, New York

www.cambridge.org
Information on this title: www.cambridge.org/9781108021104

This edition first published 1833
This digitally printed version 2010

ISBN 978-1-108-02110-4 Paperback

J. Pollard del. *P. Gauci lithog.*

STRATA of TILGATE FOREST, in SUSSEX.

Printed by Graf & Soret.

THE GEOLOGY OF THE SOUTH-EAST OF ENGLAND.

BY

GIDEON MANTELL, F.R.S.

FELLOW OF THE ROYAL COLLEGE OF SURGEONS;
AND OF THE LINNEAN AND GEOLOGICAL SOCIETIES OF LONDON;
MEMBER OF THE BRITISH ASSOCIATION OF SCIENCE;
HONORARY MEMBER OF THE ACADEMY OF NATURAL SCIENCES OF PHILADELPHIA;
OF THE PHILOMATHIC SOCIETY OF PARIS;
OF THE PHILOSOPHICAL SOCIETY OF YORKSHIRE; ETC. ETC.

"Geology, in the magnitude and sublimity of the objects of which it treats, undoubtedly ranks next to Astronomy in the scale of the sciences." — SIR J. F. W. HERSCHEL.

LONDON:
LONGMAN, REES, ORME, BROWN, GREEN, & LONGMAN,
PATERNOSTER-ROW.
1833.

TO

HIS MOST EXCELLENT MAJESTY

WILLIAM THE FOURTH,

THIS WORK

ON THE

GEOLOGY OF THE SOUTH-EAST OF ENGLAND,

IS,

WITH HIS MAJESTY'S GRACIOUS PERMISSION,

MOST HUMBLY INSCRIBED,

BY

HIS MAJESTY'S FAITHFUL AND DEVOTED

SUBJECT, AND SERVANT,

THE AUTHOR.

Castle Place, Lewes;
April, 1833.

PREFACE.

THE discovery of an unknown fossil reptile in the strata of Tilgate Forest, induced me to lay before the Geological Society of London, a Memoir on the Organic Remains of the Wealden which had been collected since my last publication on the geology of this district.

By permission of the President and Council, that memoir was withdrawn; and it was suggested to me that, as my former works were out of print, a volume which should combine the most interesting portions of the "Illustrations of the Geology of Sussex," with an account of the recent discoveries, might be acceptable both to the natural philosopher, and the general reader.

Agreeably to this suggestion the present work has been written; and the kind and liberal manner in which the proposals for its publication have been received by so many distinguished and learned characters, who have thus testified an interest in my researches, has afforded me the highest gratification and encouragement.

Engaged in the same responsible and anxious duties as at the period of my former publications, I solicit the indulgence of the reader for the imperfections and omissions of this volume, which, like its predecessors, has been composed during short and uncertain intervals of leisure, after active and laborious professional exertions.

Castle Place, Lewes;
April, 1833.

CONTENTS.

CHAP. VI.

CHAP. VII.

CHAP. VIII.

CHAP. IX.

CHAP. X.

CHAP. XI.

ERRATA ET ADDENDA.

Page 30.	line 7.	from the top,	for *its*	read *their.*
— 30.	— 14.	———	*Bignor*	— *Bognor.*
— 122.	— 12.	———	*genuine*	— *generic.*
— 124.	— 4.	———	*spirous*	— *spinous.*
— 204.	— 14.	———	*these contain*	— *the soil is composed of.*
— 206.	— 22.	———	*some hundred*	— *many thousand.*
— 228.	— 16.	———	*no*	— *the.*
— 286.	— 8.	———	*horn*	— *horns.*
— 180.	-	- -	Chap. X.	— Chap. VII.
— 232.		- - -	Chap. XI.	— Chap. VIII.

*** *Lepisosteus Fittoni*, p. 252. Since the notice of this ichthyolite was written, Mr. Martin, of Pulborough, has discovered, in western Sussex, a splendid specimen of the head and anterior part of the body, which he has, with great liberality, submitted to my examination. By this interesting example we have been able to ascertain that the mouth was rounded, and the jaws were very strong and thick; there are no teeth of detention, but those in the front of the mouth are more pointed and elongated than those which occupy the middle.* There are eleven or twelve teeth remaining in the lower and ten in the upper jaw. The skull is so much crushed, that neither its form, nor the situation of the orbits, can be determined.

* Mr. Webster's fig. 5. pl. vi. "Geological Transactions," second series, vol. ii., represents a portion of the jaw, with some of the teeth of the front row, of a young individual. Mr. W. mentions that he has found similar teeth in the Purbeck limestone.

INTRODUCTORY OBSERVATIONS.

"If we look with wonder upon the mighty remains of human works, such as the columns of Palmyra, broken in the midst of the desert; the temples of Pæstum, beautiful in the decay of twenty centuries; or the mutilated relics of Greek sculpture in the Acropolis of Athens, or in our own museums, as proofs of the genius of artists, and power and riches of nations, now passed away; with how much deeper feeling of admiration and astonishment must we consider those grand monuments of nature which mark the revolutions of the globe: continents broken into islands; one land produced, another destroyed; the bottom of the ocean become a fertile soil; whole races of animals extinct, and the bones and exuviæ of one class covered with the remains of another, and upon the grave of past generations — the marble or rocky tomb, as it were, of a former animated world — new generations rising, and order and harmony established, and a system of life and beauty produced out of chaos and death: proving the infinite power, wisdom, and goodness, of the GREAT CAUSE of all things."

Sir Humphry Davy's Last Days of a Philosopher.

To the mind which is wholly unacquainted with the nature and results of geological enquiries, and has been led to believe that the globe we inhabit is in the state in which it was originally created, and that, with the exception of the effects of an universal deluge, its surface has undergone no material change, many of the phenomena described in the following pages will appear marvellous and incredible, and the inferences drawn from their investigation be considered as the vagaries of the imagination, rather than the legitimate deductions of sound philosophy. If, therefore, it be absolutely necessary, as it unquestionably is, that, in the pursuit of knowledge of any kind, before experience itself can be used with advantage, we must dismiss from the mind all prejudices, from whatsoever source they may arise; this mental purification becomes the more indispensable in a science like Geology, in which we meet, at the very threshold, with facts that disturb all our preconceived opinions of the nature of the globe on which we live, and teach us that, though man be, as it were, but the creature of yesterday, the earth has teemed with countless forms of animal and vegetable existence, myriads of ages before the creation of the human race.

Dismissing from his mind all prejudices of opinion, the geological enquirer must be prepared to learn, that the earth's surface has been, and still is, subject to incessant fluctuation and movements; and that, as the land has been the theatre of perpetual mutation, the sea can alone be considered as having undergone no change; the apostrophe of the noble bard to the ocean is therefore as philosophical, as it is highly poetical and sublime:—

> "Unchangeable, save to thy wild waves' play,
> Time writes no wrinkle on thine azure brow —
> Such as creation's dawn beheld, thou rollest now!"
>
> BYRON.

To render this volume more intelligible to the general reader, we shall offer a brief notice of the leading principles of modern geology; referring, for further information on this sublime and important science, to Mr. Bakewell's Introduction, as one of the most popular, lucid, and highly interesting elementary works on the subject; and to Mr. Lyell's Principles of Geology, as presenting the most able, comprehensive, and philosophical view of the science, that has hitherto appeared.

The superficies of our planet is computed to contain 190 millions of square miles, of which three fifths are covered by the ocean; another large portion by polar ice, eternal snows, and vast bodies of fresh water: the space habitable by man and terrestrial animals being thus reduced to scarcely more than one fifth of the whole.* The strata of which the earth is composed, have been examined through a thickness nearly equal to eight miles, calculating from the summits of the highest mountains to the greatest natural or artificial depths: but as the earth is nearly 8000 miles in diameter, the entire series of strata that has been explored, is but as the paper which covers a globe 12 inches in diameter; and the highest mountains, and the deepest valleys, may be compared to the inequalities and fissures in the composition with which the surface of such an instrument is coated. To render this more obvious, let us assume that the height of this page of letter-press is equal to the radius of the earth; then this letter, I, will represent a thickness of 100 miles, and a mere line, —, the greatest elevations in the world. It is therefore clear that disturbances in the earth's surface to ten times the depth to which the researches of man can extend,

* Bakewell's Introduction to Geology, 3d edit., p. 6.

may have taken place, without affecting, in any sensible degree, the whole mass of the globe. If these facts be duly considered, the mind will no longer refuse its assent to the interpretation which modern geology offers of those natural records of its physical history, that are to be found in every known region of the earth; and it will be prepared to believe, that the rocks composing the highest mountains have been formed in the depths of the ocean, and raised to their present situations by earthquakes, at various, and, in many instances, comparatively very recent, periods. At every step the geological enquirer will find, in the displacements and fractures of the strata, evidence that, from the remotest eras, the earthquake and the volcano have been in active operation; and in the beds of gravel, and other accumulations of water-worn materials, equally decisive proofs of diluvial agency.

The mineral masses of which the crust of the globe is composed, may be separated into two principal divisions; namely, the *primary* and the *secondary*. The *primary* are destitute of organic remains, and occupy the lowermost place in the order of superposition of the strata; yet, having manifestly been injected from below, they also very generally form the summits of the highest peaks and mountain chains in the world. Those which are decidedly of igneous origin, are *granite*, *sienite*, *porphyry*, *basalt*, &c. They were called primary, because it was supposed, from the absence of fossils, that they were formed before the creation of animals and vegetables; but it is now well known that granite and its associated rocks are, in fact, ancient lavas of various ages; and it is certain that granite has been erupted, even since the period when the chalk was deposited. The other primary rocks appear, on the contrary, to be sedimentary deposits altered by the effects of high temperature under great pressure; such are *gneiss*, *mica slate*, &c.

The *secondary* rocks contain the fossilised remains of animals and vegetables, are generally stratified, and have evidently been deposited by water: to this class all the formations of the south-east of England belong. These strata, for the convenience of study, are subdivided into the *secondary*, strictly so called, which comprise all the sedimentary rocks from the primary to the chalk inclusive: and the *tertiary*, under which division all the beds, from the chalk to the alluvial deposits of the modern epoch, are placed.

The organic remains entombed in the sedimentary strata, afford conclusive evidence of the former existence of a state of animated nature widely different to the present; and furnish data by which we can determine the comparative ages of the various formations, and even calculate the relative periods when the existing mountain chains were lifted up. Nay, more; by these relics, these medals, as they have been aptly termed, struck by nature to commemorate her revolutions, we learn the physical mutations which the surface of the earth has undergone, and the temperature of the climate of various regions, in periods far beyond all human history and tradition; and, by bringing to our assistance the sciences of anatomy and botany, we can even restore anew the forms of the animals and vegetables which flourished on the earth, when our present continents were engulfed beneath the depths of the ocean.

The interest and importance of this branch of natural philosophy are now so highly appreciated by all, save those minds which are alike destitute of all capacity and relish for intellectual pursuits, that I feel it to be wholly unnecessary for me to offer one remark on its practical utility, or on the lofty and sublime pleasures which its investigations afford. An eminent astronomer has, however, cited, as an example of the value of physical knowledge in teaching us to avoid attempting impossibilities, so remarkable an instance in which ignorance of the first principles of geology led to an expensive and abortive undertaking, in a part of Sussex which we shall hereafter have occasion to describe, that I am induced to subjoin the relation in the words of the distinguished author. "It is not many years since an attempt was made to establish a colliery at Bexhill, in Sussex. The appearance of thin seams and sheets of fossil wood and wood coal, with some other indications similar to what occur in the neighbourhood of the great coal beds in the north of England, having led to the sinking of a shaft, and the erection of machinery on a scale of vast expense; not less than 80,000*l.* are said to have been laid out in this project, which, it is almost needless to add, proved completely abortive, as every geologist would at once have declared it must: the whole *assemblage of geological facts being adverse to the existence of a* *regular coal bed in the Hastings strata;* while this on which Bexhill is situated, is separated from the *coal measures* by a series of interposed beds of such enormous thickness, as to render all idea of penetrating through them absurd. The history of mining operations is full of similar cases, where a very moderate acquaintance with the *usual order of nature,* to say nothing of theoretical views, would have saved many a sanguine adventurer from utter ruin."*

The sequence of physical events which the geological phenomena of the south-east of England establish, will be fully elucidated in the following pages; and, as the strata of this part of our island form a series from the most recent deposits to the *Oolite,* the investigation of their characters and relations will be found highly interesting, and may be considered as one of the best practical lessons for the student in British geology.

* Sir John Herschel's Discourse on the Study of Natural Philosophy.

GEOLOGY

OF THE

SOUTH-EAST OF ENGLAND.

CHAPTER I.

PHYSICAL GEOGRAPHY OF THE COUNTY OF SUSSEX.

SUSSEX, the ancient kingdom of the South Saxons, and the southernmost maritime province of England, is bordered on the west by Hampshire, on the north by Surrey, on the east and north-east by Kent, and on the south by the British Channel. It is about seventy-six miles in length, and twenty-eight in average breadth; containing 1461 square miles.*

The strata of which it is composed form three principal groups, each possessing characters that materially affect the geographical features of the country, and present a striking instance of the intimate relation that exists between the physical

			Of Greenwich.
* " Northernmost point	*Black-Corner,*	N. lat. 51° 9′;	48′ W. long.
Southernmost -	*Selsey Bill,*	N. lat. 50° 43′;	47′ W. long.
Easternmost -	*Kent Wall,*	N. lat. 50° 56′;	49′ E. long.
Westernmost -	*Stansted Park,*	N. lat. 50° 53′;	58′ W. long.

—*Dallaway's History of the Western Division of the County of Sussex,* 4to. 1815, vol. i. p. 5.

appearance of the surface of the earth, and its geological structure. The popular division of this tract, into the DOWNS, WEALD, and FOREST-RIDGE, may therefore be considered as sufficiently correct and comprehensive for our present purpose, since it is descriptive of the external characters of the district, and is agreeable to the natural arrangement of the strata.

The Downs* are a chain of hills covered with a fine verdant turf, possessing in a striking degree that smoothness and regularity of outline, for which the mountain masses of the chalk formation are so remarkable. Commencing with the bold promontory of Beachy Head, they traverse the county in a direction nearly east and west, and pass into Hampshire, near Compton. Their length is between fifty and sixty miles, their greatest breadth seven miles, and their mean altitude about five hundred feet above the level of the sea. Their northern escarpment is in general steep and abrupt, but on the south they descend by a gentle declivity, and unite almost imperceptibly with the low lands of the coast.

From Beachy Head to Brighton they present an immediate barrier to the sea, forming a bold and precipitous line of coast; but, proceeding westerly, they extend inland in an oblique direction, and occupy the centre of Western Sussex.

* "Though I have now travelled the Sussex Downs upwards of thirty years, yet I still investigate that chain of majestic mountains with fresh admiration, year by year. This range, which runs from Chichester east, as far as Eastbourn, is about sixty miles in length. and is called the South Downs, properly speaking, only round Lewes." —*Natural History of Selborne*, 1802, p. 276.

From this circumstance, a considerable difference exists in the geological relations of the eastern and western divisions of the county; the latter being characterised by a range of chalk hills in the centre, with a maritime district formed of clay and gravel on the south, and a weald composed of sand and clay on the north.

Throughout its whole extent this chain exhibits decisive manifestations of the action of water. Not only are the ridges and summits of the hills rounded and even, but their surface is every where furrowed by coombes, or narrow undulating ravines: these, uniting, terminate in valleys, that intersect the downs in a direction nearly north and south, and form extensive outlets for the rivers that flow from the interior of the country into the British Channel. The course of the smaller excavations or coombes is exceedingly various, but their general bearing is east and west; they gradually increase in breadth as they descend, and their opposite sides have corresponding angles and sinuosities: this appearance, however, is not observable in the principal valleys.

The chalk hills of Sussex are separated into five distinct masses, by the following rivers; viz. the *Arun*, the *Adur*, the *Ouse*, and the *Cuckmere*.

The first is situated in Western Sussex: it rises in the forest of St. Leonard, near Horsham, and, taking its course to the westward for a few miles, turns suddenly to the south, passes through the chalk near Arundel, and falls into the sea to the west of Little Hampton.

The Adur constitutes the western boundary of the South Downs, properly so called: like the former, it has its origin in St. Leonard's forest, and, passing by Steyning and Bramber, enters the British Channel at New Shoreham.

The Ouse, which is the principal river in the south-eastern part of the county, rises by two branches; the one has its source in St. Leonard's forest, and the other in the forest of Worth, north of Cuckfield. The river formed by the confluence of these streams pursues a tortuous course to the southward, and, passing to the east of Lewes, which it separates from the adjacent town of the Cliff, flows through the flat alluvial tract of Lewes Levels, and discharges itself into the sea at Newhaven harbour.

The Cuckmere has its source near Warbleton, and being augmented by numerous tributary streams, in its course by Hellingly, Arlington, Alfriston, &c., falls into the British Channel at the haven which bears its name.

By these rivers the drainage of the country is effected; and it is worthy of observation, that they invariably flow from an older through a newer country; or, in other words, that the strata forming the district from whence they take their rise are of anterior formation to the chalk valleys by which they empty themselves into the ocean.

The WEALD* of Sussex is an extensive vale that occupies the centre of the south-eastern part of the county, and, running parallel with the

* "Opposite to the South Downs, on the north, are the Surrey hills, falling abruptly southward, and sloping gradually to the north, and between these two lines of hills is the Weald of Sussex and Surrey." —*Young's Agricultural Survey.*

Downs, forms their northern boundary. It was anciently an immense forest, (called by the earlier colonists *Coid Andred*, by the Romans *Silva Anderida*, and by the Saxons *Andreadswald*,) which, even in the time of Bede, was a mere retreat for deer and swine: the greater part is now in an excellent state of cultivation. It consists of various beds of clay, sand, and limestone, and is comparatively of low elevation: its breath is from five to ten miles, and its length from thirty to forty miles; it is estimated to contain 425,000 acres. The surface is intersected by numerous valleys, which generally occur at the outcrop or basseting edges of the harder strata, and form channels for the numerous streams that are tributary to the rivers in their vicinity. The whole tract rises with a gradual sweep from the foot of the Downs, and unites with the higher lands of the Forest-Ridge.

The FOREST-RIDGE constitutes the north-eastern extremity of the county. It is composed of the more elevated portions of the sands and sandstones; and, from the rocky and abrupt termination of its ridges, which are for the most part either crested with forests or overgrown with underwood, forms a tract of country remarkable for its romantic and picturesque scenery. The principal heights in this range are Wych Cross, Brightling Down, Dane's Hill, Fairlight Down, and Crowborough Beacon: the last-mentioned is the highest and most central eminence, and is 804 feet above the level of the sea.

"The climate in the western part of the maritime

division is very warm, and highly favourable to the powers of vegetation. The Downs fronting the south-west are bleak, being exposed to violent winds, which are impregnated with saline particles, occasioned by the spray beaten against the sea-beach; and this influence affects the animals as well as vegetables indigenous to the hills. In the Weald the due circulation of air is greatly impeded by the forests and thick hedges, and the climate is in consequence cold and damp."*

Such are the geographical features of the masses which compose the county of Sussex; but as, in the course of our investigations, we shall have occasion more immediately to refer to the south-eastern division, it will be necessary to point out with greater precision the course and position of the chalk hills of that district, and more especially of those in the vicinity of Lewes and Brighton.

The South Downs, strictly so called, are that portion of the Sussex range which lies between Eastbourn and Shoreham. They are twenty-six miles long, about seven miles in breadth, and are divided by the intervention of rivers into four groups.

The easternmost rises with a gentle slope near Eastbourn proceeds inland as far as Folkington, and is separated from the middle division by the Cuckmere. The southern escarpment composes a rocky and precipitous range of cliffs, extending eastward along the coast from the embouchure of Cuckmere River to Beachy Head, where it rises to the altitude of 564 feet.

* Dallaway's Western Sussex, p. 6.

The middle group is bounded on the east by the line of separation above mentioned, on the west by Lewes Levels, and on the south by cliffs which reach from Cuckmere Haven to Seaford Point, from whence to Newhaven harbour it is skirted by a low marshy coast: the northern margin is formed by the elevated ridge of Firle Hills.

The western division embraces the most considerable extent of down in the county. The Adur forms the natural limits of this chain on the west, and the Ouse on the east; the southern slope is washed by the British Channel, except towards the south-west, where a flat maritime district, extending from near Brighton to Shoreham harbour, intervenes, and separates it from the sea-shore. The ridge by which it is bounded on the north, presents a steep escarpment to the Weald, and is the highest land in the county; Ditchling Beacon, the centre of this line, being 864 feet above the level of the sea. Eastward of the beacon lies Plumpton Plain, an elevated platform, commanding an extensive view of the rich scenery of the Weald on the one hand, and of the Downs and British Channel on the other: it is celebrated in history as the field where Henry III. was defeated by the barons, under Sir Simon de Montfort. The prospect from this spot is equal to many in the finest parts of Europe, extending thirty miles towards the sea, and forty miles inland to Surrey.

Brighton and Lewes, two of the principal towns in the county, are situated in this division of the South Downs. The former lies nearly in the centre of the southern edge, on the margin of an extensive bay, comprehended between Beachy Head and Selsey Bill, and is sheltered by a range of hills on the east, north, and north-east: the peculiarities of its site, and the structure of the cliffs in its vicinity, will be hereafter particularised.

Lewes is delightfully situated on the eastern extremity of this range. It lies 50° 52′ north latitude, and is distant fifty miles south from London. The Downs form an amphitheatre of hills to the east and west of the town; but the northern and southern slopes are skirted by the Levels.

The CLIFF HILLS, near Lewes, constitute the last division of the South Downs: they are a small insulated group, separated from the central and western chains by the intervention of Lewes Levels. The edge of this range runs parallel with the road from Southerham to Glynd and Glyndbourn, passes near Ringmer in its course westward, and terminates at Old Malling, near the banks of the Ouse. The south-eastern angle is formed by Mount Caburn, and the western escarpment is deeply indented by the steep valley of the Coombe.

" The soil of the Downs is subject to considerable variation. On the summit it is usually very shallow; the substratum is chalk, and over that a layer of chalk rubble, and partially rolled chalk flints, with a slight covering of vegetable mould. Along the more elevated ridges there is sometimes merely a covering of flints, upon which the turf grows spontaneously. Advancing down the hills, the soil becomes deeper, and at the bottom is constantly found to be of very sufficient depth for ploughing: here the loam is excellent, generally ten or twelve inches thick, and the chalk rather broken, and mixed with loam in the interstices." *

Some parts of the South Downs are converted into arable, but in general they are reserved for pasturage, and support a breed of sheep, equal, if not superior, to any in the kingdom. †

LEWES LEVELS, which have already been mentioned as intervening between the western and central divisions of the South Downs, form a

* Young's Agricultural Survey of Sussex, 8vo. p. 5.

† The sheep fed on the South Downs amount to nearly 200,000; and, as there are no natural springs on the chalk hills, the flocks are supplied with water from large circular ponds, made on the summits of the Downs; the bottoms of these excavations are covered with a layer of ochraceous clay, to prevent the water from percolating through the chalk, and they are seldom known to fail even in the hottest summers. The late Mr. White considered this circumstance as very remarkable; and has particularly noticed it, in his interesting volume on the Natural History of Selborne. " To a thinking mind, few phenomena are more strange, than the state of little ponds on the summits of chalk hills, many of which are never dry in the most trying droughts of summer: on chalk hills, I say, because, in many rocky and gravelly soils, springs usually break out pretty high on the sides of elevated grounds and mountains; but no person acquainted with chalky districts will allow that they ever saw springs in such a soil but in valleys and bottoms, since the waters of so pervious a stratum as chalk all lie on one dead level, as well-diggers have assured me again and again.

" Now, we have many such little round ponds in this district; and one, in particular, on one sheep-down, three hundred feet above my house; which, though never above three feet deep in the middle, and not more than thirty feet in diameter, and containing perhaps not more than two or three hundred hogsheads of water, yet never is known to fail, though it affords drink for 300 or 400 sheep, and for at least twenty head of large cattle beside."—*White's Nat. Hist. of Selborne*, p. 206.

What, however, appears to me still more remarkable, is the fact, that, soon after a new pond has been made, and has received a partial supply of water from a few passing showers, it becomes inhabited by various kinds of freshwater plants and shell-fish, and even frogs and lizards; although it may be remote from any other pond, and at an elevation of 400 or 500 hundred feet above the level of the surrounding country.

marshy alluvial plain, through which the Ouse winds its way to the British Channel. This tract consists of silt, clay, and peat, and is nearly ten miles long; its breadth varying from half a mile to two miles and a half. Towards the north-western confines of this plain are two remarkable oval mounds or hillocks of chalk marl, situated at a short distance from each other, near the borough of Southover. They bear the name of *rhies;* a provincial term, derived from the Saxon *hryg*, a heap, or longitudinal projection *, and are about seventy feet high, and from two to three furlongs in length.

* History of Lewes, 8vo. 1795, p. 416.

CHAP. II.

GEOLOGICAL STRUCTURE OF SUSSEX.

The investigation of the geology of this district is attended with considerable difficulty. The displacement and disintegration which many of the strata have sustained; the excess of soil and vegetation with which, in most places, their basseting edges are covered at the line of junction; and the absence of sections in those situations where the relative position of the rocks is involved in obscurity, present numerous, and in some instances insuperable, obstacles to accurate examination. Under such circumstances, induction and analogy must supply the place of actual observation; but, the relative position of the principal masses having been correctly ascertained, whatever errors may have originated from the causes alluded to are of minor importance; since they chiefly relate to the geographical extent of the strata, and cannot affect the geological deductions that may be drawn from these researches.

For the information of the general reader, it may be necessary to observe, that in Sussex, as in every other part of England, the strata maintain a certain order of superposition; and that, however great the displacement or interruption they may have sustained, this order is never inverted. To illustrate this remark, we may observe, that the *firestone* which separates the *grey marl* from the *gault*, at Southbourn, is altogether wanting at Hamsey and several other places; but in these instances the *grey chalk marl* reposes immediately upon the *gault*, the relative position of the masses remaining unaltered by the absence of the intervening deposit.

Before proceeding to a more particular examination of the several formations, we present the following tabular arrangement of the strata, which will render our subsequent remarks more intelligible to the general reader, and, with the assistance of the map and sections, convey a correct idea of the stratification of this county, and of the adjoining parts of Surrey Kent, and Hampshire.

The Strata arranged according to their Order of Superposition, commencing with the uppermost or newest deposit.

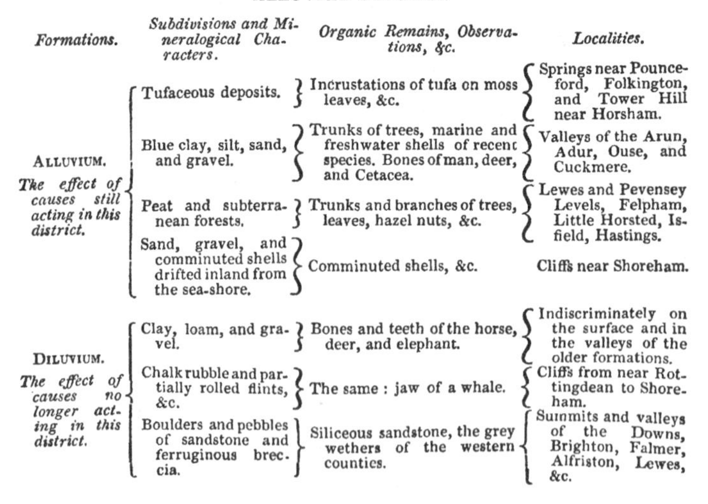

ALLUVIAL DEPOSITS.

Formations.	*Subdivisions and Mineralogical Characters.*	*Organic Remains, Observations, &c.*	*Localities.*
ALLUVIUM. *The effect of causes still acting in this district.*	Tufaceous deposits.	Incrustations of tufa on moss leaves, &c.	Springs near Pounceford, Folkington, and Tower Hill near Horsham.
	Blue clay, silt, sand, and gravel.	Trunks of trees, marine and freshwater shells of recent species. Bones of man, deer, and Cetacea.	Valleys of the Arun, Adur, Ouse, and Cuckmere.
	Peat and subterranean forests.	Trunks and branches of trees, leaves, hazel nuts, &c.	Lewes and Pevensey Levels, Felpham, Little Horsted, Isfield, Hastings.
	Sand, gravel, and comminuted shells drifted inland from the sea-shore.	Comminuted shells, &c.	Cliffs near Shoreham.
DILUVIUM. *The effect of causes no longer acting in this district.*	Clay, loam, and gravel.	Bones and teeth of the horse, deer, and elephant.	Indiscriminately on the surface and in the valleys of the older formations.
	Chalk rubble and partially rolled flints, &c.	The same: jaw of a whale.	Cliffs from near Rottingdean to Shoreham.
	Boulders and pebbles of sandstone and ferruginous breccia.	Siliceous sandstone, the grey wethers of the western counties.	Summits and valleys of the Downs, Brighton, Falmer, Alfriston, Lewes, &c.

TERTIARY FORMATIONS.

(*Partly Marine and partly Freshwater.*)

Formations.	*Subdivisions and Mineralogical Characters.*	*Organic Remains, Observations, &c.*	*Localities.*
Plastic Clay.	Clay, sand, and gravel.	Potamides, Cyclades, Ostreæ, Cyrenæ, teeth of fishes, &c. Leaves of terrestrial vegetables; cone of an unknown plant. Subsulphate of alumine, gypsum, surturbrand, &c.	CastleHill,nearNewhaven. Chimting Castle,near Seaford. Falmer, Lewes, &c. Insulated patches in many other localities.
London Clay.	Blue Clay.	Ampullariæ, Turritellæ, Venericardiæ, and other marine shells peculiar to the London clay. Vertebræ, teeth, and palates of fishes.	Bracklesham Bay, Stubbington, &c.
	Grey calcareous sandstone.	Pectunculi. Vermiculariæ, Ampullariæ, Nautili, Pinnæ, teeth of fishes, &c.	Bognor, Barn, and Mixen Rocks.

SECONDARY FORMATIONS.

(*Marine Deposits.*)

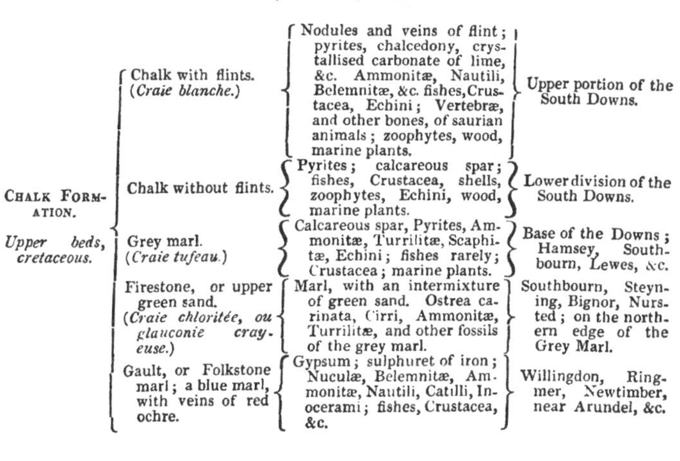

Formations.	*Subdivisions and Mineralogical Characters.*	*Organic Remains, Observations, &c.*	*Localities.*
Chalk Formation. *Upper beds, cretaceous.*	Chalk with flints. (*Craie blanche.*)	Nodules and veins of flint; pyrites, chalcedony, crystallised carbonate of lime, &c. Ammonitæ, Nautili, Belemnitæ, &c. fishes, Crustacea, Echini; Vertebræ, and other bones, of saurian animals; zoophytes, wood, marine plants.	Upper portion of the South Downs.
	Chalk without flints.	Pyrites; calcareous spar; fishes, Crustacea, shells, zoophytes, Echini, wood, marine plants.	Lower division of the South Downs.
	Grey marl. (*Craie tufeau.*)	Calcareous spar, Pyrites, Ammonitæ, Turrilitæ, Scaphitæ, Echini; fishes rarely; Crustacea; marine plants.	Base of the Downs; Hamsey, Southbourn, Lewes, &c.
	Firestone, or upper green sand. (*Craie chloritée, ou glauconie crayeuse.*)	Marl, with an intermixture of green sand. Ostrea carinata, Cirri, Ammonitæ, Turrilitæ, and other fossils of the grey marl.	Southbourn, Steyning, Bignor, Nursted; on the northern edge of the Grey Marl.
	Gault, or Folkstone marl; a blue marl, with veins of red ochre.	Gypsum; sulphuret of iron; Nuculæ, Belemnitæ, Ammonitæ, Nautili, Catilli, Inocerami; fishes, Crustacea, &c.	Willingdon, Ringmer, Newtimber, near Arundel, &c.

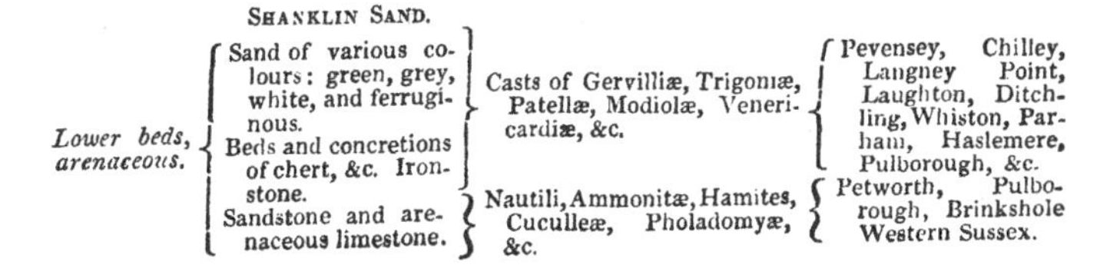

Formations.	*Subdivisions and Mineralogical Characters.*	*Organic Remains, Observations, &c.*	*Localities.*
Lower beds, arenaceous.	Shanklin Sand. Sand of various colours: green, grey, white, and ferruginous. Beds and concretions of chert, &c. Ironstone.	Casts of Gervilliæ, Trigoniæ, Patellæ, Modiolæ, Venericardiæ, &c.	Pevensey, Chilley, Langney Point, Laughton, Ditchling, Whiston, Parham, Haslemere, Pulborough, &c.
	Sandstone and arenaceous limestone.	Nautili, Ammonitæ, Hamites, Cucculleæ, Pholadomyæ, &c.	Petworth, Pulborough, Brinkshole Western Sussex.

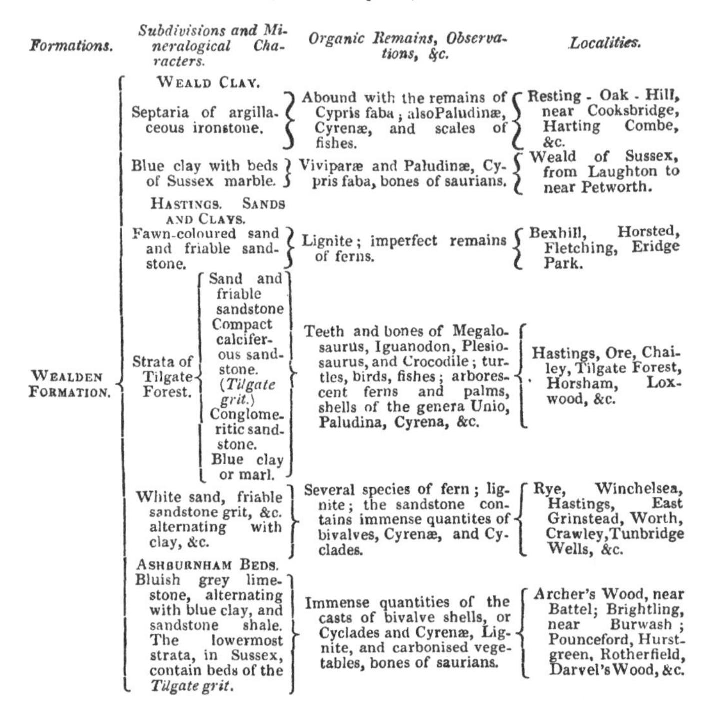

(*Fluviatile Deposits.*)

Formations.	*Subdivisions and Mineralogical Characters.*	*Organic Remains, Observations, &c.*	*Localities.*
WEALDEN FORMATION.	WEALD CLAY.		
	Septaria of argillaceous ironstone.	Abound with the remains of Cypris faba; also Paludinæ, Cyrenæ, and scales of fishes.	Resting-Oak-Hill, near Cooksbridge, Harting Combe, &c.
	Blue clay with beds of Sussex marble.	Viviparæ and Paludinæ, Cypris faba, bones of saurians.	Weald of Sussex, from Laughton to near Petworth.
	HASTINGS. SANDS AND CLAYS.		
	Fawn-coloured sand and friable sandstone.	Lignite; imperfect remains of ferns.	Bexhill, Horsted, Fletching, Eridge Park.
	Strata of Tilgate Forest. — Sand and friable sandstone. Compact calciferous sandstone. (*Tilgate grit.*) Conglomeritic sandstone. Blue clay or marl.	Teeth and bones of Megalosaurus, Iguanodon, Plesiosaurus, and Crocodile; turtles, birds, fishes; arborescent ferns and palms, shells of the genera Unio, Paludina, Cyrena, &c.	Hastings, Ore, Chailey, Tilgate Forest, Horsham, Loxwood, &c.
	White sand, friable sandstone grit, &c. alternating with clay, &c.	Several species of fern; lignite; the sandstone contains immense quantites of bivalves, Cyrenæ, and Cyclades.	Rye, Winchelsea, Hastings, East Grinstead, Worth, Crawley, Tunbridge Wells, &c.
	ASHBURNHAM BEDS. Bluish grey limestone, alternating with blue clay, and sandstone shale. The lowermost strata, in Sussex, contain beds of the *Tilgate grit.*	Immense quantities of the casts of bivalve shells, or Cyclades and Cyrenæ, Lignite, and carbonised vegetables, bones of saurians.	Archer's Wood, near Battel; Brightling, near Burwash; Pounceford, Hurstgreen, Rotherfield, Darvel's Wood, &c.

The general inclination of the beds in Sussex is towards the south-east, consequently, a line drawn from the coast, through the interior of the country, would pass over the basseting edges of the strata in regular succession. On the surface the occurrence of a new formation is indicated by the intervention of a rivulet or valley, by a difference in the physical appearance of the country, and a corresponding change in the nature of the soil and its productions.

Alluvium.

The geological changes effected by agents which are still in active operation form the subject of this section: these appear to have proceeded, with but little modification, from the period when our continents and islands assumed their present form. The encroachments of the sea upon the shore, by which the cliffs are undermined, and their ruins swept away; the overflowing of low districts, and their conversion into marshes subject to periodical inundations; or, on the contrary, the formation of downs along the coast by the drifting inland of sand and gravel; the production of alluvial tracts by rivers, and the accumulation of estuaries at their mouths; and the disintegration of the exposed surface of the strata, by atmospherical causes; are the principal changes which are now taking place in the strata of the south-east of England.

1. *Blue Clay or Silt, deposited by Rivers, Lakes, &c.*

The numerous tributary rills and streams which flow through the sand and clay districts into the rivers, bring with them particles of mud, sand, &c.: these are carried on by the current, towards the sea; but, as the motion of the waters becomes less rapid, the larger particles subside, and, by degrees, the greater portion is deposited at the bottom of the river or on its banks. It is clear that, by a process of this kind, Lewes and Arundel levels have been produced. If the operation be continued without interruption, the bed of the river becomes more shallow, and the water overflows its banks, till, by degrees, an extensive lake is formed. The levels in the vicinity of Lewes constitute an extensive marshy tract, through which the Ouse winds its way to the British Channel. Tradition, ancient records, and the names of several hamlets situated upon its borders, testify that in distant ages it was covered by an arm of the sea, which extended up the country far beyond the town of Lewes; the site of the Cliff being buried beneath its waters. During the last century, and before the present improved state of the navigation of the Ouse, the levels were annually exposed to extensive inundations, from the overflowing of the banks of the river; the *rhies* and other eminences forming islands in the midst of the lake. These levels offer an instructive explanation of the manner in which the change above alluded to is effected. A section of the soil presents the following deposits:—

1. Vegetable mould.
2. Peat: a mass of decayed vegetable matter, with leaves, and occasionally trunks of trees; from three to five feet thick.
3. Dark blue clay, provincially termed *silt*, containing freshwater shells in the upper part; and an intermixture of sand and marine shells in the lower: from five to twenty-five feet.
4. Pipe clay; the detritus of the chalk-marl basin, in which the preceding beds are deposited: from one to two feet.

The shells of the silt are, in the upper part, *Cyclas cornea*, *Succinea amphibia*, *Planorbis carinatus*, *Planorbis corneus*, *Limnea stagnalis*, *Lim-*

nea palustris, Limnea limosa, Valvata piscinalis, Paludina impura ; in the lower part, *Lutraria compressa, Tellina solidula, Cardium edule, Turbo ulva,* and the *Indusia*, or cases of the larvæ of *Phryganeæ*, in abundance, with minute shells of the genera *Planorbis* and *Limnea* adhering to them, the intermediate layers containing an intermixture of both. Hence it appears, that after the catastrophe which broke through the chalk hills, and thus formed the transverse valleys of the South Downs, the basins of the chalk were filled with salt water; the currents of fresh water flowing from the interior brought down clay, silt, and decaying vegetables, and soon occasioned an intermixture of lacustral testaceæ, and at length so far changed the nature of the element, as to render it fit for the habitation of freshwater shell-fish only.* The transition of the ancient lake into a narrow river has probably been occasioned partly by natural, and partly by artificial causes; within the last fifty years the levels were covered with water, during several months in the winter season; and even now it requires all the resources of art, to confine the river within the limits of its own bed.†

* This conclusion naturally results from the occurrence of marine shells in the lower beds only, and of freshwater in the upper, the two being intermixed in the intermediate layers, since the experiments of M. Beaudant have shown, that if freshwater mollusca be suddenly introduced into sea water, they die in a very short time; but if the fresh water be gradually impregnated with salt, they will live in it when of the strength of sea water without any injury: the same experiments repeated on freshwater mollusca gave similar results.—Vide *Annal. de Chim. et Physique*, ii. 32.

† The alluvial deposits above described are clearly of very remote antiquity; as is evident from the superficial situation in which ancient

2. *Subterranean Forests.*

The occurrence of large trees beneath the surface of the earth, with their leaves, roots, and even fruits, more or less preserved, attracted the attention of philosophers at a very early period. These subterranean forests have been noticed in almost every part of England, and various conjectures offered in explanation of the catastrophes by which they have been overwhelmed. In some instances they appear to have been torn up by a sudden irruption of the sea, and overwhelmed by the sand and mud which it dashed over them; in others they seem to have been submerged by a subsidence of the land. The subject has been ably treated by Mr. Parkinson*, to whose work the reader is referred for an interesting account of the most remarkable examples.

The trees are chiefly oak, hazel, fir, birch, yew, willow, and ash; in short, almost every kind that is indigenous to this island occasionally occurs. The trunks, branches, &c. are dyed throughout of a deep ebony colour; the wood is firm and heavy, and sometimes sufficiently sound for domestic use;

---

coins, &c. have been found. On the west side of Glynd Bridge, a paved Roman causeway was discovered, lying three feet beneath the turf, upon a bed of silt twenty feet thick; and near it was found a large brass coin of *Antoninus Pius*. In forming a road across the levels, from Ranscomb to Beddingham, a coin of Domitian was discovered immediately beneath the surface of the soil. This formation of the levels was therefore antecedent to the Roman advent: and we may infer, that since that period it has not received any material addition.

* Organic Remains of a former World, vol. i.

in Yorkshire it is employed in the construction of houses. Several accumulations of this kind have been discovered on the coast of Sussex, occupying low alluvial tracts, that are still subject to periodical inundations.

At Felpham, near Bognor, on the 25th of October, 1799, a submarine forest was laid bare by a north-east hurricane. It was situated about five feet beneath the surface, but neither its thickness nor extent could be ascertained; notwithstanding there can be no doubt that it pervades the Felpham Levels, probably as far as the village of Barnham. Large portions of the trunks of trees, and heaps of reeds, oak-leaves, &c. matted together, were observed, permeated throughout with a bituminous stain. This storm also exposed on the strand, at low water, upwards of forty large oak trees, lying with their heads towards the south-east. The body of the largest measured four feet in diameter; the wood was extremely black, and emitted a strong sulphureous smell during combustion. Trees of this kind have often been observed by the inhabitants of Bognor after a north-east storm; and, doubtless, may again be witnessed under similar circumstances by any curious enquirer.*

In Pevensey Levels the trunks of large trees have often been observed, imbedded in a mass of decayed vegetables. The substratum is an inferior peat, with an intermixture of sand, reposing upon a thick bed of blue alluvial clay, containing marine

* Communicated by the late Rev. J. Douglas, F.A.S. of Preston.

shells of the same species as those that occur in Lewes Levels. In that division of the marsh called *Hoo Levels*, a submarine forest was discovered a few years since. It lies in the western extremity of Bexhill parish, just above low water mark, adjacent to a manor farm of the Duke of Dorset, called *Conden*, nearly midway between Hastings and Eastbourn. The following description, from an anonymous correspondent, was published in the Gentleman's Magazine for 17—.

"In this place there are the remains of 200 or more trees, which are firmly rooted to the soil, now become sand, and still retain their perpendicularity, and original position. Some of the trees are four or five feet above the surface; others have been cut down, or rather, I conjecture, worn away by the continued flux and reflux of the water. The ramifications, &c. of the roots are very perfect. The trees are of the same species as those of which our Sussex woods are composed, being principally oak and birch. At high water this spot is covered by the sea to the depth of ten or twelve feet; so that it is evident that the earth must here have experienced some grand convulsion, as it is utterly impossible, under present circumstances, that any other than marine vegetation could thrive or even exist there.

"The adjacent country, inland, is a marsh from which the sea has been expelled, and is now kept out with great difficulty, and at a vast expense, and there is no woodland nearer than four miles, on the hill adjoining these levels."

The marsh called the Wish, near Eastbourn,

consists chiefly of peat, of the same character as that of Pevensey, containing leaves, nuts, branches of trees, &c., and the bones of ruminants * : and at Isfield, in sinking the well near the paper-mill, a bed of similar materials was passed through; it is nearly twenty feet thick, and contains oak-leaves, nuts, branches of trees, &c.

3. *Calcareous Tufa, deposited by Springs.*

The deposition of calcareous earth from water flowing through beds of limestone, is a fact so well known as to require but little comment.

Springs of this kind occur in many parts of England, particularly in Derbyshire, where the incrustations they form are generally considered as petrifactions, although certainly having no claim to that title. The chemical changes which give rise to the phenomena in question admit of an easy explanation.

At the temperature of 60°, lime is soluble in 700 times its weight of water; and if to this solution a small portion of carbonic acid be added, a carbonate of lime is formed, and precipitated in an insoluble state. † If, however, the carbonic acid be in such quantity as to supersaturate the lime, it is again rendered soluble in water; and it is thus that carbonate of lime, held in solution by an excess of fixed air, not in actual combination with the lime, but contained in the water, and acting as a menstruum, is commonly found in all waters.

* In 1817, Thomas Smith, Esq. F.R.S., discovered in this alluvial bed the bones of a species of *Bos*.

† Organic Remains, vol. i. p. 373.

Hence it is obvious, that a deposition of carbonate of lime from water may be occasioned either from an absorption of carbonic acid, or from the loss of that portion which exists in excess.

Of the incrusting springs that occur in Sussex, one of the most powerful has its source in the beds of limestone of the Wealden formation; it is situated in a wood at Pounceford, between Heathfield and Burwash. It forms an inconsiderable cascade over a rock of sandstone, and pursuing a tortuous course, deposits carbonate of lime on every extraneous body that lies in its channel; converting the mosses and other vegetables within reach of its waters, into masses of calcareous tufa. The specimens in my possession consist of incrustations of mosses, Equiseta, small branches of trees, leaves, &c.: some of them are composed of a porous friable calcareous earth; and others of a compact carbonate of lime of a subcrystalline structure, perfectly resembling the tufaceous depositions of Derbyshire. When recently collected, the moss on the surface was green and flourishing, and had evidently continued to vegetate, although the roots, &c. were completely imbedded in the stone.

The water has not been analysed, but is evidently possessed of very considerable lapidescent powers, and might doubtless be applied to the same ingenious purposes as the waters of Tivoli, and the baths of St. Phillip, in Tuscany.*

* At the baths of St. Phillip, in Tuscany, a manufactory is established, where casts of medals and bas-reliefs are formed of calcareous tufa. The water is propelled from a considerable height into a large vessel, and being interrupted in its fall by a wooden cross, is separated

Incrusting springs exist also at Tower Hill, near Horsham, and at Folkington, near Ratton.

4. *Sand, &c. drifted inland.*

In Sussex, the effects of this operation are unimportant: a few low banks along the sea-shore, and a ridge of sand and comminuted shells near the entrance of Shoreham Harbour, being the only instances worthy of notice: but in other parts of England they have produced extraordinary changes on the surface of the country, covering extensive tracts, and burying churches, and even whole villages, beneath mountains of sand, which have subsequently been converted into compact sandstone.*

ENCROACHMENTS OF THE SEA.

The destruction and removal of extensive tracts of land on the Sussex coast, by the inroads of the sea, have been noticed in the earliest historical records of the county. At the present time, the ocean, silently, but incessantly, is carrying on the

---

into a fine spray, and dashed against the moulds of the medals, which are placed round the sides of the vessel: by this means excellent impressions are produced.— Vide *Org. Rem.* vol. i. p. 363.

The waters of the Ouse also contain a considerable proportion of calcareous earth. A wooden pipe, which had for several years been used for the conveyance of water from the paper-mill at Lewes, had its interior coated wlth a very compact carbonate of lime, nearly 0·4 inch thick. The presence of a large proportion of calcareous matter in almost all the springs in Sussex, renders the *goître*, or Derbyshire neck, very common among the female population.

* Vide a most interesting account of a recent formation of sandstone, on the northern coast of Cornwall, by Dr. Paris; Trans. Geological Society of Cornwall, vol. i. p. 4. *et seq.*

work of destruction, almost along the whole line of coast. In a few centuries, even without the aid of any violent inroads, the patches of tertiary strata still remaining on the Sussex chalk cliffs must entirely disappear: within the memory of the writer, the plastic clay near Newhaven, the cliffs at Brighton, and the sandstone of Bognor, have visibly diminished.

These encroachments of the sea along the coast of Sussex have continued incessantly, from time immemorial; and when so considerable as to have occasioned sudden inundations, or overwhelmed fertile or inhabited tracts, have been noticed in our historical records. In the " *Taxatio Ecclesiastica Angliæ et Walliæ, auctoritate P. Nicholas,* (A. D. 1292), and *Nonarum inquisitiones in curia scaccarii* (A. D. 1340), the following notices occur, of the losses sustained by the action of the sea, between the years 1260 and 1340; a period of only eighty years.

At *Pett,* marsh land overflowed by the sea; the tithes of which were valued at two marks per annum.

Iklesham and *Ryngermersh,* lands of which the tithes were 49*s.* 8*d.* per annum.

Thornye, 20 acres of arable, and 20 acres of pasturage.

Selseye, much arable land.

Felpham, 60 acres of land.

Middleton, 60 acres.

Brighthelmston, 40 acres.

Aldrington, 40 acres.

Portslade, 60 acres.

Lancing, land, the tithes of which were 44*s*. 6*d*. per annum.

Siddlesham and *Westwythering*, much land.

Houve, 150 acres.

Terringe, land, the tithe valued at 6*s*. 8*d*. per annum.

Bernham, 40 acres.

Heas, 400 acres.

Brede, great part of the marsh called *Gabberghes*.

Salehurst and *Udimer*, land, the tithes of which were valued at 40*s*. per annum.

At Brighton, the inroads of the sea have been very extensive. The whole of the ancient town was situated on the spot which is now covered by the sands, and the present cliffs were then behind the town, like those of Dover. The sea, as Mr. Lyell remarks, has therefore merely resumed its position at the base of the cliffs, the site of the old town having been a beach which had for ages been abandoned by the ocean.* In the year 1665, twenty-two tenements *under* the cliff had been destroyed, among which were twelve shops, and three cottages, with land adjoining them. At that period, there still remained, *under the cliff*, 113 tenements; and the whole of these were overwhelmed in 1703 and 1705. Since that time, an ancient fort called the *Block-house*, with the *Gun garden*, wall, and gates, have been completely swept away, not the slightest trace of their ruins having been perceptible for the last fifty years.†

* Lyell's Principles of Geology, vol. i. p. 279.

† Lee's Hist. of Lewes and Brighton.

At the present time, the whole line of coast, between the embouchure of the Arun, and Emsworth Harbour, is visibly retreating, and the means adopted for its prevention have hitherto been attended with but little success.*

The process by which this destruction of the coast is effected, is sufficiently obvious. By the incessant action of the waves, the cliffs are undermined, and at length fall down, and cover the shore with their ruins. The softer parts of the strata, as chalk, marl, clay, &c., are rapidly disintegrated and washed away; while the flints, and more solid materials, are broken and rounded by the continual agitation of the water, and form those accumulations of sand and pebbles that constitute the beach, and which serve, in some situations, to protect the land from farther encroachments. But when the cliffs are entirely composed of soft substances, their destruction is very rapid, unless artificial means be employed for their protection; and even these, in many instances, are but too frequently ineffectual.†

* Dallaway's Western Sussex, vol. i. p. 55.

† In November, 1824, a violent storm swept over the southern coast of England; and its effects on the Sussex shores were in some places very considerable. I was at Brighton during the greatest violence of the gale, and at the height of the tide; the waves rolled over the towers of the chain pier, and dashed with violence on the Steine; many large masses of the cliffs were thrown down. At Seaford, the bank of shingle, the accumulation of centuries, and the only bar against the ocean, was swept away, and the town and all the adjacent low country inundated: when I visited Seaford, a few days afterwards, the road was annihilated, and I had to make my way over shingle and heaps of sea-weed: another bank has since gradually accumulated.

CHAP. III.

DILUVIUM.

Beds of partially rolled flints appear immediately beneath the turf on the summits, and also in some of the valleys of the Downs; and loam, clay, sand, gravel, and other diluvial débris, are spread over the surface of the regular strata throughout the interior of the country, obscuring their outcrop, and forming the immediate subsoil of the district. These accumulations of water-worn materials have clearly resulted from the destruction of the more ancient deposits; the flints and gravel from the disintegration of the upper strata of the chalk; and the loam, clay, &c. from that of the sands and clays of the Weald and Forest-Ridge. Large blocks of a siliceous sandstone, of which no regular bed now exists in this country, and of the ferruginous breccia of the tertiary formations, are also of frequent occurrence. It is in these accumulations of diluvial débris, that the bones and teeth of elephants, horses, and other quadrupeds, are discovered. We shall not, however, in this place, dwell upon the inferences which may be deduced from these striking facts, but confine our remarks to a brief notice of the most interesting examples.

The bed of flints, slightly rolled, which occurs on the Downs, immediately beneath the turf, contains rounded masses of chalk, blocks of crystallised carbonate of lime, sometimes stalagmitical, ferruginous breccia, scoriaceous ironstone, a coarse grit containing angular fragments of quartz, and flattened oval pebbles of siliceous sandstone. The flints are more or less broken, have suffered but little from attrition, and are so abundant as to form a constant supply for repairing the roads in the south-eastern part of Sussex. This bed has clearly been formed by the destruction of the upper portion of the chalk; and it is equally evident, that the cause which produced the disintegration of the superior strata, was as transient as it was effective; since, although the chalk in which the flints were imbedded has been entirely destroyed, the latter have sustained but very little injury, and must have been elevated above the reach of the waters so soon as the chalk which had invested them was worn away.

Descending into the valleys, accumulations of chalk rubble and ochraceous clay are again seen lying upon the basseting edges of the solid strata; and the slopes of the hills are generally composed of similar materials. Examples of this kind occur in almost every locality of the South Downs.

The gravel-pits (as they are called) of Barcombe are part of a ridge of broken chalk flints, slightly rolled, resting upon an eminence of the Weald clay. The flints are of various shades of yellow, brown, and carnelian. The colour, which in all probability results partly from decomposition, and partly from an impregnation of metallic oxides, pervades the substance of the flint, but is much paler towards the centre than on the surface. These flints are not reduced to the state of pebbles, much less of gravel, but are merely broken, and the sharpness of their angles worn away: they offer one of the few examples of a bed of partially rolled chalk flints lying at a distance from the chalk escarpment.

At Isfield, Little Horsted, Barcombe, Wellingham, &c., the surface of the Weald clay, Iron sand, and Green sand, is covered with beds of gravel, composed of water-worn fragments of *sandstone* and *ironstone*, which in some instances are consolidated into a coarse aggregate, and are evidently the detritus of the upper beds of the Hastings sand formation. A considerable bed of it occurs in the parish of Barcombe, near the Anchor; at Hamsey, on the estate of the Rev. Geo. Shiffner; and at Wellingham, near the seat of Mr. John Rickman.

At Ringmer, and Laughton Place, a layer of loam and ochraceous clay is distributed over the surface of the Galt, and frequently contains belemnites and other organic remains, that have been washed from the upper beds of that deposit.

But the most considerable and important diluvial deposits in Sussex, are those forming the cliffs at Brighton; and which possess characters so remarkable, as to require particular notice in this place: they will, in all probability, hereafter be found to belong, if not to the Crag, to an extensive formation*, which, though of modern origin as

* On the summits of the chalk cliffs at Dover, I have observed a capping of a similar stratum.

compared with the chalk on which it reposes, is referable to the period when elephants lived in our climate, and were contemporary with the horse and the deer.

BRIGHTON CLIFFS.

BRIGHTON CLIFFS, from Kemp Town, looking towards Rottingdean. The diagram in p. 31. explains its geological structure.

The town of Brighton is situated on an immense accumulation of water-worn materials, which fills up a valley, or hollow, in the chalk. This diluvial deposit is bounded on the north-west by the South Downs; on the east it extends to Rottingdean, and is there terminated by the chalk; on the west it may be traced more or less distinctly to Bignor; on the south it is washed by the sea, and forms a line of cliffs from 70 to 80 feet high; these exhibit a vertical section of the strata, and enable us to ascertain their nature and position.

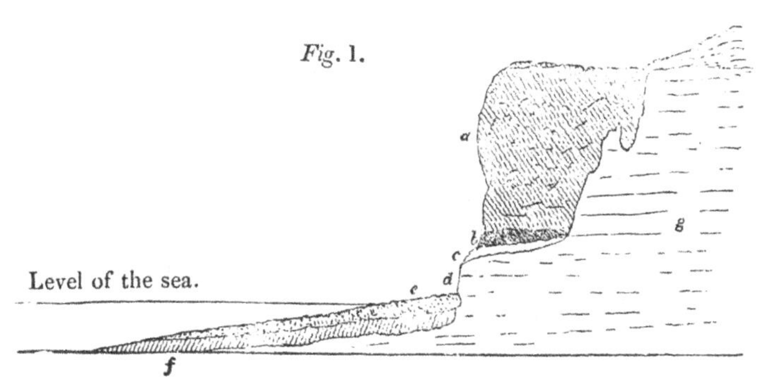

a. Elephant bed.
b. Ancient shingle.
c. Ancient sand.
d. Base of the cliffs of chalk rock.
e. Modern beach.
f. Modern sand.
g. Chalk rock.

A vertical section of the cliffs, about half a mile east of Brighton, is represented above. The lowermost bed is,—

1. The Upper, or flinty chalk; which constitutes about six or eight feet of the lower part of the cliff, and, dipping southward, extends to an unknown distance into the sea. The continuation of the chalk behind the calcareous bed is marked *g* in the sketch; and is introduced to show the relative situation of the masses, but without any regard to proportion.

2. Bed of fine sand, from three to four feet.

3. *Shingle bed*, from five to eight feet.

4. *Elephant bed**, formed of the ruin of the chalk strata, with an intermixture of clay; the harder masses are provincially termed *Coombe rock;*—from 50 to 60 feet.

* This name is given to distinguish it from the other calcareous beds, which do not contain remains of elephants.

The *chalk* presents its usual characters; and in various parts of its course is traversed by vertical and oblique veins of flint.

The *sand* is very fine, varying from pure white to a light reddish brown colour. It disappears about a mile to the east of Kemp Town, where the succeeding deposit lies immediately upon the surface of the chalk.

The *shingle* bed consists of pebbles, formed, like the present beach, of broken chalk flints rounded by attrition. It contains also water-worn blocks of granite, porphyry, slate, limestone, and of tertiary sandstone, breccia, &c. It occasionally envelopes masses of broken shells. The upper part of this bed is cemented together by calcareous spar, of a light yellow or amber colour, forming a coarse conglomerate of a very singular appearance.*

The *Elephant* bed is composed of broken chalk, with angular fragments of flint, imbedded in a calcareous mass of a yellowish colour, constituting a very hard and coarse conglomerate. It is not stratified, but is merely a confused heap of alluvial materials; where it forms a junction with the shingle bed, a layer of broken shells generally occurs: they are too fragile to extract whole: they appear to belong to the genera modiola, mytilus, nerita, &c. It varies considerably in its appearance and composition, in different parts of its course. In the inferior portion of the mass, the chalk is reduced to very small pieces, which

* A specimen of this mineral is figured in Sowerby's British Mineralogy.

gradually become larger in proportion to their height in the cliff: at length fragments of flint appear; and these increase in size and number as they approach the upper part of the bed, of which they constitute the most considerable portion. These flints are more or less broken, and resemble those of our ploughed lands that have been long exposed to the action of the atmosphere.

In some parts of the cliff, irregular masses occur of an extraordinary hardness; these have been produced by an infiltration of crystallised carbonate of lime. Large blocks of this variety may be seen on the shore, opposite to the *New Steine*, where they have for years resisted the action of the waves.

This bed also contains water-worn blocks of siliceous sandstone, and ferruginous breccia. Small nodular masses, composed of carbonate of iron in lenticular crystals, interspersed with brown calcareous spar, have occasionally been found at the depth of ten or twelve feet from the summit of the cliff. The organic remains discovered in this deposit are the bones and teeth of the ox, deer, horse, and of the Asiatic elephant*; these occur but seldom, and are generally more or less water-worn†; but, in some instances, they are quite

* In April, 1822, a large molar tooth of the Asiatic elephant was discovered in Lower Rock gardens, in a well fifty feet deep; and four very fine and perfect ones were dug up by the workmen employed on the foundation of the walls for the esplanade, at the Chain Pier, in 1831.

† I have specimens of the teeth, found in a well fifty yards inland, at the depth of forty-six feet, in the *Coombe rock*, and immediately above the bed of shingle.

entire, and cannot have been subject to the action of the waves.

The wells in the less elevated parts of the town pass through the calcareous bed, shingle, and sand, in succession; upon reaching the chalk, springs of good water burst forth, and these are said to be influenced by the tides.*

Such are the leading features of these remarkable beds, in the immediate vicinity of Brighton; in their course eastward towards Rottingdean, other characters are exhibited, which we shall now proceed to examine.

About a mile to the east of Brighton, veins of tabular flint traverse the chalk in an oblique direction, and terminate with the chalk, immediately beneath the shingle bed, *fig.* 2. To avoid repe-

STRATA EAST OF KEMP-TOWN; *fig.* 2.

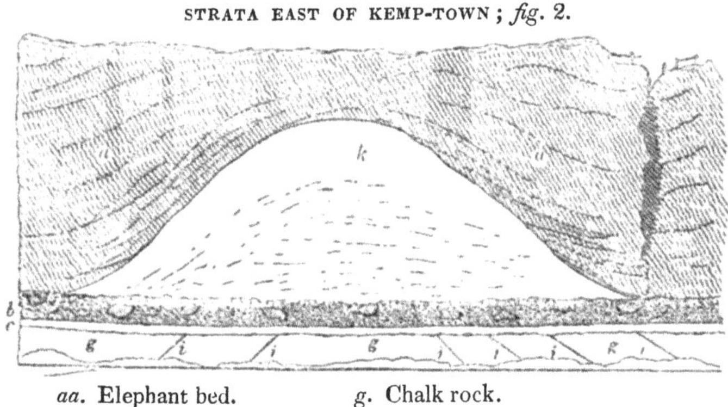

aa. Elephant bed.	*g.* Chalk rock.
b. Shingle bed.	*i.* Flint veins.
c. Ancient sand.	*k.* Chalk in a state of ruin.

* Some wells at Tetney (a village on the coast of Lincolnshire) that are sunk in the chalk, are also affected by the tide; the wells overflowing with a greater flux at the time of high water, and particularly at spring tides; showing that the water in the chalk communicates with the sea."—*Geolog. Trans.* vol. iii. p. 394.

tition, it may be proper in this place to remark, that the veins of flint, so numerously distributed both horizontally and vertically throughout the chalk, are *invariably confined to that formation, and in no instance whatever appear either in the shingle bed, or in the calcareous bed above it.** The shingle bed is perfectly horizontal, and contains boulders of chalk, druid sandstone, and ferruginous breccia. In the *Elephant bed*, the proportion of chalk is so great, that the cliff at a distance assumes the appearance of a regular stratum; but, upon closer examination, it is evident that the chalk at some remote period has been broken, and displaced; and having fallen upon the shingle, previously to the formation of the *Elephant bed*, has subsequently been covered by that deposit.

In the above sketch, the chalk, traversed by oblique veins of flint, is seen forming the base of the cliff. The shingle bed succeeds; and immediately above it, is a heap of chalk in a state of ruin; the latter is invested by the *Elephant bed*, of which the upper part of the cliff is composed. This appearance is curious, but the manner in

* An opinion having been expressed (by a gentleman well known in the scientific world), that the flint veins traverse not only the shingle bed, but also the calcareous deposit, and have been formed "subsequently to the accumulation of an alluvial bed, by the attrition of agitated water," and that the cliffs at Brighton are to be regarded as "two very distinct chalk formations †," I carefully repeated my examination of the strata in question, but could not discover any appearance to support such a hypothesis.

* Vide Royal Institution Journal, No. VIII. p. 227. et seq.; Phillips's Outlines, Edit. 1822, p. 106.

which it has been produced is easily explained, by a reference to those natural operations that still continue in full activity on our coasts. Were a bed of calcareous rubble to be deposited over the ruins of the chalk cliffs that are scattered along the shore, a collection of materials would be formed, corresponding in every respect with those above described; and a vertical section would exhibit an appearance precisely similar, namely, a stratum of solid chalk at the base; then a layer of sand and of shingle; and, lastly, a heap of displaced chalk, surrounded by calcareous diluvium. In corroboration of this opinion, it may also be remarked, that while, in general, the variations observable in the colour and composition of the *Elephant bed* are nearly horizontal in the circumstances under discussion, they are no longer conformable to the subjacent deposit, but rise over the heaps of chalk rubble, as in the section page 34. These interspersions of pure chalk are frequent in other parts of the bed, but the present example is one of the most remarkable.

Proceeding eastward, at the distance of two miles and a half from Brighton, the cliff is composed of the Upper chalk, to the extent of three hundred yards. This remarkable change in the structure of the cliff has evidently been occasioned, partly from the destruction of the diluvial deposits by the inroads of the sea, and partly from a projection of the chalk, which formed their ancient boundary; for there appears to have been but little correspondence in the sinuosities of the ancient and modern shores. An abrupt recess marks this alteration in the face of the cliff; and here the calcareous bed rises suddenly to the summit of the chalk, over which it is continued in a layer of inconsiderable thickness, *fig.* 3. The shingle

CLIFFS BETWEEN KEMP-TOWN AND ROTTINGDEAN; *fig.* 3.

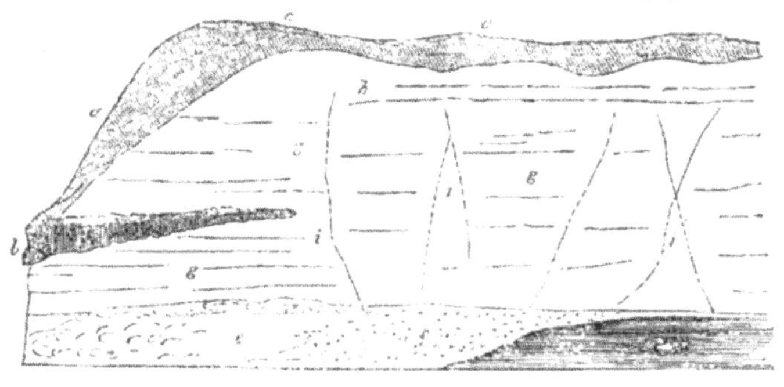

a. Elephant bed.
b. Ancient shingle.
e. Modern beach.
f. Modern sand.
g. Chalk rock.
h. Horizontal flint veins.
i. Vertical and oblique flint veins.

bed, which, at a short distance to the west, contains large masses of chalk, here suffers a remarkable contraction, and is divided by thin seams of sand and fine rubble. At the curvature of the recess, the shingle diminishes very abruptly, and soon entirely disappears. Along the face of the chalk, slight traces of it are here and there perceptible; and in these situations, the vertical flint veins that traverse the cliff *invariably pass behind, and are concealed from view,* by the insular patches of shingle.

The face of the chalk is remarkably even; it is not, however, vertical, but forms a precipitous slope. In the upper part, the chalk is much broken, and contains two horizontal veins of tabular flint: the inferior strata are more regular. It is particularly necessary for the reader to bear in mind, that although the chalk, with its horizontal flint veins (vide section above), is *higher* than the insular portions of the shingle bed, it is not situated *perpendicularly above* them. The cliff, as before mentioned, forms an inclined plane, its summit receding considerably from the shore: consequently a vertical section would cut off all traces of the shingle.*

On the eastern extremity of the recess, the chalk is traversed by numerous veins of marl, but in other respects presents nothing worthy of observation. At the termination of the chalk, a bold projection of the cliff occurs, in which the shingle and calcareous bed appear in their usual position and proportions

Towards Rottingdean the cliffs increase in altitude, but the calcareous bed diminishes considerably in thickness, and, wherever a vertical section is exposed, is seen lying upon the shingle, in contact with a sloping bank of broken chalk; the latter being evidently the ruin of the ancient chalk cliffs, the flints it contains presenting no appearance of having suffered either from attrition or exposure.

* It was probably from want of attention to this circumstance, that the respectable writer previously alluded to was led to adopt the opinion, that the shingle bed was situated between two distinct beds of flinty chalk.

CLIFFS WEST OF ROTTINGDEAN; *fig.* 4.

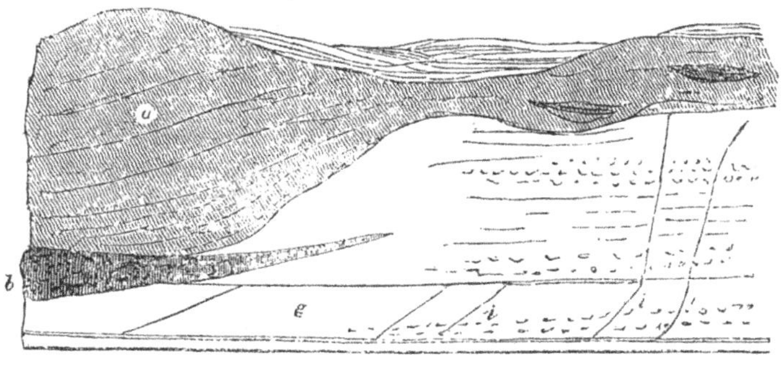

a. Elephant bed.
b. Shingle bed.
g. Chalk rock.
i. Flint veins.

To the west of Rottingdean, the cliffs are chiefly composed of the regular chalk strata, containing, as usual, horizontal beds of siliceous nodules, and veins of tabular flint. Veins of marl are also very numerous; and there is one of remarkable extent, which appears beneath the shingle, and, extending in a horizontal direction to within a short distance of Rottingdean, re-appears on the eastern bank of the landing place.

In that portion of the cliffs we are now describing, the shingle terminates as represented above; and, by a singular coincidence, a bed of flint nodules commences immediately beneath it, and, pursuing a horizontal course in the chalk, resembles at a distance a continuation of that bed. It is scarcely necessary to observe, that this apparent identity is a mere illusion; a bed of *rounded pebbles lying upon* the chalk, cannot readily be mistaken for a stratum of *perfect chalk flints, still occupying the cavities in which they were originally formed.*

On the west side of the landing place at Rottingdean, the cliff is low, and its upper part occupied by a mass of chalk rubble, analogous in some of its characters to the *Elephant bed*, of which it may possibly be a continuation. It is strongly marked with undulating lines, of an ochraceous yellow colour. The chalk, on both sides of the gap, is more or less disturbed, and the veins of tabular flint are broken and contorted; this is remarkably the case with those on the eastern bank (*fig.* 5.),

LANDING PLACE, ROTTINGDEAN; *fig.* 5.

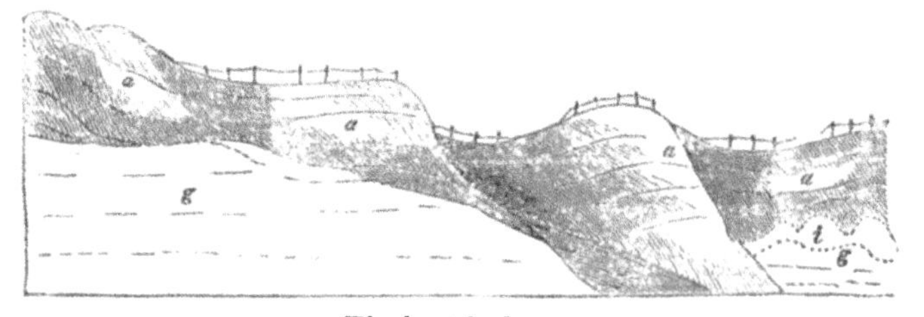

a. Elephant bed.
g. Chalk rock.
i. Tortuous flint veins.

in which the fragments of flint are detached from each other. The beach near this place contains semi-translucent pebbles of agate, and chalcedony, of a bluish grey colour. These are collected by visiters, and when cut and polished are used for bracelets and other ornamental purposes: they are usually called "*Rottingdean pebbles.*"

In concluding this description of the cliffs between Brighton and Rottingdean, I would observe, in the words of Mr. Webster, "that it is impossible to view them, without immediately perceiving that they do not owe their existence to original stratification, but are simply the section of an immense heap of fragments of chalk and flints, mixed with clay and sand, the whole having, at some distant period, been subjected to the action of water, and deposited upon the solid chalk stratum."

ORGANIC REMAINS OF THE DILUVIAL DEPOSITS.

It is always in deposits of this kind, that is, in diluvial beds spread over the surface of plains, or accumulated in the bottoms of vallies, that the teeth and bones of mammalia have been discovered in various parts of England. In Sussex, however, these remains but very rarely occur; the bones and teeth of the horse, ox, deer, and elephant, being the only examples at present known.

J. Hawkins, Esq., of Bignor Park, informed me, that about sixty years since, the bones of an elephant were dug up in Burton Park, near Arundel, but no satisfactory account of the circumstances attending the discovery was preserved.

In the brick-loam at Hove, near Brighton, a fragment of a bone resembling the femur, and a grinder of a large size, were found at the depth of about six feet; the tooth was decidedly that of the Asiatic elephant.

At Peppering, near Arundel, the bones, and several grinders of elephants, have been found in a bed of gravel, on the estate of John Drewett, Esq., of Peppering, who kindly favoured me with the following remarks concerning them.

"The remains in question were found in a bed of gravelly loam, situated near the foot of the Downs, and reposing upon the chalk, at an elevation of about eighty feet above the level of the Arun. They were lying very superficially, the first fragment of bone that attracted our notice being scarcely three feet beneath the surface. The specimens collected consist of a tusk, four grinders, and several fragments of other bones, apparently portions of the skull; the body appeared to lie beneath a bank of earth of considerable thickness, and could not have been removed without much labour. The tusk was lying upon its convex part, and notwithstanding every precaution, broke into several pieces, upon our attempting to remove it. It measured four feet and a half long, and from twenty-two to twenty-four inches in circumference; but neither the base nor point was perfect. The largest grinder of the lower jaw weighed six pounds four ounces; its upper surface being three inches and a half wide, and seven inches long; one of the molares of the upper jaw was broken in two, and the pieces detached from each other."

The grinders and bones of the elephant, horse, ox, and deer, occur in the calcareous bed at Brighton, as previously mentioned; and the jaw of a whale was found in the shingle bed.

The antlers and bones of the red deer are said to have been discovered in a bed of loam, in sinking a well near the barracks, a mile to the north-east of Brighton. The remains of a deer were found in the diluvium at Copperas Gap, near

Hove, by the Rev. H. Hoper; also in digging a well near the Western road.

In a diluvial clay, or loam, in a valley near Hastings, a tooth of the Asiatic elephant has been discovered: and the men employed in levelling a low bank near the road at Patcham, dug up several molar teeth, and some portions of the bones of an elephant, which are in the select collection of Richard Weekes, Esq. of Hurstperpoint.

CHAP. IV.

TERTIARY FORMATIONS.

In the deposits previously described, but little order or regularity was perceptible; their varied contents being, for the most part, indiscriminately mingled, and bearing incontestable proofs of having been produced by the action of water in a state of agitation, on the more ancient strata; but those which form the subject of the following sections, will be found to present a certain and constant order of superposition; particular fossils will be seen to occur in some of the strata, and to be wanting in others; yet, even in these formations, traces of extensive diluvial action appear in the pebbles, and other water-worn materials, of which some of the strata are almost entirely composed.

The class of deposits to which these belong, were but imperfectly known, till the researches of MM. Cuvier and Brongniart, in the environs of Paris. The publication of their masterly delineation of the *Géographie Minéralogique*, of that district, excited universal attention, and attached to the investigation of these strata a high degree of interest and importance. The inquiry was pursued with equal zeal and success, in our own country, by Mr. Webster, who discovered in London, Hampshire, and the Isle of Wight, a series of beds, corresponding in their characters, and geological position, with those of the neighbourhood of Paris. Insular portions of these strata have subsequently been noticed in numerous localities of the English chalk; and the facts already known, are sufficient to warrant the conclusion, that contemporaneous deposits are extensively distributed throughout the globe.*

Of the strata which exist above the chalk, traces are perceptible in many localities of Sussex, and the adjoining counties; so numerous, indeed, are the proofs of their having formerly extended very far beyond the limits generally assigned to the Isle of Wight basin, that it seems evident, either that the basins of London and Hampshire were united, previously to the last displacement of the chalk, and its superincumbent strata, or that there were many hollows or basins on the surface of the chalk, which received the sedimentary deposits of the tertiary seas. The flat maritime district, which extends from near Worthing to Bracklesham Bay, and from thence into Hampshire, is composed of clay, sand, brick-earth, gravel, &c.; and at Chimting Castle, near Seaford, and Castle Hill, near Newhaven, outlying portions of the same series of beds remain in situ. These strata are supposed to have belonged to the Isle of Wight basin. To obtain a

* Even in North America, tertiary beds are found over a vast extent of country, and their fossil shells vie in beauty, numbers, and interest, with those of Europe. Dr. Morton, Mr. Say, Mr. Conrad, and other American savans, have pursued the inquiry with so much zeal, talent, and success, that even a work expressly devoted to the tertiary shells of North America has just appeared. It contains beautiful and accurate figures; and will be of the highest interest and importance to the European geologist. It appears in monthly numbers; and may be obtained of Mr. O. Rich, American bookseller, Red Lion Square, Holborn.

clear idea of their geological relations, it will therefore be necessary to take a brief view of the extent, and characters, of that depression of the chalk.

The district comprehended by the Isle of Wight basin, is about 100 miles in length, and at its greatest breadth does not exceed twenty miles. The southern side is formed by the highly inclined chalk, extending from the Culver Cliffs, at the east end of the Isle of Wight, to White Nose, in Dorsetshire, five miles west of Lulworth; the north side by the South Downs, that pass from Beachy Head, to Dorchester, in Dorsetshire. The strata of which these hills are composed, dip generally from 15° to 5° to the south; the inclination varying in different places. The south side of the basin must therefore have been extremely steep, while the slope of the north side was very gentle. The western margin cannot be distinctly traced, and the eastern is now entirely destroyed, the sea flowing through the opening.*

The annexed sketch will illustrate this description, and show the connection between the outlying fragments of these beds in Sussex, with those of the Isle of Wight.

* Mr. Webster on the Strata overlaying the Chalk, Geolog. Trans. vol. ii. p. 170. *et seq.*

The strata contained in the Hampshire or Isle of Wight basin, form five principal divisions, viz.

1. Lowest marine formation over the chalk, including the plastic clay and sand, together with the London clay.
2. Lowest fresh water formation.
3. Upper marine formation.
4. Upper fresh water formation.
5. Alluvium.

The remains of the tertiary formations that occur in Sussex, admit of the following arrangement; it must, however, be remarked, that from the ruin and displacement to which they have been exposed, it is scarcely possible in every instance accurately to determine their geological positions.

TERTIARY FORMATIONS IN SUSSEX.

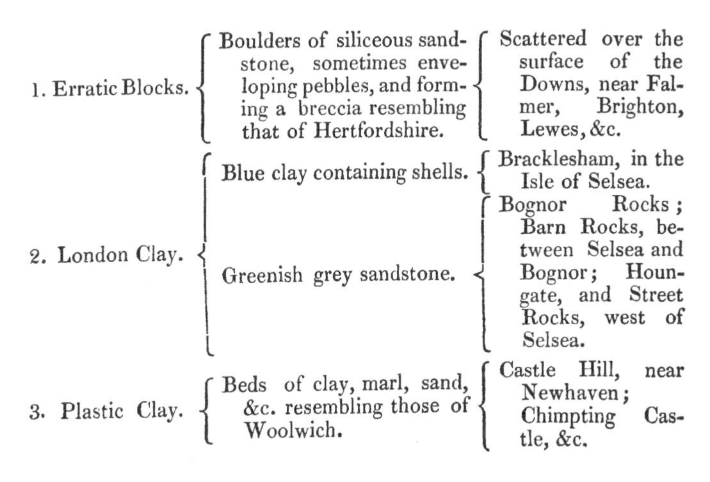

Formation	Description	Localities
1. Erratic Blocks.	Boulders of siliceous sandstone, sometimes enveloping pebbles, and forming a breccia resembling that of Hertfordshire.	Scattered over the surface of the Downs, near Falmer, Brighton, Lewes, &c.
2. London Clay.	Blue clay containing shells.	Bracklesham, in the Isle of Selsea.
	Greenish grey sandstone.	Bognor Rocks; Barn Rocks, between Selsea and Bognor; Houngate, and Street Rocks, west of Selsea.
3. Plastic Clay.	Beds of clay, marl, sand, &c. resembling those of Woolwich.	Castle Hill, near Newhaven; Chimpting Castle, &c.

ERRATIC BLOCKS OF SANDSTONE.

IMMENSE blocks and boulders of siliceous sandstone, composed of granular quartz, and occasionally enveloping chalk flints, and other extraneous bodies, lie scattered over the Downs, and on the ploughed lands, near Brighton, Falmer, and other places. This sandstone is perfectly analogous to that which occurs in Berkshire and Wiltshire, where it is distinguished by the provincial term of "*Grey Weathers.*" Of this substance, Stonehenge, and other druidical monuments are composed,—a circumstance that has given rise to its present geological appellation. The cement of the beautiful conglomerate or puddingstone of Hertfordshire, agrees in its characters with the druid sandstone, and from that breccia also occurring in detached blocks above the chalk, it is now generally supposed that they are both of contemporaneous origin; the siliceous deposition, when it did not envelope any foreign substance, forming the rock called the "*Grey Weathers,*" and when it fell among pebbles of any kind, composing a breccia or *puddingstone.**

The puddingstone is exceedingly rare in Sussex, but specimens sometimes occur; and several examples have been found in the vicinity of Newhaven, that could not be distinguished from the Hertfordshire breccia.

Examples of the siliceous sandstone may be seen on the hill near Lewes Race-course; at Bormer;

* Geological Transactions, vol. ii. p. 225.

in Stanmer Park; and on the ploughed lands near Hogshrove farm. At Falmer, the pond that supplies the village with water is surrounded with large blocks, and there are a considerable number in Stanmer Park, the seat of the Earl of Chichester; and also around Brighton. These boulders have their edges rounded and even, and exhibit incontestable proofs of long exposure to the action of the waves. They are of various sizes, some of them exceeding nine feet in length; their colour is either white, or of different shades of grey and reddish brown. Their texture is sub-crystalline; the white varieties, when recently broken, much resembling lump sugar. In a few instances, they enclose chalk flints slightly worn, and small fragments of a dark green substance, the nature of which is unknown.

Boulders of druid sandstone also occur in the shingle bed, and calcareous deposit, at Brighton, and may be observed lying on the sea-shore in considerable numbers after a recent fall of the cliffs.

Upon comparing the sandstone of Stonehenge with that of Sussex, no perceptible difference can be detected; and in this county, as well as in Wiltshire, it has been employed by the earlier inhabitants, as landmarks to denote the boundaries of towns*, and villages, or to commemorate the

* The frequent occurrence of large smooth blocks of stone, on the boundary line of villages and parishes, in the south-eastern part of Sussex, must have been noticed by many of my readers. A large boulder of druid sandstone, placed at the corner of *Ireland's Lane*, in St. Ann's parish, forms the western boundary of the borough of Lewes.

site of battles; as sepulchral stones, to perpetuate the memory of their chiefs; and as altars, on which to sacrifice to their gods.

No regular stratum of the druid sandstone has yet been discovered in this country, and its geological position is still undetermined.

LONDON CLAY.

This formation consists chiefly of a dark blue clay, which in some localities includes beds of grey limestone and sandstone.

Both the clay and limestone occur in Sussex: the former constitutes the flat maritime district of the south-western part of the county; the latter composes groups of rocks on the coast.

Blue Clay.

This deposit forms the line of coast from Worthing to Christchurch, in Hampshire, extending from the latter place inland, by Ringwood, Romsey, and Fareham; and passing a mile or two south of Chichester, terminates near Worthing, from whence to Brighton, the surface of the chalk near the coast, is covered with beds of loam, clay, brick-earth, gravel, &c.*

Similar stones are not unfrequent in the large tumuli on the Downs; several may be seen near Lewes Race Course. It seems probable that the ancient Britons regarded this sandstone with superstitious veneration; for besides employing it in the construction of their temples, kist-vaens, &c., they converted the pebbles and smaller stones into amulets and beads.

* Phillips's Geology, p. 32., edit. 1822.

In some parts of its course in Sussex, it contains an abundance of the organic remains for which it is so remarkable. Emsworth and Stubbington, on the confines of the county, have been noticed by Mr. Webster, as abounding in fossil shells. Bracklesham, near Selsea, is equally productive; and if I may judge from the liberal contributions of my kind and excellent friends Mr. and Mrs. Hawkins, of Bignor Park, will almost rival the celebrated cliffs of Hordwell.

On the coast westward of Selsea, near Thorney and Bracklesham, vast quantities of fossil shells are washed out of the clay and deposited on the shore, by the action of the waves, particularly after severe storms. This bed of clay is, however, only accessible at low water, and even then but for a very short period.

Below the beach at Bracklesham, in the parish of East Wittering, the clay envelopes the trunks, roots, and branches of trees.

In the second volume of the Geological Transactions, Mr. Webster has enumerated the fossils discovered by him at Bracklesham. The interesting collection from that place, collected by my friends, has enabled me to add very considerably to their number; the various species are enumerated in the systematic catalogue at the end of this volume.

SANDSTONE, OR ARENACEOUS LIMESTONE OF BOGNOR.

The sandstone rocks of Bognor are the ruins of a deposit once very extensive, and which, even within the memory of man, formed a line of low cliffs along the coast; at present, a few groups of detached rocks, covered by the sea at high water, are all that remain, and the period is not far distant, when even these will be swept away by the encroachments of the ocean. The lowermost part of the rocks is a dark grey limestone, in some instances passing into sandstone: the upper part is siliceous. The *Barn* rocks between Selsea and Bognor, the *Houndgate* and *Street* rocks on the west, and *Mixen* rocks on the south of Selsea, are portions of the same bed. The fossils enclosed in these strata are similar to those which occur in the London clay.

These beds are decidedly analogous to the *Calcaire grossier* of Paris; the correspondence in their geognostic situation, and in the nature of their materials and organic remains, sufficiently evinces their identity.

The sandstone is of a grey colour inclining to green; and varies considerably in hardness and composition. The shells are generally white and friable, consisting of a soft calcareous earth, but they also occur in a good state of preservation.

They consist of various species of the genera *Nautilus*, *Rostellaria*, *Lingula*, *Turritella*, *Pyrula*, *Pinna*, *Pectunculus*, *Pholadomya*, &c., — wood perforated by teredines, &c. *Septaria* sometimes occur with numerous *turritellæ*, the shells of the latter being converted into a porcelain-like carbonate of lime: polished slabs of this kind are very beautiful. Teeth and bones of fishes are also found.

PLASTIC CLAY.

In conformity with the nomenclature of MM. Cuvier and Brongniart, a principal division of the tertiary formations, consisting of various beds of sand, clay, marl, and gravel, is distinguished by the name of Plastic Clay (*Argille Plastique*). An attentive examination of the general points of resemblance in the physical characters, and organic remains, of these irregular alternations above the chalk, leaves no doubt of their being members of a series of nearly contemporaneous depositions, intermediate between that formation and the London clay; but the propriety of the separation of these beds from the latter is not very obvious.

A fine series of strata belonging to this division, occurs on the western side of Newhaven harbour, lying upon the chalk cliffs, which are there about fifty feet high. The summit of the hill is broken and rugged, and its appearance differs so remarkably from the smooth rounded surface of the surrounding downs, that the geologist, even at a distance, would suspect the existence of strata very dissimilar to any that exist in the vicinity.

Castle Hill is the Eastern extremity of the chalk cliffs, that extend from Rottingdean to Newhaven harbour, and is about a mile to the south of the town, on the west of the embouchure of the Ouse, eight miles from Lewes. Its summit bears marks of an ancient entrenchment, which, when first constructed, must have embraced an area of twelve or fourteen acres; although, from the encroachments of the ocean, scarcely a third of it remains. The hill is about 180 feet high, and presents a ruinous aspect towards the sea: near the edge of the cliff, the strata are in so crumbling a state, and the fissures so numerous, as to render their examination difficult and dangerous. From the decomposition and destruction of the strata which are constantly going on, their surface seldom exposes an instructive section, except after recent falls of the cliff. The thickness of the strata, exclusively of the chalk, does not exceed seventy or eighty feet; and they reach along the coast to about a mile west of the signal house, gradually becoming less, till a thin cap of clay and sand marks their termination.

Commencing with the uppermost bed, the cliff is composed of,

1. *Sand and pebbles*, — from 10 to 15 feet.

2. *Coarse argillaceous rock;* chiefly composed of oyster shells; with a few cerithia, &c., — *5* feet.

3. *Foliated blue clay*, with an immense quantity of shells; the upper part contains shells of the genera Cytherea and Cyclas; the lower division Cerithia, and teeth of sharks *, — 10 feet.

4. *Reddish brown marl*, more or less slaty, bearing impressions of leaves, wood, cones of a plant of the palm tribe? and casts of Potamides and Cyclades, — from 4 to 5 inches.

* A few shells of the genus *Cyrena* have been discovered in it.

SECTION OF THE STRATA AT CASTLE-HILL, NEAR NEWHAVEN HARBOUR.

5. *Surturbrand or lignite*,—from 4 to 6 inches.

6. *Blue clay*, with marl of a sulphur colour, including gypsum both crystallized and fibrous, — 20 feet.

7. *Sand* of various shades of yellow, green, ash colour, &c., without fossils, — 20 feet.

8. *Breccia* of pebbles, broken flints, and green sand, strongly impregnated with iron, forming a hard conglomerate, — 1 foot.

9. *Ochraceous clay*, containing *hydrate* and *sub-sulphate* of *alumine*, with gypsum, &c., — 1½ foot.

10. Chalk with flints, — about 100 feet.

Of the strata above enumerated, Nos. 3, 4. 6. are considered by Professor Buckland as analogous to the plastic clay beds of Woolwich (Nos. 7, 8. Pl. xiii. *Geolog. Trans.* vol. iv.), which contain also the same species of *Potamides* and *cyclas*.

No. 7. of Castle Hill, is the ash-coloured sand of Woolwich, in diminished thickness. The *breccia*, No. 8., corresponds with the Reading oyster bed, which " though inconsiderable in thickness, seems constantly to occur immediately above the chalk; although organic remains have been noticed in it only at Reading."

The ochraceous clay (No. 9.) contains the substance that has rendered Castle Hill so interesting to the mineralogist — the *subsulphate* of *alumine*. As this mineral is peculiar to Sussex, a particular description of it is subjoined.

(*Hydrate* and *subsulphate* of alumine. *Brit. Min.* Tab. 499. *Annals of Philosophy*, vol. ii. p. 238.)

This substance is imbedded in a layer of ochraceous clay that lies immediately upon the chalk. The bed is situated nearly midway between the summit of the cliffs and the sea-shore, and therefore cannot be examined without much difficulty, and exposure to considerable danger.*

The first specimen of the subsulphate was discovered by the author among some gravel that

* Specimens may generally be found among the ruins of the cliffs that lie scattered on the shore, from half a mile to a mile west of Newhaven harbour.

had been brought from Newhaven, and was lying in a wharf near Lewes.* A few months afterwards, Mr. Webster, in a geological excursion along the Sussex coast, collected a specimen at Newhaven. This was analyzed by Dr. Wollaston, and found to consist of alumine, in combination with sulphuric acid, and a small proportion of silex, lime, and oxide of iron.

This mineral occurs massive, in veins, and in tabular and tuberose masses; the former frequently attaining several feet in length, and the latter exceeding three or four pounds in weight. It appears to have been of stalactitical origin, and is supposed to result from the decomposition of iron pyrites, and the reaction of other substances. As the superincumbent strata contain all the elements necessary for its production, it probably has been introduced into its present situation by infiltration.† When pure, it is perfectly white, but is generally more or less discoloured by an intermixture of yellow clay. It is dull and opaque, with an earthy fracture, and yields to the knife. It is infusible at 166° of Wedgwood‡, but fuses rapidly when exposed to the stream of the hydro-oxygen blowpipe: the result is a pearl white translucent enamel, a partial combustion taking place during its fusion.§ According to Stromeyer it consists of—

* Vide Mr. Sowerby's description of this substance in British Mineralogy.

† Professor Buckland on the Plastic Clay. Geological Transactions vol. iv. p. 294.

‡ Kirwan.

§ History of the Gas Blow-pipe, by E. D. Clarke, LL.D. 8vo. 1819, p. 56. The experiments of my brother gave similar results.

Alumine	-	30
Sulphuric acid,		24
Water	-	45*

Crystals of gypsum are frequently disseminated through the masses of alumine, and the two substances enter into various states of combination, sometimes giving rise to specimens that are semi-translucent. Chalk flints, indurated ochraceous clay, and other extraneous bodies, are also occasionally enveloped.

From the experiments of the late Dr. Clarke, it appears that the purer masses of aluminite are destitute of sulphuric acid, and consist simply of water and aluminous earth. Hence a suspicion has arisen that the sulphuric acid, in the examples analyzed by Stromeyer and Dr. Wollaston, may have originated from the presence of gypsum: this, however, is not the case; since in many specimens it is evident that the sulphate of lime has been decomposed, and the sulphuric acid, entering into combination with the alumine, has formed a true sub-sulphate.

The hydrate occurs in friable masses, of the colour and consistence of magnesia: it adheres to the tongue, and may be reduced to powder between the fingers. In this respect it differs from the subsulphate, which possesses considerable hardness, and is susceptible of a fine polish.†

* Phillips's Mineralogy, 2d edition, 1819, p. 111.

† In the elegant compendium of geology, inserted in Professor Brand's Manual of Chemistry, are the following observations on this subject: —

"In the cliffs at Newhaven, on the Sussex coast, a very curious series of changes is going on. A stratum of marl containing decom-

The flints or pebbles composing the *breccia* (No. 8.) are characterised by their green and ferruginous crusts.

This appearance is so peculiar, that it frequently serves to identify the situations formerly occupied by the breccia, even where the stratum itself has been broken up. These pebbles are scattered over the ploughed lands on the summits and slopes of the Downs, near Tarring, Piddinghoe, Falmer, Stanmer, Bormer, and many other places in the vicinity of Lewes. I have also detected them in the diluvium of the interior of the country. Waterworn fragments of the breccia occasionally occur in similar situations; some of considerable magnitude may be observed lying bare in the fields near Brighton church, Goldstone Bottom, and Falmer Hill.*

---

posing pyrites, lies upon the chalk, which gives rise to the formation of sulphate of alumine: this is decomposed by the chalk; and aluminous earth, selenite, and oxide of iron, are the results." (*Manual of Chemistry*, 3 vols. 8vo. 1821. Vol. iii. p. 312.)

In the Annals of Philosophy, for August 1820, Mr. Cooper, of the Strand, gives a description of an aluminous chalybeate spring, situated on the coast between Newhaven and Rottingdean.

" The spring is situated, as I understand, about midway between Newhaven and Rottingdean, at an elevation of about 15 or 16 feet above the level of the sea at high water mark. It issues from between the cliffs or fissures of the chalk in small streams, and these, when united, pour forth from twenty to twenty-five gallons in the hour. The chalk about the place is every where tinged with an ochreous deposit. Its temperature as it issues is 65° Fahr. and remains constantly the same. When I received it, there was a deposit of a brownish colour, which proved on examination to be oxide of iron. Its specific gravity, at the temperatnre of 60° Fahr. was 1·076: it is slightly acidulous, changing the colour of litmus paper both before and after boiling, by which operation it deposits a further portion of oxide of iron, and also

The selenite or crystallized gypsum (of No. 6.) occurs in flattish crystals, from six to eight inches long, which are generally in the form of oblique parallelopipeds, or of rhomboidal prisms. The fibrous gypsum is deposited in veins in the marl; the foliated variety occurs in large tabular masses, composed of thin laminæ, and is frequently coated with a coaly substance.

The *surturbrand* or *coal* (No. 5.) appears to be analogous to that of the Paris Basin, Corfe Castle, and Alum Bay: it also resembles the surturbrand of Iceland; some specimens are exactly similar to the Bovey coal.

---

a little lime. Reagents show it to contain the following substances in solution: —

Oxide of iron,	Lime,
Alumina,	Carbonic acid,
Muriatic acid,	Soda.
Sulphuric acid,	

"This last substance I will not be quite certain of; but I expect shortly to be able to make a more perfect analysis, and to give a better account of its situation, which is of some importance, as I expect it is not far distant from the spot where the native alumina or subsulphate is found."

* The boulders of this breccia, like those of the siliceous sandstone, were used in distant ages as sepulchral stones. Beneath one of those, near Brighton church, an urn of high antiquity, containing human bones and ashes, was discovered by the late Rev. J. Douglass, F.A.S.

An immense block of this kind is situated in Hove parish, near the Shoreham road, and is vulgarly called Goldstone, "from the British word *col*, or holy-stone; it is evidently a tolmen of the British period. This stone is in a line to the south of Goldstone Bottom, at the end of which, close to the rise of the hill, is a dilapidated *cirque*, composed of large stones of the same kind. On the farm of Thomas Read Kemp, Esq. opposite Wick, are two dilapidated *kist-vaens*, formed of similar materials; and on each side of the British trackway, leading to the *Devil's Dyke*, blocks of the same substance may also be observed." *Extract of a Letter from the late Rev. J. Douglass to the Author, dated May*, 1818.

Rolled masses of this substance are frequently found on the shore at Brighton, and were formerly so abundant as to be used for fuel by the poorer inhabitants.* They are provincially called *strombolo,* a corruption of *strom-bollen, stream* or *tide balls;* the name given them by the Flemings, who formerly settled in that town.

The use of this substance was prohibited, on account of the very offensive smell emitted during its combustion. It was employed by the late Dr. Russell as a fumigation in certain glandular complaints, and it is said with decided benefit.

The organic remains found in these deposits are enumerated in the catalogue: they consist of dicotyledonous wood; impressions of leaves; a coniferous fruit; shells of the genera Potamides, Helix, Cyclas, Cyrena, Cytherea, Ostrea, &c.

Between Castle Hill and Seaford, a flat alluvial tract intervenes, through which the Ouse flows into the British Channel. To the north-east of this marshy plain, the truncated terminations of the Downs are covered with a cap of fawn-coloured and greenish sand, with rolled blocks of chalk, and flint pebbles. An excavation on the side of the hill, near the road leading from Newhaven to Seaford, exhibits a good section of these deposits. The rolled pebbles and sand occupy about fifteen feet of the upper part of the bank, and lie in a hollow or basin of chalk rubble; and wherever the chalk is accessible to observation along this margin of the Downs, it is invariably in a broken and ruinous state.

* Lee's History of Lewes and Brighthelmstone, 8vo. 1795, p. 554.

At Chimting Castle, about a half mile to the east of Seaford, the upper part of the cliffs is composed of a bed of sand, about fifty feet thick. Here a stratum of the ferruginous breccia, previously mentioned, is seen in situ, lying beneath the sand, and immediately upon the chalk. The sand is of a fawn-colour, passing into olive green; it contains numerous irregular veins, and concretions of mammillated ironstone. The puddingstone, or breccia, is precisely similar to that of Castle Hill, with which, there can be no doubt, it was once continuous. The flints that compose it present the same characters; some being rolled, others angular, and all of them either of a dark green or yellow colour externally. The bed of breccia, in some places, is nearly four feet thick; from this stratum the blocks distributed through the diluvial deposits were, no doubt, derived. The strata above named extend eastward about half a mile, and disappear near the Signal House; they dip to the west at an angle of from 10° to 20°.

Eastward of this place, the chalk has only a covering of ochraceous clay, and vegetable mould, and, with the exception of the blocks of breccia at Brighton, &c., previously alluded to, and a few insular patches of olive green sand in hollows of the chalk at Piddinghoe, I am not aware of the existence of any other decided examples of the Plastic clay in the south-eastern part of Sussex.

In the western division of the county, Professor Buckland observed a red variety of Plastic clay, in a small valley, at the village of Binstead, three miles west of Arundel; and also on the declivity of the hill by which the Binstead and Chichester road descends into Arundel.

The country around Chichester has a foundation of chalk, with a subsoil of fine red gravel, and pebbles, mixed with sand, loam, and chalk rubble. Furrows and wells in the chalk, filled with these materials, are commonly observable in the quarries near that city.

From the immense quantity of marine, and the almost entire exclusion of freshwater shells, it is manifest that of the strata of Castle Hill, the beds below the blue clay, with cerethites, can alone be considered as analogous to the freshwater formation of the Paris Basin. In the *Geographie Minéralogique des Environs de Paris*, the Plastic clay formation is divided as follows:—

Premier terrain d'eau douce. { Argile plastique. Lignites. Premier gres.

From the description of the strata around Paris, in the celebrated work above referred to, it appears, that although its illustrious authors consider the Plastic clay as a freshwater formation, yet, in its upper division, marine shells predominate. That, generally, in the beds resting immediately upon the chalk, organic remains do not occur*; that, in the middle portion, freshwater shells (Planorbis, Lymnæa, Paludina, Cyrena, &c.) and vegetable remains, commonly prevail; and that if marine

* "C'est ordinairement dans les parties inferieures que se trouve la *veritable argile plastique;* l'argile pure, infusible, ne renfermant aucun debris organique."—*Géog. Min.* p. 261.

exuviæ occur in this division, or in the lowermost bed of clay, upon the chalk itself, we may conclude that the inferior or freshwater deposits are wanting, and that the uppermost division reposes immediately upon the chalk. Such appears to be the case at Castle Hill; and we must therefore refer the lignite, red marl with vegetable remains, and the few freshwater shells which accompany them, to the *middle* of the series; and the clay, with cerithia and other marine shells that form so large a proportion of the cliff, to the *upper* division; the lowermost bed of the French series, the pure *argille plastique*, being absent. In the London clay at Bracklesham (the *Premier terrain marin* of M. Cuvier and Brongniart), a more striking analogy with the corresponding strata in France is observable; and the fossils are identical with those of the *Calcaire grossier* of the environs of Paris.

On the opposite coast of France, strata corresponding with those of Castle Hill, and Chimting Castle, occupy the same relative position.

In the perpendicular cliffs, under the lighthouse of St. Margaret, to the west of Dieppe, the following beds occur:—

1. Chalk.
2. Sand and sandstone in thick beds, containing concretions of the same substances.
3. Plastic clay, impure, and containing lignite much charged with pyrites. Also oysters, cerithia, &c., both in beds, and irregularly disseminated.
4. Alluvium.

These deposits M. Brongniart considers as identical

with the beds of the Plastic clay formation in many other parts of France; particularly at Marly, and in the Soissonnois, where similar organic remains occupy strata disposed in the same manner, and identical with those near Dieppe.

In the preceding sketch of the strata of the *Plastic clay* of Sussex, the geologist will immediately recognise the usual characters of that formation, which, " viewed on an extended scale, is composed of an indefinite number of sand, clay, and pebble beds, irregularly alternating; the distribution of the organic remains, like the alternation of the strata, being exceedingly variable: sometimes they occupy the clay; at other times the sand or pebbles; and very frequently are altogether wanting in both."*

* Geological Transactions, vol. iv.

CHAP. V.

THE CHALK FORMATION.

THIS formation includes the following deposits; namely,—

1. Upper Chalk, with Flints ..... } *Craie Blanche* of the French
2. Lower Chalk, without Flints .. } Geologists.
3. Chalk Marl .... *Craie Tufeau.*
4. Firestone, or Upper Green Sand .. *Glauconie Crayeuse.*
5. Galt, or Folkstone Marl.
6. Shanklin, or Lower Green Sand .. *Glauconie Sableuse.*

---

1. UPPER OR FLINTY CHALK.
2. LOWER CHALK.

THESE deposits form by far the most considerable and important divisions of the chalk formation, and constitute the most striking features in the geology of Sussex. As their investigation is highly interesting, we shall endeavour to elucidate the subject, by subjoining a brief notice of the course of the chalk through the south-eastern part of England, and the Continent.

We are informed by Mr. Townsend, " that the chalk hills are bounded by a line which stretches from south-west to north-east, and that within these limits they form three principal mountain ranges. The first, leaving Berks, runs north through Bucks, Bedfordshire, and Hertfordshire, into Cambridgeshire, by Dunstable, Hitching, Baldock, and Royston, to Gogmagog Hills, near Cambridge. The second, passing from Berkshire eastward, stretches through Surrey, where it forms the Hog's Back, that beautiful ridge which extends from Farnham to Guildford, and then appears at Boxhill. This branch forms the hilly country and the downs north of Reigate, Bletchingley, and Godstone. It enters Kent to the north of Westerham, and extends by Riverhead to Wrotham, south of Dartford, Rochester, Lenham, and Canterbury, to Folkstone and Dover. One division of this ridge is continued to the north coast of Kent, by Feversham, near Sheppey, Margate, and the North Foreland to Ramsgate.

"The third range, leaving Wiltshire and Berkshire, enters Hants, and to the south passes round Petersfield, then, stretching to the east, forms a barrier against the sea along the coast from Chichester, constituting the South Downs; and ranges from Maple-Durham, Houghton, Steyning, and Lewes, as far as Beachy Head."*

Insular parts occur in the Isle of Thanet and Isle of Wight.†

In France, the chalk prevails on the skirt of the western boundary of Mount Jura, extending nearly in a direction from S.E. to N.W., and covering a space of at least 210 miles long and 50 broad.

* Townsend's Moses, vol. i. page 142.

† Smith's Strata. For a more particular account of the range and extent of the chalk formation, vide Phillips's Geological Outlines, edition of 1822, p. 77.

Chalk also occurs in Ireland, Spain, Denmark, Sweden, Germany, and Poland.*

The thickness of the chalk formation varies considerably in different parts of its course. Near Royston, it attains an elevation above the sea of 481 feet; south of Dunstable, it is 994 feet; south of Shaftesbury, 941 feet; between Lewes, in Sussex, and Alton, in Hampshire, various parts of the range rise to the height of between 800 and 900 feet; and between Alton and Dover, between 700 and 800 feet.† In the Sussex range, Ditchling Beacon, which is the highest point, is 856 feet above the level of the sea.

* Dr. Berger, Geological Transactions, vol. i. p. 14.

It is supposed that the white chalk does not exist in North America, although the discovery of extensive beds of marl and sand, in which the same genera of fossil shells prevail as in the chalk formation of Europe, leaves no doubt that these deposits belong to the Cretaceous group. To Dr. S. G. Morton, Secretary to the Academy of Natural Sciences of Philadelphia, science is indebted for this important and highly interesting discovery. This gentleman has figured and described, in Professor Silliman's American Journal of Science, Belemnites, Hamites, Nautili, Ammonites, Echini, &c. which an European geologist would immediately recognise as inhabitants of the chalk. The specimens in my collection, and for which I am indebted to the liberality of Dr. Morton, leave no doubt on the subject. The fossils from the limestone superimposed on the marls of New Jersey more closely resemble those of Maestricht; there are more littoral shells, and fewer pelagic, than in the inferior beds. It is much to be desired that Dr. Morton should favour the world with a more particular account of his invaluable researches. Since the above was written, Mr. Constable, of Dover's Green, near Reigate, has obliged me with an extract from his journal made during a tour through North America many years since, in which he mentions having observed white chalk in the wilderness in Alabama. Mr. Constable's account is so interesting, that I have transmitted it to America, in the hope that Dr. Morton will be able to examine the locality, and determine the question.

† Phillips's Outlines, 2d edition, 1816.

The mountain ranges formed by the chalk are characterised by their smooth and unbroken outline, and are generally covered with a short verdant turf.*

The earlier inhabitants of this island, either from choice or necessity, fixed their settlements on the elevated ridges and platforms of this formation; and vestiges of their sepulchral mounds are still visible, scattered here and there over the Downs. Stonehenge and other druidical temples are situated upon it, being composed of immense blocks of the siliceous sandstone, that occurs in the form of boulders on various parts of its surface.

The description of the South Downs, inserted in a former part of this work, will sufficiently explain the range and extent of the Upper and Lower Chalk of Sussex. Varying in altitude from 300 to upwards of 800 feet, this chain of hills extends from Beachy Head along the coast to Brighton, from whence it stretches through the centre of Western Sussex into Hampshire. On the north it presents a precipitous escarpment to the Weald, but its southern side descends with a gentle slope, and on the south-west is lost beneath the beds of the Isle of Wight basin; while the south-eastern part forms a line of chalk cliffs of considerable extent.

* "In Champagne, in France, there are immense plains of chalk absolutely destitute of vegetation, except where patches of the *Calcaire grossier* occur as islands, or oases, in the midst of these deserts. Many parts of this tract have, perhaps, not been visited for ages by any living being, no motive existing that could induce any one to wander there. This chalk is said to contain 11 per cent. of magnesia, to which the barrenness of the soil is supposed to be owing."—*Geological Transactions*, vol. ii. p. 175.

The general dip or inclination of the strata of the Sussex range is to the south-east; in many instances, however, the influence of local causes has occasioned exceptions; and the beds which flank the transverse valleys generally diverge from the openings, as if they had once formed an anticlinal axis.

The face of the chalk is marked with fissures or wells, &c. and scooped into deep hollows, furrows, and basins, which are more or less filled with tertiary sand and gravel. Masses of stalactitical and stalagmitical carbonate of lime, which must have originally been formed in caverns or grottoes of the chalk, also occur in the most elevated parts of the Downs. In numerous places on the sides and at the base of the Downs, quarries have been opened, and kilns erected, for converting the chalk into lime, of which immense quantities are annually consumed by the Sussex agriculturists. These partial sections in the interior, together with the line of coast from Brighton to Beachy Head, afford ample opportunities for the examination of the geological structure of this interesting chain.

We now proceed to a more particular survey of the deposits included in the present section; but, before entering on their investigation, it may be necessary to offer a few remarks, upon the substance of which they are principally composed.

CHALK* is a mineral too well known to require

* Various conjectures have been offered respecting the probable origin of chalk, and the mode of its formation. Patrin† supposed that it was the production of three different causes:—

† Dict. d'Histoire Naturelle, tom. vi. p. 472.

description, yet its characters are such as could not fail to excite attention, if less frequently presented to our notice.

The Sussex chalk varies in colour from pure white to a bluish grey, and differs considerably in its coherence and composition. It has an earthy fracture, is meagre to the touch, and adheres to the tongue; it is dull, opaque, soft, and light, its specific gravity being about 2·3. It is composed of lime and carbonic acid, and contains an inconsiderable proportion of silex and iron.

The harder varieties of this substance were formerly in great request for building, and, when protected from the influence of the atmosphere by a thin casing of limestone or flint, proved very durable. The ruins of the priory of St. Pancras, near Lewes, which have stood nearly 800 years, afford a remarkable instance of this kind; the interior of many of the walls are six feet thick, and are entirely formed of chalk, the outside having a facing of Caen-stone and squared flints. At pre-

---

1. Animal earth, proceeding from the decomposition of organic bodies.
2. Calcareous lava ejected by submarine volcanoes.
3. Detritus of calcareous mountains.

Delametherie imagined it to have been deposited by water in a state of great agitation.*

In Ireland, the chalk acquires a degree of hardness equal to that of compact limestone. In its geological position, and in the nature of its fossils, it corresponds, however, with that of England, with which it is considered to be entirely identified. In many places it is covered by basalt, and its hardness is probably attributable to the high temperature to which it has been subjected from this cause. It contains *echinites*, *terebratulæ*, *ammonites*, and *belemnites*.—*Geological Transactions*, vol. iii. pp. 129. 169.

* Journal de Physique, tom. lxxx. p. 37.

sent, chalk is seldom used in architecture, except in the construction of vaults, cellars, and other subterranean works.

The upper portion of the chalk formation in Sussex is naturally separated into two divisions; viz. the CHALK with FLINTS, and the CHALK without FLINTS.* The lower strata also form two well-marked deposits: the CHALK MARL, which contains a large proportion of argillaceous earth; and the FIRESTONE, or malm-rock, that principally differs from the marl in having a considerable intermixture of green and grey sand. The Gault or Galt, (the blue marl of Folkstone), and the Shanklin or Lower Green Sand, may likewise be ranked as subordinate members of the chalk formation, notwithstanding the dissimilarity of their mineralogical characters; for their organic contents correspond in so many particulars with those of the marl, firestone, and chalk, as to warrant the conclusion that they were deposited by the same ocean, and that the entire series constitutes but one geological epoch or formation.† We proceed to take a

* Even this character, so constant and perfect as it is in the south of England, is not universally maintained. " At Havre on the opposite coast, the lower chalk contains an abundance of flint and chert nodules where it passes into the upper green sand."—*Manual of Geology, by Mr. De la Beche.*

† See the excellent observation on the relative value of the mineralogical and zoological characters of rocks, by M. Brongniart, Essai Minéralogique; note to pp. 327. et seq.

Messrs. Conybeare and Phillips have separated not only the galt and firestone from the chalk, but also the chalk marl, including the whole under the general name of " Chalk Marl." These gentlemen, however, remark, that, in applying the term formation to these subdivisions, they merely use it as a convenient designation for a large assemblage of similar strata.

rapid view of the characters and distribution of these deposits, premising that the white chalk forms the Downs; the argillaceous and arenaceous strata emerging from their base along their northern escarpment, and constituting the low banks and terraces of the district which intervenes between the sand and the chalk hills, except in the west of Sussex, where the Shanklin sands rise into hills, equal in altitude to the neighbouring chalk.

I. CHALK with FLINT.

From the situation occupied by this division of the chalk, it has suffered most extensively from the effects of those catastrophes to which we have already alluded. In almost every part of the interior of the country, its ruins may be seen in the form of beds of gravel, or partially rolled flints. What its original thickness may have been, cannot be ascertained; but the occurrence of large blocks of stalagmites and stalactites on the summits of the highest hills proves that chalk caverns formerly existed above the most elevated points of the South Downs.

The chalk of this subdivision is generally of a purer white, and of a softer texture, than the inferior strata; but in other respects, presents no sensible difference. It is regularly stratified, and partakes of the general inclination of the other divisions of the series. It is separated by horizontal

---

The strata of Sussex thus constitute three grand formations; viz.

1st, *Partly freshwater and partly marine.* The strata above the chalk.

2d, *Marine.* The Chalk and Shanklin sand inclusive.

3d, *Freshwater.* The Weald clay, Hastings sands, and Ashburnham beds.

layers of siliceous nodules into beds that vary from a few inches, to several feet in thickness, and which, in some localities, are traversed by obliquely vertical veins of tabular flint, that may be traced for many yards without interruption. These are sometimes disposed horizontally, and form a continuous layer of thin flint, of considerable extent.

The nodular masses of flint are very irregular in form, and variable in magnitude: some of them scarcely exceeding the size of a bullet, while others are several feet in circumference. Although thickly distributed in horizontal beds or layers, they are never in contact with each other, but every nodule is completely surrounded by the chalk. Their external surface is composed of a white opaque crust, consisting of an intermixture of chalk and silex, probably formed by a combination of the outer surface of the nodule with its investing matrix, while the former was in a soft state. Internally they are of various shades of grey, inclining to black, and often contain cavities lined with chalcedony and crystallised quartz.

When first extracted from the quarry, flint is brittle, has a conchoidal fracture, and feeble lustre; thin fragments are translucent. Its specific gravity is 2·594. According to the analysis of Klaproth, it consists of

Silex -	98	Oxide of iron	0·25
Lime -	0·5	Water -	1·
Alumine -	0·25		

It is infusible, but upon being submitted to a great heat, becomes white and opaque. By exposure to the atmosphere, it undergoes considerable change, and assumes a yellow or ferruginous colour, an appearance commonly exhibited by the flints of our ploughed lands. When in contact with ochraceous clay, or sand containing iron, it frequently attains a dark carnelian colour externally, the interior being of a lighter shade: of this kind, numerous beds occur in the parish of Barcombe.

Flints so commonly enclose the remains of sponges, alcyonia, and other zoophytes, that some geologists are of opinion that the nucleus of every nodule was originally an organic body.* That this has been the case, in most instances, is very evident; and in Sussex, there are comparatively but few flints that do not possess traces of zoophytal organization. These nodules oftentimes exhibit not only the outline of the original zoophyte, but also its internal structure, preserved in the most delicate and beautiful manner that can be conceived. In some examples the zoophyte has undergone decomposition, and the space it occupied been partially filled with an infiltration of agate, chalcedony, and crystallised quartz.

Although, even in the present advanced state of chemical science, we are unacquainted with the process by which silex may be dissolved in water, yet that its solution was formerly effected by natural causes on a very extensive scale, the siliceous nodules, whose history is the subject of these remarks, afford the most conclusive evidence. At the pre-

* "So far as my observation extends, zoophytes appear universally to have formed the nuclei of nodulated and coated flints."—*Townsend's Character of Moses.*

sent moment, nature, in her secret laboratories, is still carrying on a modification of the same operation; of which we have remarkable instances in the boiling springs of the Geyser, in Iceland ‡, and of Carlsbad, in Bohemia.† Nor is a high temperature absolutely essential to the solution of silex in water, since this earth occurs in a large proportion in the mineral waters of our own island*, and also enters into the composition of the epidermis of various plants of the bamboo tribe, and of the English reeds and grasses. The epidermis of the Equisetum hyemale, or Dutch rush, consists almost entirely of silex. §

There is scarcely a single fact in geological science, which has given rise to so many unsatisfactory conjectures as the formation of the siliceous nodules of the chalk. Upon this interesting subject, I shall content myself with offering a condensed view of the theory proposed by Dr.

* The deposition of siliceous tufa, or chalcedony, formed by the boiling springs of the Geyser in Iceland, are well known: these waters contain 31.38 of silex per gallon. Vide *Travels in Iceland*, by Sir George Stewart Mackenzie, Bart. 4to. Edinburgh, p. 389.

† According to the experiments of Klaproth, the spring at Carlsbad contains 25 grains of silex in 1000 cubic inches of water. Mr. De la Beche observes, " We find silex is held in solution by thermal waters, which also, as in the case of those of St. Michael's in the Azores, may contain carbonate of lime. No springs, or system of springs, that we can imagine, are likely to have produced this great deposit of chalk so uniformly over a large surface. But although springs, in our acceptation of the term, could scarcely have caused the effects required, we may, perhaps, look to a greater exertion of the agents which now produce thermal waters for a possible explanation of the observed phenomena."—*Geol. Manual*, 1st edition.

‡ The mineral waters of Bath contain twenty grains of silex in ten pints and a half. *Nicholson's Journal*, vol. iii. p. 403.

§ Organic Remains, vol. i. p. 328.

Buckland, which, if not satisfactory, is at least more worthy attention than any other hypothesis that has been suggested.

"It does not," he remarks, "appear possible that the flints could have been formed by infiltration into pre-existing cavities, according to the theory of Werner, like the regularly disseminated geodes of the trap rocks; since this hypothesis, in the case of chalk, would imply the anomaly of there having once existed uniformly over many hundred square miles, as many strata of air bubbles as there are of flint, alternating with the chalk; and of which air-holes not one was left empty, or partially filled; whilst, on the other hand, many of the nodules could not have been formed in such air-holes, as they entirely derive their shape from some extraneous bodies affording a nucleus to the silex that has incrusted them.* Assuming that the mass which is now separated into beds of chalk and flint, was, previously to its consolidation, a compound pulpy fluid, and that the organic bodies now enveloped in the strata were lodged in the matter of the rock, before the separation of its calcareous from its siliceous ingredients, the bodies thus dispersed throughout the mass would afford nuclei, to which the flint, in separating from the chalk, would, upon the principle of chemical affinity, have a tendency to attach itself. The chalk and flint proceeded through a contemporaneous process of consolidation; the separation of the siliceous from the calcareous ingredients having been modified by attractions, which drew to certain centres

* Geological Transactions, vol. iv. p. 422.

the particles of the siliceous nodules, as they were in the act of separation from the original compound mass. The distances of the siliceous strata must have been regulated by the intervals of precipitation of the matter from which they are derived: each new mass, as it was discharged, forming a bed of pulpy fluid at the bottom of the then existing ocean, which, being more recent than the bed produced by the last preceding precipitate, would rest upon it as a foundation similar in substance to itself, but of which the consolidation was sufficiently advanced, to prevent the ingredients of the last deposit, from penetrating or disturbing the productions of that which preceded it." *

That the beds of chalk and flint were deposited periodically, cannot admit of the slightest doubt. Specimens are not unusual, in which angular fragments of black flint, that could not possibly have been originally formed in their present state, are imbedded in chalk. An example of this kind, in my possession, contains several portions of flint which are as sharp and translucent as if recently broken, and entirely destitute of the external opaque crust invariably seen in the perfect nodules; these are imbedded in, and separated from each other by, the chalk. It is sufficiently obvious that the nodule from which these pieces were derived, must have been displaced and broken, subsequently to its original formation, and the fragments afterwards enveloped in another and more recent stratum of chalk. In this country the chalk very rarely contains traces of older deposits: the only instances of extraneous rocks that have come under

* Geological Transactions, vol. iv. p. 420.

my observation are pebbles of quartz, and some fragments of green schist.

The line of coast from Brighton to Beachy Head exposes an interesting vertical section of the upper chalk, exhibiting almost every variety of character hitherto observed in the beds of that deposit.

At Brighton, the cliffs, as before remarked, are composed of an accumulation of diluvial substances, resting upon the solid chalk, which there constitutes the sea shore, and continues to Rottingdean. From thence to Newhaven the cliffs are nearly perpendicular, and on the western side of the harbour rise into an irregular elevation, called Castle Hill, the upper part of which is composed of the numerous beds of the plastic clay formation previously described: the lowermost consisting of the flinty chalk. On the opposite side of the river, a low mound of chalk, capped with a bed of plastic clay and ferruginous breccia, appears at Chimting Castle. Proceeding eastward towards the Signal-house, near Seaford, the chalk rises to a considerable height, and forms a majestic line of cliffs from thence to the embouchure of Cuckmere river; from this place they extend eastward, and terminate in the magnificent promontory of Beachy Head, which is nearly six hundred feet above the level of the sea* (*vide Vignette of the titlepage*). Along this

* The following circumstance is too singular to be omitted. One of those prodigious falls of the chalk cliffs, which make a residence near them frequently so dangerous, occurred at Beachy Head a few years since. The clergyman of East Dean was walking on the brink of the precipice, when he perceived the ground to be sinking from under him, and although he had the presence of mind instantly to rush from the

line of coast, Ammonites of a large size, Plagiostoma, Terebratulæ, Echinites, and other productions of the chalk, may be obtained.

The sections in the interior of the country are entirely artificial: of these, the following are the most interesting that occur in the south-eastern part of the county.

Holywell quarry, near Eastbourne, contains Echinites, Plagiostoma, Inocerami, Terebratulæ, the remains of Fishes, &c.

Alfriston chalk-pit; remarkable for crystallised carbonate of lime of considerable purity.

Cliff Hills. The pits formed on the sides of this insulated portion of the chalk hills, produce a great variety of fossil shells and zoophytes, the remains of fishes, and the vertebræ and bones of unknown animals. In some of these quarries, after a recent fall, the chalk presents a remarkable appearance; the newly exposed surface is of a brown colour, and uniformly marked with fine vertical striæ, giving to the mass a fibrous appearance. Small conical portions of the chalk sometimes partake of the same character, and some specimens closely resemble calcareous fossil wood. In every instance, however, this structure is confined to the surface, and does not affect the interior of the chalk. In all probability it has been produced by a subsidence of the strata, which caused

---

impending danger, a deep chasm had formed at some distance from the edge of the cliff, over which he had escaped but a few moments, before the mass of chalk upon which he had been standing, to the extent of 300 feet in length, and eighty in breadth, fell with a tremendous crash into the sea. Geological Transactions, vol. ii. p. 191.

them to slip over each other, before they were entirely consolidated.* In a quarry, now disused, and which was one of the most productive in fishes and zoophytes, there is a remarkably deep well, or conical cavity, filled up with the reddish sand and coarse sandstone of the tertiary beds.

South-street pit, near Lewes, affords a fine section of the flinty chalk, exceeding two hundred feet in height. An irregular canal or dike, varying from two to eight feet in diameter, traverses this quarry, in an oblique direction. It was noticed, many years since, at the northern extremity of the pit; and subsequent falls of chalk have from time to time exposed its course towards the centre, from whence it now appears to proceed easterly; a section of it is still perceptible at an elevation of a few feet. In some parts this cavity was almost empty, and in others nearly filled by sand, clay, and ochre of a light chocolate colour. This canal or dike has probably been formed by a subterranean current of water, the substances it contains being evidently alluvial. South-street pit is also remarkable, as being the only known locality of the detached octaëdral sulphuret of iron. It contains the scales, teeth, &c. of fishes, and numerous shells and corals.

Beddingham pit is situated on the side of the Downs, about a mile distant from the village. In ascending the hill, the grey marl, lower chalk, and flinty chalk, are passed over in succession.

* An appearance somewhat similar occurs in the limestone beds of Derbyshire, and is provinciallv termed "*slickensides.*"

The pit is between twenty and thirty feet high, and consists of, —

1. Vegetable mould intermixed with chalk rubble, — 1 foot.
2. Chalk rubble, — 3 feet.
3. Flinty chalk, — 20 feet.

The chalk is stratified in horizontal beds from two to four feet thick, and these are separated in some instances by layers of flints, and in others by chalk rubble; flints are also irregularly disposed throughout the mass. This spot is peculiarly interesting, from the vertical fissures which are every where observable, being partially filled with broken chalk and flint, cemented together by crystallised carbonate of lime, of a light amber colour. The sides of the fissures are incrusted with the same substance, which has insinuated itself into the crevices of the surrounding chalk, and also forms irregular concretions in the cavities of the flints. The surface of these stalactitical depositions of calcareous spar is frequently covered with delicate undulations, as if the water had been suddenly congealed, while in a state of agitation.

Piddinghoe. This pit lies on the road-side, near the village of the same name; it is remarkable for the purity and softness of the chalk, and for the numerous vertical and oblique veins of tabular flint by which it is traversed. These veins are of a most extraordinary character; for although the flint retains its original form and situation, yet, upon examination, it is found to be cracked and shivered in every direction. The fractured flint falls to pieces upon being removed from the chalk, but in some instances it is held together by sulphuret of iron, forming a conglomerate of silex and pyrites, of a very singular appearance. The phenomenon here remarked is not, however, confined to this quarry, but may be observed in several chalk-pits near Lewes and Brighton.

Sir Henry Englefield was the first who directed the attention of geologists to this subject. In a paper read before the Linnean Society, he notices several beds of shattered flints, which occur in a chalk-pit at Carisbrook, in the Isle of Wight; and, after describing their appearance and situation, proceeds to offer some conjectures upon the probable cause of their destruction. This he supposed might have been occasioned by some sudden shock or convulsion, "which in an instant shivered the flints, though their resistance stopped the incipient motion; for the flints, though crushed, are not displaced, which must have been the case, had the beds slid sensibly." *

Offham pit is nearly two hundred feet high, and exhibits a good section of the Sussex chalk. It contains the large fibrous bivalve, the fragments of which are so frequently met with in every locality; teeth and palates of fishes, and numerous zoophytes. It is the only locality near Lewes in which the *Marsupites* have been discovered. South of this place, in a bank on the road-side, the chalk is covered by a bed of ochraceous clay, and, where in contact with the latter, the chalk and

* Linnean Transactions, vol. vi p. 108.

flints are marked with regular stripes of yellow, bluish grey, and brown. This singular appearance extends into the substance of the chalk, but does not penetrate beyond the external crust of the flints: similar specimens sometimes occur in the pit in South-street.

Clayton pit. This locality produces Inocerami, Nautili, Plagiostoma, Terebratulæ, Marsupites, &c.

Falmer. An excavation made on this side of the road, leading from the village, towards Bormer, is particularly interesting from the evident proofs it exhibits of the changes the strata have suffered since their original deposition. The pit is about twenty feet deep, and contains the following beds, beginning with the lowermost:—

1. Chalk with horizontal layers of large flints, —6 feet.

2. Chalk much broken, containing interspersed flints,—10 feet.

3. Ochraceous clay and flint pebbles,—from 2 to 4 feet.

From the upper part of the pit, several fissures of an irregular shape, and from three to six feet in diameter, extend through the broken chalk to the more solid beds beneath. Some of these cavities are of an inversely conical form, and others are nearly cylindrical. They are filled with ochraceous clay, rolled flints, and rounded masses of a conglomerate, consisting of pebbles and fragments of chalk, held together by a ferruginous cement. An appearance somewhat analogous is observable on the north side of the chalk-hill on which the church of St. John *sub Castro*, in Lewes, is situated.* The broken chalk in Falmer pit is in very small pieces, the angles of which are perfect; a proof that, although minutely divided, it has not suffered by attrition. The sides of the valleys of the South Downs are universally composed of chalk, of a character precisely similar; an appearance which, in all probability, has resulted from the ruin of the chalk cliffs having accumulated at their base.

Brighton pits. There are several chalk-quarries in the vicinity of this celebrated watering-place; but of these, one only is particularly worthy of notice.† The pit alluded to is situated near the church, and affords an excellent example of that fractured state of the chalk, which has been previously mentioned. It is thus described by Sir H. Englefield:—"The upper part of this chalk is in separate masses, not perfectly rubble, but with all their tender angles sharp, exactly as if just broken to pieces to put into the limekiln, and quite clean, nearly of a size, and almost without any chalk powder mixed with them." Some remarkable veins of shattered flints occur in this quarry.‡

Preston. The quarry is extensive, and lies im-

* A similar fact is mentioned by MM. Cuvier and Brongniart. These naturalists remark, that in the beds of the lower marine formation, and particularly in those of Liancourt, natural wells of considerable size are sometimes found, filled with ferruginous and sandy clay, and waterworn siliceous pebbles. Geolog. Trans. vol. ii. p. 208.

† This chalk is very pure; a specimen, of the specific gravity 2·34 was composed of,

	Carbonic acid	43·4
	Lime - -	56·0
	Silica - -	0·6

‡ Linnean Transactions, vol. vi. p. 108.

mediately behind the village; it formerly produced numerous remains of fishes, palates, teeth, &c., but is now seldom worked. It is, however, deserving of attention, on account of several thin veins of pure flint that fill up vertical fissures in the chalk, and which, to use the language of Sir H. Englefield, "appear exactly as if the flint, not being quite hard when the fissures took place, had been squeezed out of the beds, and had run into the fissures as soft pitch would do: I do not mean that this was the case, but merely to describe the appearance."*

Steyning chalk-pits. These produce belemnites, plagiostoma, dianchoræ, teeth, palates, &c. The sulphuret of iron found in these quarries is of a very singular form, being cylindrical, with a small projection at both extremities. Chalk-pits near Arundel abound in fishes, palates, teeth, marsupites, and others of the most interesting organic remains.

MINERALS.

In the upper chalk the minerals are but few in number, and, like the lower chalk, it contains but one metalliferous ore.

1. Crystallised quartz: this is of frequent occurrence in the cavities of siliceous nodules, shells, &c. The form of the crystals is that of a six-sided pyramid, their colours varying from a reddish brown to a light blue, amber, grey, and white.

* Linnean Transactions, vol. vi. p. 108.

2. Chalcedony is often found occupying the hollows of flints, and is either mammillated, botryoidal, or stalactitical. It sometimes forms the constituent substance of corallines, alcyonites, and other zoophytes, displaying in the most delicate manner the complicated structure of the originals. Its colour is of various shades of grey, azure, and pearl white, and in many examples it is beautifully translucent: specimens are not uncommon in which the surface of the mammillated chalcedony has received an investment of crystallised quartz.

The stalactitical and botryoidal varieties are confined to those nodules which retain a part of the original zoophyte. In some instances the flint passes insensibly into chalcedony; in others the line of separation is most distinctly marked; but in all, there is sufficient evidence that the chalcedony and quartz were deposited by infiltration, and must have passed through the substance of the flint.

On this subject it has been remarked, that "although, in the present compact state of the matter of flint, it is not easy, though possible, to force a fluid slowly through its pores, yet it is probable that before its consolidation was complete, it was permeable to a fluid whose particles were finer than its own; and that the particles of chalcedony, whilst yet in a fluid state, being finer than those of common flint, did thus pass through the outer crust to the inner station they now occupy; where they also allowed a passage through their own interstices to the still purer siliceous matter, which is often crystallised in the form of quartz in the centre of the chalcedony, and is so entirely surrounded by it, that it could have no access to its present place, except through the substance of the chalcedony, and the flint enclosing it."*

3. Calcareous spar. This mineral is abundant in the fissures and hollows of the chalk, and forms the constituent substance of the shells and echinites. It is of various shades of amber colour, brown, and pearl white; the variety into which the shells and echinites are converted is opaque, and has an oblique fracture. The other modifications generally possess some degree of transparency; in some of the larger bivalves, of the genus Inoceramus, the structure is fibrous.

The crystals of carbonate of lime are of various forms: the most usual are the rhomboidal, columnar, and acicular. The first occurs abundantly in cavities in the chalk, immediately beneath the turf, on Plumpton Plain; and it is worthy of notice, that the hollows it occupies have manifestly been formed subsequently to the consolidation of the chalk. In Western Sussex, branched cavities in the chalk, apparently occasioned by the decay of ramose zoophytes, are incrusted by this variety of calcareous spar.†

Of the columnar crystals, some fine specimens were brought to view by the tremendous fall of the cliffs near Beachy Head, that happened a few years since. These occurred in large masses of a yellowish colour, and the crystals when detached were semitransparent; Plumpton Plain, Alfriston

* Geological Transactions, vol. iv. p. 419.

† From the correspondence of J. Hawkins, Esq.

chalk-pit, and some other localities, have produced similar examples.

Obtuse rhomboidal crystals, of great beauty, have been found in a chalk-pit near Alfriston: their colour is of a delicate pearl white, and in their general appearance they resemble the double-refracting spar of Iceland, except in their inferior degree of transparency. The cavities of echinites are sometimes lined with rhomboidal crystals of carbonate of lime, disposed in lines parallel with the sections formed by the areæ of the shell; and the inner surfaces of the terebratulæ are frequently frosted over with drusy crystals of the same substance.

4. Sulphuret of iron, or iron pyrites, in subglobular and irregular masses, is very common in the upper chalk. The external surface of the specimens is invested by crystals of a pyramidal, octaedral, or cubo-octaedral form; and their interior exhibits a radiated structure, possessing a brilliant metallic lustre. When broken, and exposed to the action of air and moisture, they undergo decomposition with great rapidity; and even in cabinets, frequently form an efflorescent sulphate of iron, and crumble into dust. This mineral occasionally encloses flints, shells, echinites, &c., and frequently fills up the cavities of the latter. A specimen in my possession exhibits, on the upper side, a sharp cast of the interior of a spatangus; and its base is covered with an elegant group of quadrangular pyramids, evidently the terminations of octaedrons, with their inferior angles concealed.

The lower beds of the flinty chalk, in South-street, contain detached crystals of sulphuret of iron, remarkable for their neatness and elegant figure. They are usually regular octaedrons, having their planes studded with small quadrangular pyramids; but some examples occur in which the solid angles are replaced by quadrangular planes, forming a crystal with fourteen sides.

LOWER CHALK.*

The absence of siliceous nodules, and the superior hardness of the chalk, distinguishes this deposit from that which lies above it.

Its colour is of a light grey, enclosing masses of pure white. It forms the low elevations at the foot of the Downs; and, as the situations it occupies are generally easy of access, a considerable number of quarries have been opened in different parts of

* In some parts of England the chalk admits of a more minute division. The cliffs in the vicinity of Dover, described by Mr. Phillips (Geological Transactions, vol. v. p. 18.), are separated by that gentleman into the following, viz.: —

I. Chalk with numerous flints, 350 feet thick; which is subdivided into —
  1. A bed with few organic remains.
  2. Chalk with interspersed flints, consisting chiefly of organic remains, in which numerous flints of peculiar forms are interspersed, and a few beds of flint.

II. Chalk with a few flints: this stratum is about 130 feet thick.

III. Chalk without flints, 140 feet thick, consisting of—
  1. A stratum containing very numerous and thin beds of organic remains, 90 feet thick.
  2. A bed 50 feet thick, with few organic remains.

IV. Grey chalk, estimated at 200 feet in thickness.

its course. It is regularly stratified, the lines of separation being composed of a softer chalk, that in some places contains so great a proportion of argilla as to form veins of marl. The latter substance also occurs in transverse and vertical veins, in which the remains of fishes are more frequent than in the more solid strata.

The general inclination of the beds is towards the south-east, at an angle of from 5° to 15°. Their total thickness has not been determined, but is probably not less than 200 feet. A well sunk on the side of the hill, near Glyndbourn, passed through 120 feet of the lower chalk only. The lowermost beds were of a deeper grey than the upper, but presented no other material variation. The cliffs that extend from near Beachy Head to Southbourn expose this bed at their base, and afford considerable facility for its investigation.

Near Lewes, the lower chalk occurs in the quarries at the foot of Malling Hill, Southerham, Glynd, Glyndbourn, Swanborough, Plumpton, &c.; and in other parts of the county, along the northern edge of the Downs, reposing immediately on the grey chalk marl.

The quarry at Southerham is remarkable for the inclination and direction of its beds: it is situated on the east side of the road, on the south-western extremity of Cliff Hills. It is about thirty feet high, and contains from eight to ten layers of chalk, the latter varying in thickness from one to eight feet, being separated from each other by intervening seams of friable chalk marl. The strata exhibit decided proofs of having suffered considerable displacement; they are inclined obliquely toward the north, at an angle of from 20° to 30°, their planes being depressed towards the west. Northward from this spot, at the distance of about 300 yards, the upper chalk is exposed in the pit of Messrs. Hillman, in South-street, and here the strata are slightly inclined to the south. The hill in which both these quarries occur presents a smooth unbroken outline, conveying no indication of the changes that have taken place beneath its surface.

MINERALS.

1. Sulphuret of iron is the only metallic substance that occurs in the lower chalk; and of this mineral some elegant crystals, of a reddish or yellowish brown colour, have been discovered, in the quarries at the foot of Malling Hill.

They consist of nine or ten quadrangular columns, formed of octaëdrons piled upon each other: these proceed from one common centre, and each terminates in a quadrangular pyramid.

The lower chalk, near Beachy Head, contains small cylindrical masses of pyrites of a steel grey colour, that possess a very brilliant lustre; their surface is generally invested with pyramidal crystals, having their solid angles replaced by quadrangular planes.

CHAP. VI.

ORGANIC REMAINS OF THE UPPER AND LOWER CHALK.

THE organic remains of the chalk strata are very numerous; but, notwithstanding the considerable and important additions which modern discoveries have made, the fossil productions of these extensive deposits are still but imperfectly known.

The fossils of the French chalk have been described by MM. Cuvier and Brongniart *; those of the English, by Mr. Parkinson †, W. Phillips, and others; and the organic remains of the chalk of Yorkshire are admirably figured in Mr. John Phillips's highly interesting geology of that county.‡

The contents of the Sussex beds will be found to differ in some respects from those here mentioned; while many species of fossils, described by

* Essai sur la Géographie Mineralogique des Environs de Paris, Par MM. G. Cuvier et Alex. Brongniart, p. 11.

† Geological Transactions, vol. i. p. 344.

‡ Illustrations of the Geology of Yorkshire, 4to. 1829.

Miss Benett's beautiful and scientific "Catalogue of the Organic Remains of Wiltshire" must not be omitted: it contains a list of all the species of the Wiltshire chalk. It is much to be regretted that the amiable and highly talented authoress reserved it for private distribution.

Mr. Sowerby's beautiful and faithful representations of the shells of the chalk, in his Mineral Conchology, are too well known to require mention.

the authors above named, are unknown in this district. In their mode of preservation, however, a perfect correspondence exists in the productions of different localities. They are for the most part remarkably entire, the delicate coverings of the crustacea, the spines of the shells, &c. remaining unbroken; in short, their appearance, as Mr. Parkinson justly remarks, "warrants the conclusion, that they have been enveloped by the chalk, while living in their native habitats, and that this was effected in the tranquil depths of a profound ocean."

In every instance the shells, echinites, madreporites, and encrinites are converted into calcareous spar, their cavities being filled with chalk, flint, or sulphuret of iron.

The remains of the softer zoophytes occur in the form of chalky casts, tinged with a yellowish or reddish oxide of iron: this appearance, which facilitates the separation of the fossils from the chalk, results from the decomposition of pyrites. The vertebræ and bones are soft and friable; but the teeth and palates are finely preserved, and have the natural polish of the enamel, heightened by an impregnation with iron. The scales and fins of fishes, and the coverings of the crustacea, are changed into a brown substance, which is exceedingly brittle, and fades upon exposure to the air.

The zoological characters of the chalk prove unquestionably that almost the whole of the animals and vegetables, whose remains it entombs, were inhabitants of the sea; and the proportion of those shells which are pelagian, or, in other words, inhabited deep waters, is so great as to show that the strata, as before observed, were formed at the bottom of a profound ocean. The abundance of the Ammonites, Hamites, Turrilites, Nautili, and other chambered shells (at once the habitations and swimming apparatus of the mollusca to which they belonged), confirms this conclusion. The Belemnites, Echini, Terebratulæ, Plagiostoma, Inocerami, Spongiæ, Alcyonia, and other related zoophytes; Caryophylleæ, Turbinoliæ, and other simple stellular corals; Marsupites, and a few of the Crinoideæ, compose the principal fossil remains of the population of the once extensive ocean of the chalk. The species already ascertained in the upper and lower chalk exceed 300; those from the Sussex beds are enumerated in the tabular arrangement at the end of this volume; and we shall only notice, in a cursory manner, in this place, the most interesting of the organic remains that have been discovered in the chalk of the South-East of England.

VEGETABLE REMAINS.

Of marine plants, traces of *Confervæ*, and *Fuci*, occur in the lower chalk; a specimen of *Confervites fasciculata* (of M. Adolphe Brongniart) has been noticed in flint; and a fine species of Fucus in chalk, which I have named in honour of the distinguished author of the *Végétaux Fossiles*, *Fucoides Brongniarti*.

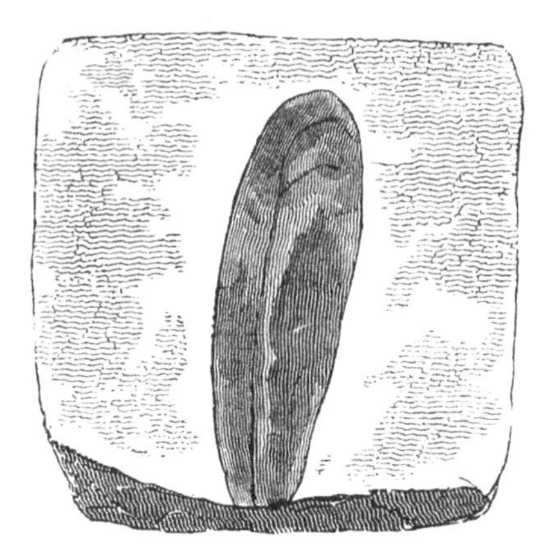

Fucoides Brongniarti.

Dicotyledonous wood is found imbedded in flint, and appears waterworn, as if drifted: it is commonly marked with perforations from teredines and fistulanæ. It also is enclosed in chalk, and in that state is very friable, and of a deep reddish brown colour; but the more compact specimens have the appearance and texture of Bovey coal.

ZOOPHYTES.

Although the remains of this class of organized beings are very abundant in the chalk, they are referable, comparatively, to but few genera and species. The stony corals are for the most part transmuted into calcareous spar; while the softer spongeous zoophytes are either enveloped in flint or chalcedony, or form chalky ferruginous masses.

On the echini and shells several species of Cellipora and Flustra occur, but they have not been examined with the attention and discrimination necessary to determine their specific characters. I have also very beautiful examples of a small *Millepora* and *retipora*.

The spongeous, inarticulated, porous zoophytes are by far the most abundant; and, in fact, are so numerous, that scarcely a flint is broken in which traces of their structure may not be detected. Many of these are decidedly branched sponges; others approximate to the Alcyonia; but those which in the "Fossils of the South Downs" are referred to the new genus Ventriculites, have characters so well defined, that notwithstanding some respectable writers have amused themselves with either giving them new names, or arranging them as spongiæ, alcyonia, &c., I shall still consider them as distinct; and as they are among the most numerous and characteristic of the chalk zoophytes, a description with figures, is subjoined.

Ventriculites. — A zoophyte of a funnel shape, having the base or stirps furnished with radical fibres or processes of attachment; the external surface reticulated, the inner covered with minute perforated papillæ. The original substance spongeous or gelatinous. The zoophyte capable of contraction and expansion? The common species, *V. radiatus*, has the external integument formed of subcylindrical, anastomosing fibres, which radiate from the centre to the circumference: the papillæ on the inner surface are formed by the open extremities of short transverse tubuli.

So numerous are the accidental varieties of form assumed by the fossil remains of this species, that it is difficult to distinguish them correctly, without the assistance of an extensive suite of specimens. This circumstance is partly attributable to the various stages of growth, or states of expansion and contraction, in which the originals were introduced into the mineral kingdom, and partly to the mode in which their remains are preserved.

The specimens enveloped in flint, are usually of a cyathiform, or turbinated shape (*No.* 8.), while

FLINTS DERIVING THEIR FORMS FROM VENTRICULITES; *No.* 8.

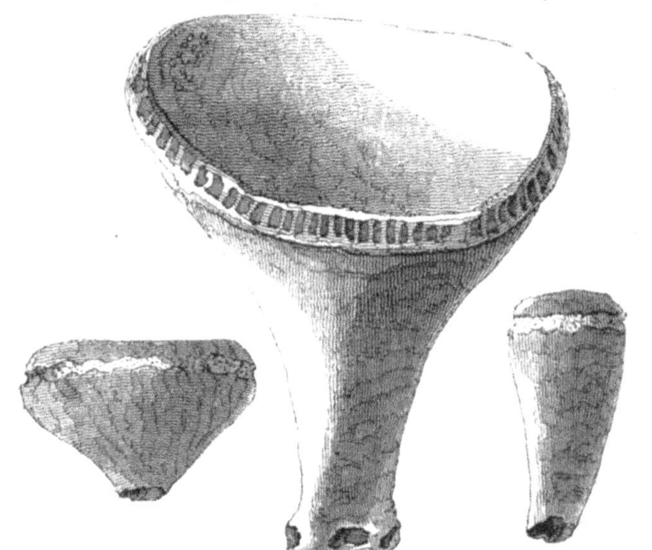

those imbedded in chalk are more frequently expanded into a broad circular disk. The external surface is composed of cylindrical fibres, that extend in a radiating manner from the centre or base to the outer margin, and by frequently subdividing and anastomosing, constitute a reticulated integument capable of very considerable contraction and expansion.

The fibres are solid, and when viewed through a lens exhibit a porous structure, bearing considerable resemblance to dried sponge. The meshes, or interstices between the fibres, are narrow and elongated in the specimens that are expanded, but very irregular in those which are corrugated by contraction. In some instances, slender transverse filaments extend from one fibre to another, by which the entire plexus is more firmly connected together. The surface of the interior, or funnel-shaped cavity, is studded with small perforated tubercles, or papillæ, the open extremities of short, straight, cylindrical tubes, that arise between the fibres of the external integument, and passing in a transverse direction, terminate on the inner surface. Siliceous casts of these tubuli are frequently observable in flints deriving their form from ventriculites. The base forms an elongated stem or stirps, and terminates in diverging root-like processes, by which the original was fixed to other bodies.

This zoophyte, when contracted into a cylindrical form, is from one to six inches in length; when expanded, its diameter occasionally exceeds nine inches: the thickness of its substance is seldom more than 0·2 inch.

The siliceous specimens exhibit no traces of organization, except at the margin and base; the outer surface of the original being obscured by the silex, in which it is imbedded. In some examples, however, the enclosed zoophyte may be separated from the surrounding flint by a well directed blow on the margin, and very delicate casts and impressions may be thus obtained.

The former exhibit the external integument changed into a white friable carbonate of lime; the latter form conical cavities covered with numerous interrupted ridges, disposed in a radiated manner.

The casts of the funnel-shaped cavity are solid cones; their surface exhibiting numerous minute papillæ that have been moulded in the open extremities of the tubuli. A chalk specimen of this kind is figured by Lhwyd, No. 176.*: but it is drawn in an inverted position.

VENTRICULITES RADIATUS,
(Showing the formation of the siliceous specimens.)

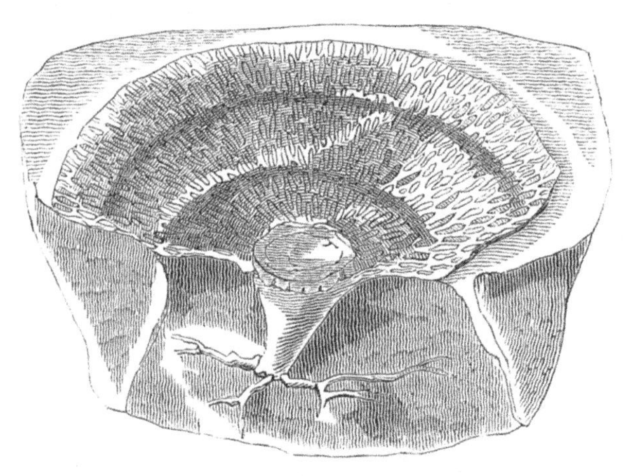

The fossil above represented is remarkably interesting, since it elucidates the formation of those

* It is thus described:—"Astroitæ congener *radularia* cretacea. E puteis cretaceis juxta *Aston Rowant* in agro Oxoniensi."—*Lhwyd, Lith. Britt.*

before mentioned, and establishes the identity of the chalk and flint specimens.

A turbinated flint fills up the lower portion of the funnel-like cavity, and is surrounded by the impression of the external surface of the upper portion; several radical processes proceed from its base. The dissimilarity in the size and shape of the flints, represented page 98., is purely accidental, arising from a greater proportion of silex having been deposited in the one instance than in the other: if, in this example, the quantity of silex had been sufficient to have filled the entire cavity, the flint thus formed would, in every particular, have resembled figure 1, instead of figure 3, *No.* 8.

Among the singular forms assumed by the siliceous specimens of ventriculites, none are apparently more difficult of explanation than the broad annular flints occasionally found on the ploughed

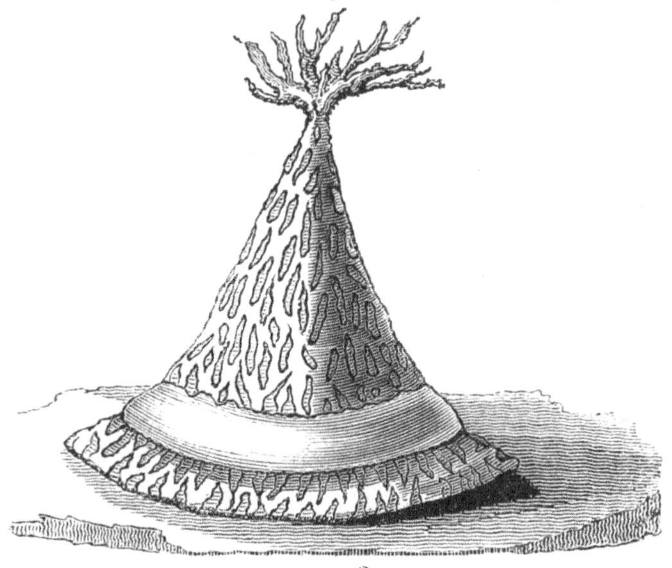

lands of the Downs, and which bear considerable resemblance to a quoit; their origin is, however, satisfactorily illustrated by the specimen figured p. 101. In this example the ventriculite is inverted, and is attached to the chalk by its inner surface, the outer integument forming a narrow zone round the annular flint which, in the perfect state of the fossil, encircled the stem.

The appearance of this specimen seems to warrant the conclusion, that at the period of its mineralization the silex was in the state of a thick viscid fluid, otherwise it is difficult to understand why it should not have extended to the margin of the zoophyte, instead of being consolidated in its present situation. A cyathiform flint, in my cabinet, might also be adduced in support of such an opinion, in which the silex not only fills the cavity of the ventriculite, but is elevated considerably above the margin, as if a pulpy or glutinous fluid had been gradually poured in till the cup-like cavity was overflowing.

In concluding this description, it may be proper to offer a few remarks on the probable economy of the recent animal; and, from the facts that have been presented to our notice, endeavour to illustrate the nature of the original.

From a careful examination of a numerous and interesting suite of specimens in my possession, the structure of the recent ventriculite may be readily understood. The general form of the animal appears to have been that of a hollow inverted cone, having numerous ramose fibres proceeding from the base, by which it was attached to other bodies. Externally it was composed of a reticulated integument, which seems to have been capable of expanding and contracting according to the impressions it received; and, internally, it possessed a surface covered with the apertures of numerous tubuli, in all probability the openings of vessels, by which nutrition was effected.

These inferences naturally present themselves, even upon a slight inspection of the fossils above described. It has already been shown that the specimens occur in every intermediate form, between that of a simple elongated cone, and a flat circular disk; the thickness of the parietes of the cone being considerable when short, thinner when more extended, and thinnest when completely expanded: hence it seems obvious, that the substance of the original must have been soft and elastic, susceptible of spontaneous expansion and contraction, or it could not have accommodated itself to such a variety of shapes, without fracture or laceration. The fibres composing the external integument are nearly straight in the expanded specimens, but are corrugated and moniliform in those which are contracted; the thickness of the latter is also much greater than in the former examples;—circumstances that strongly corroborate the opinion here advanced.

The expanded state of the animal might have been favourable for the discovery of the substances destined for its nutriment, and which, by the subsequent contraction, would be imprisoned in the funnel-like cavity. Whatever may have been the aliment, it must have undergone a certain degree of digestion and assimilation before it was fitted for support; the nutritious particles were no doubt absorbed by the openings so numerously distributed on the inner surface of the ventricular cavity.

Whether the recent ventriculites were confined to one spot, or possessed a certain degree of locomotion, and by detaching their radical processes, were able to change their situation by floating in the water, cannot with certainty be determined; but it seems more probable, that, like the *alcyonia* and *actiniæ*, they were permanently fixed to the rock upon which they grew.

The annexed figures will perhaps serve to render the subject more intelligible.

Group of Ventriculites

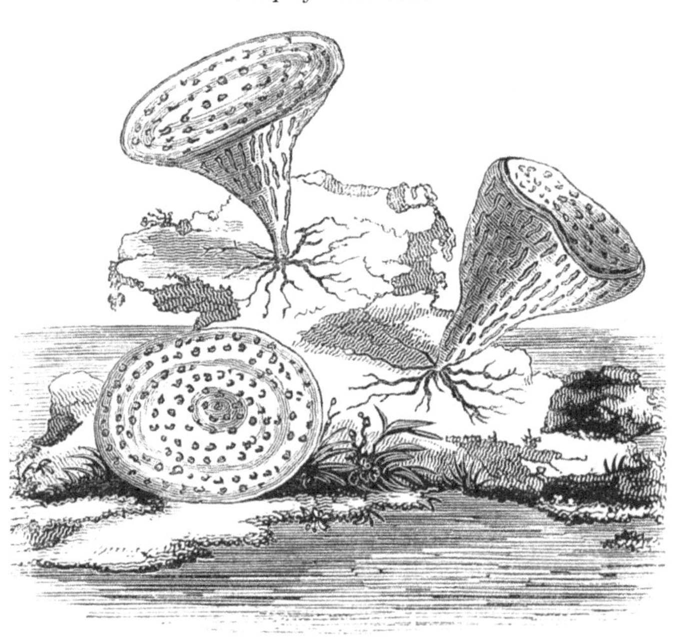

a c b

Fig. *a*. A ventriculite in an expanded state, showing the inner surface.

Fig. *b*. A specimen partially contracted, exhibiting the external integument.

Fig. *c*. A ventriculite more expanded, and exposing the internal cavity.

From what has been remarked, we may therefore conclude that the ventriculites were more nearly related to the actiniæ than to the alcyonia, and, like the former, were capable of contraction and expansion.

CHOANITES.

Another zoophyte deserving particular mention is the spherical or globular species, to which the name of Choanites, is given in the Fossils of the South Downs. The fossils derived from the remains of this zoophyte are very numerous: the originals appear to have held an intermediate place between the *alcyonia*, properly so called, and the *ventriculites*. They are distinguished from the former by the central cavity in their superior part, and from the latter, by being destitute of an external reticulated integument, &c. and possessing but a slight degree of contractile power.

The *alcyonium ficus*. of Linne (*figure de substance et d'eponge et d'alcion*, of Marsilli) may be considered as the type of the genus. "It is of the form of a fig, being attached to the rocks by branches proceeding from its smaller end; the upper part is a little flattened, and has a cavity in the middle. Its colour resembles that of tobacco, and its parenchymatous substance cannot be compared to any thing better than to nutgalls when well dried."*

The fossil remains of this genus (hitherto indiscriminately placed among the alcyonia) were first noticed by M. Guettard at Verest, and at Montrichard in Tourain, and form the subject of a paper published in the *Memoirs of the Academy of Sciences* at Paris (ann. 1757). He observes, that they are of a globular form, having the base in many examples elongated into a pedicle. In the centre of the superior part is a circular opening, generally filled "with the substance in which the fossils are imbedded. This cavity is larger in its upper than in its lower part, and is continued almost to the pedicle, in some specimens appearing to penetrate it. From the circumference of the opening, lines may be traced, that not only pass over the whole of the spherical part, where they form striæ more or less distinct, but also penetrate the substance of the zoophyte. There is seldom more than one opening, but instances have occurred in which there were three." The fossils represented in Pl. ix. figs. 1. 3, 4. 6 ? 8., and Pl. xi. fig. 8. *Org. Rem.* Vol. ii. belong to this genus. The largest Sussex species is that which, in honour to Charles Konig, of the British Museum, I have named *Choanites Konigi.* It is inversely conical; externally marked with irregular fibres, some of

* Organic Remains, vol. ii. p. 81.

which penetrate the substance, and terminate in openings on the inner surface; central cavity, cylindrical, deep, narrow; base fixed by radical processes.

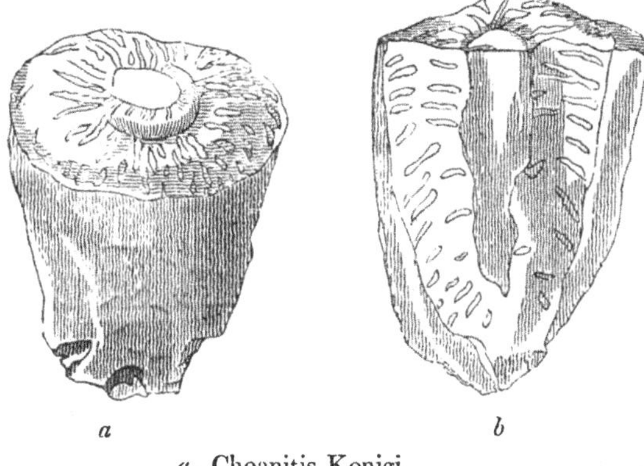

a. Choanitis Konigi.
b. Section of the same.

This species is for the most part enveloped in large irregular flints, which exhibit but slight traces externally, of the body they enclose. The superior part (*a*) presents a convex surface, with a cylindrical body in the centre, from whence interrupted fibres, slightly relieved, ramify in a radiating manner towards the margin. At the base, numerous perforations are seen, through which the radical processes pass. The vertical section (*b*) exposes the cylindrical cavity filled with flint, and the substance of the zoophyte traversed by numerous tubes; some of which appear to terminate on the outer, and others on the inner surface. In chalk specimens, this structure is also very distinctly displayed, and on the surface are markings as if of *cruciform spines*, like those with which some recent alcyonia are furnished.

POLYPOTHECIA.

The only other spongeous zoophytes which I shall notice here, are those which are so common in the flints of certain localities, particularly in those from the chalk pits near Edward Street, Brighton. These occur also very abundantly in the Wiltshire chalk, and in the beds of waterworn flints on the Downs. Miss Benett, of Norton House, near Warminster, (a lady to whom British geology is greatly indebted, and who has composed one of the best local catalogues of organic remains that has appeared in this country), has given them the name of *Polypothecia*, and has beautifully delineated several species in the elegant and scientific Catalogue of the Organic Remains of the County of Wilts, before mentioned. Miss Benett describes the following, all of which I have noticed in Sussex: —

Polypothecia obliqua.
————— clavellata.
————— fissa.
————— latissima.
————— maxima.
————— palmata.
————— infundibulum.

P. clavellata, from near Lewes, is here represented.

POLYPOTHECIA CLAVELLATA, IN FLINT.

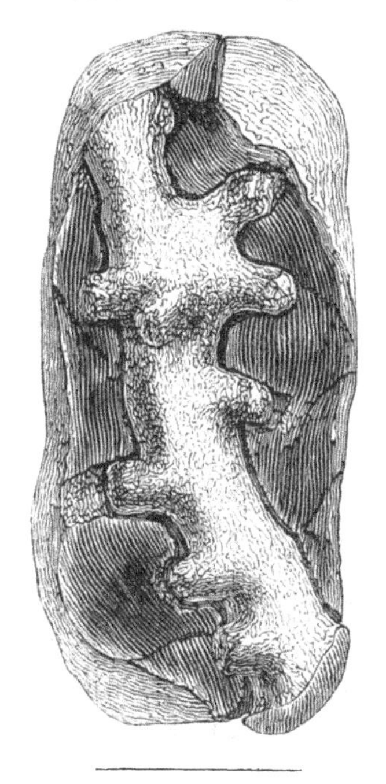

RADIARIA.

*Encrinites and Pentacrinites.**

These names have long been applied to the petrified skeletons of those zoophytes that possess a pelvis or basin, composed of an immense number of crustaceous articulated plates, and ossicula†, supported by a jointed flexible column.

* In the *encrinites*, the bones of the vertebral column are circular or elliptical; in the *pentacrinites* they are angular or pentagonal.

† Mr. Parkinson has shown, that upon a moderate calculation, the lily encrinite must have been composed of nearly thirty thousand dis-

The pelvis, which contained the viscera of the animal, is surrounded by long jointed arms or tentacula, and affixed to the vertebral column by a pentagonal plate placed in the centre of the base.

The column, in most species, is of an immense length, and consists of separate joints or vertebræ, regularly united, pierced in the centre, and having their articulating surfaces ornamented with radiating, stellular, or floriform markings. The inferior part of the column has a pedicle, or process of attachment, by which the animal was fixed to the rock.*

In the recent state, the skeleton was in all probability clothed with a fleshy or coriaceous integument; the central perforation in the vertebral column is supposed by Mr. Martin to have been filled with a medullary substance, by which sensation was conveyed to the inferior extremities of the animal†; according to Mr. Miller, it served as an alimentary canal‡; the former is the most probable supposition.

The detached vertebræ are known to collectors by the name of *trochitæ;* and when several are

---

tinct bones, and the Briaræan pentacrinite must have possessed double or treble that number; in fact, the zoophytes of this order must have more ossicula or bones in their skeletons, than any other animals.—*Org. Rem.* vol. ii. p. 181.

* For a more particular account of the natural history of this extraordinary tribe of animals, consult the 2d vol. of Parkinson's Organic Remains, and Miller's Natural History of the Crinoidea, or Lily-shaped Animals; 1 vol. 4to. 1821; a work that has been justly characterised by an eminent writer, as "a model of patient, sagacious, and successful research."

† Martin's Syst. Arrangement, p. 209.

‡ Miller's Crinoidea, p. 11.

united together, so as to form part of a column; the series is termed an *entrochite.*

The remains of this family of zoophytes so rarely occur in the chalk formation, that portions of two or three species are the only examples hitherto found in Sussex. Of these, the most perfect is the *bottle encrinite* of Parkinson (*Apiocrinites* of Miller), which we shall now proceed to examine.

A RESTORED FIGURE OF APIOCRINITES ELLIPTICUS.

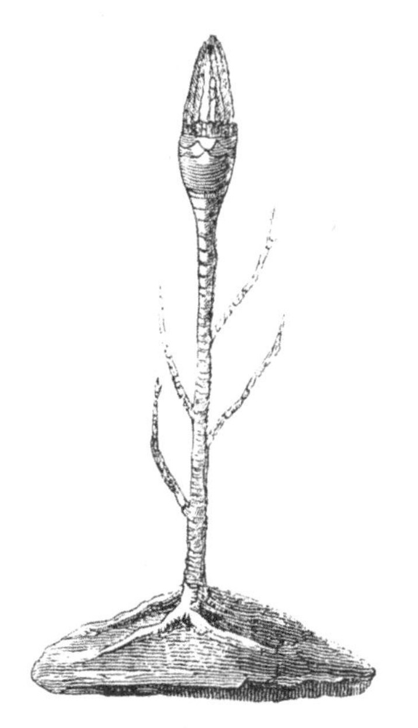

Apiocrinites *ellipticus*. *Miller's Crinoidea,* p. 34.

"A crinoidal animal, having a column composed of oval joints, articulating by a transversely grooved surface; the two upper joints of the column enlarged, sustaining the pelvis, costæ, &c. The column provided with auxiliary side arms. Base formed by numerous irregular columnar joints, sending off fibres for adhesion to other bodies."

The different parts of this animal were first described by Mr. Parkinson, under the various names of bottle, straight, and stag's horn encrinite; and have since been accurately investigated by Mr. Miller, who considers them as belonging but to one species, which he has placed in his first division of the *Crinoidea;* in the same genus with the celebrated *Pear encrinite* of Bradford.

The column of this species consists of smooth ossicula, somewhat enlarging in the middle; their articulating surfaces being elliptical, finely granulated, and having two narrow transverse ridges, in the centre of which is the small perforation supposed by Mr. Miller to contain the alimentary canal.

The pelvis, or body, is of a tumid utricular form, and is divided into separate ossicula of various shapes, to which the names of ribs, clavicles, and scapulæ, &c. have been applied by the authors above named.

Of the pentacrinites we have part of the vertebral column, consisting of eleven thin pentagonal vertebræ, with markings on their articulating surfaces, similar to those of the *Pentacrinus Caput Medusæ:* some ossicula have also been found of a quadrangular form, having the angles rounded, and the surface ornamented with figures resembling a floret of four rays, like the entrochite, No. 1170. of Lhwyd. Trochitæ of four rays are very rare; Mr. Parkinson mentions that he had seen but one (fig. 59. Tab. xiii. Vol. ii.

Org. Rem.) and that is unlike the fossil in question.

MARSUPITES.

This genus was formed by the author for the reception of a fossil that had previously been placed among the encrinites, from which, however, it differs most essentially, in being destitute of a vertebral column, and processes of attachment; hence it is obvious, that the recent animal, instead of being fixed to one spot, was capable of locomotion, and floated *ad libitum*, like the Medusæ and some other zoophytes.

Mr. Parkinson, the publication of whose work, on the *Organic Remains of a former World*, formed an important era in oryctological science, was the first author that accurately noticed this zoophyte. In his 2d vol. an admirable description is given of the pelvis of the animal, under the name of *tortoise encrinite;* and the structure of the original has since been ably illustrated by the ingenious author of *The Natural History of the Crinoidea*, who has adopted the the name by which I have been accustomed to distinguish it.

The following definition is the result of an attentive examination of more than a hundred specimens; but as the recent animal is unknown, and the fossils never occur in a perfect state, it is very probable that some of the characters which are here assumed as permanent distinctions, may hereafter prove to be only accidental varieties of form.

Marsupites *Milleri*, from near Brighton.

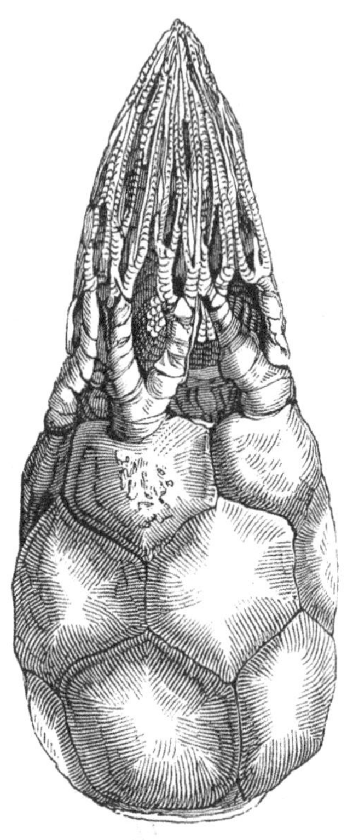

Gen. Char. Body orbicular, contained in a pelvis composed of crustaceous plates, having five articulated arms or tentacula proceeding from the margin: the opening of the pelvis covered by articulated ossicula, in the centre of which the mouth is placed.

Spec. Char. Pelvis composed of sixteen convex, radiated, angular plates: the arms dichotomous, united to the margin by a corresponding number of intermediate semilunar bones: the ossicula covering the aperture of the pelvis disposed in a proboscideal form.

The fossil remains of this zoophyte have hitherto been found only in the upper chalk of Sussex, Wiltshire, and Yorkshire, and, like most other crustaceous bodies enclosed in this formation, are transmuted into a spathose calcareous spar.

But one species is known; the following description will therefore illustrate both the generic and specific characters: —

This fossil is generally of a suborbicular form, more or less distorted, with the lower extremity closed and obtuse, and the upper, truncated and open, being filled with chalk or flint. It is composed of numerous thin angular plates, that are not united as in the echinites, but are simply held in apposition to each other, by the chalk in which they are imbedded. The name of "*cluster stones*," given them by the Sussex quarry-men, not inaptly expresses their general appearance.

The pelvis, or cavity in which the viscera of the animal were contained, is very capacious, and is composed of sixteen angular convex plates, arranged in the following manner, viz.

1. A pentagonal plate (*abdominal*) placed in the centre of the base.

2. Five pentagonal (*costal*) plates, attached to the sides of the centre.

3. Five hexagonal (*intercostal*) plates, placed between the superior angles formed by the union of the costal.

4. Five pentagonal (*scapular*) plates, filling up the angles in the superior margin of those last described, each having a semilunar depression on

their marginal edge: these form the margin of the pelvis, properly so called.

These sixteen plates are succeeded by

5. Five semilunar ossicula (*clavicles*), attached to the articulating depressions of the scapular plates.

6. Five cuneiform ossicula (*cuneiform* or *humeral bones*), attached to the clavicles. These are the first bones of the arm, and their superior edge is divided into two articulations, from which the tentacula are sent off.

7. Numerous reniform ossicula, by which the aperture of the pelvis is closed.

The plates of the pelvis are convex, sometimes umbonated in the centre, and ornamented with radiated ridges on the external surface. Their markings vary in different examples, and even in the same individual; specimens occurring in which some of the plates are nearly smooth, and others richly ornamented.

In every instance, however, the edges of the plates are more or less crenulated, and when united form a suture in the same manner as the scales of the tortoise, but they readily separate when the chalk is removed. The central or *abdominal* plate is larger, and more depressed, than the surrounding *costals;* the latter are readily distinguished by their *pentagonal*, and the intercostals by their *hexagonal* form. The *scapulæ* are generally less ornamented than the rest of the series, and are easily identified by the semilunar cavity in their upper edge; this articulating surface is traversed by a longitudinal ridge, with a minute depression in the centre, and is adapted for the reception of the clavicles.

The *clavicles* are small, and of a semilunar form externally; the upper edge is thick, nearly straight, and unites with the *humerus;* the lowermost is rounded, and corresponds with the semilunar cavity of the scapula; between these two surfaces, on the inside, is a triangular space, the use of which is not at present known.

The *cuneiform* or *humeral* bones, may be considered as the first of the arms; they have four articulations, and are attached to the clavicles by the two lowermost. Their upper margin forms two oblique surfaces, each divided by a longitudinal ridge, in the same manner as the first joint of the finger in the Bradford encrinite (*apiocrinites rotundus*). From this structure it may be inferred, that the arms were dichotomous; and probably were subdivided, and terminated in elongated tentacula, as in the *crinoidea.* On the inner surface of the humerus, a smooth space is observable, appearing like a continuation of the triangular interval, on the corresponding part of the clavicle: this may be the articulating surface for the attachment of the pectoral bones.

The *reniform ossicula,* or *pectoral* bones, are united to each other by their upper and under surfaces, both of which are divided by a ridge, into two depressions. In the only specimen, in which these bones remain, the respective parts have suffered so much displacement, that their mode of arrangement is no longer distinguishable: there is, however, reason to conclude, that in the recent animal they were attached to an epidermis extending over the cavity of the pelvis in the form of a proboscis, the mouth being placed in the centre.

From this examination of the skeleton of the marsupite, it is evident that the recent animal was nearly related to the crinoidea; but the absence of the vertebral column separates it most decidedly from that tribe.

It may, however, as Mr. Miller observes, be considered as forming a link between the *crinoidea articulata* and the *stelleridæ.*

The folds, radiating ridges, the striæ on the plates, and the lateral adhesion of plate to plate by simple sutures, plainly indicate that the whole was invested by a muscular integument; the markings on the plates being the effects of its action.*

From the rudiments of the arms, it is also equally obvious, that the recent animal was furnished with tentacula, to enable it to seize and detain its prey, in the same manner as the encrinites, &c. Its position, when floating in the water, was in all probability with the mouth downwards, like the *Medusa pulmo, M. campanulata,* and other species of that family.

The specific name is in commemoration of the valuable researches of the late J. S. Miller, Esq. A.L.S. of Bristol; a tribute of respect, which is justly due for his able investigation of the *Natural History of the Crinoidea.*

* Miller's Natural History of the Crinoidea, p. 137.

The specimen figured represents the only one in which the reniform or pectoral ossicula, and those of the first joint of the arms, remain. I have restored the arms, to convey a more correct idea of the nature of the original animal. To the liberality of the Rev. H. Hoper I am indebted for the beautiful specimen from which the drawing was made.

Specimens of the pelvis, consisting of the abdominal, costal, intercostal, and scapular plates, more or less distorted, are the only parts of the animal generally found; the clavicles, humeral bones, &c. are among the rarest productions of the chalk formation.

ASTERIA, OR STAR-FISH.

But few remains of these animals have been discovered in Sussex, although they are not uncommon in the Kentish chalk.

The specimens in my possession consist of a few detached ossicula of *pentagonaster semilunatus*, (*Org. Rem.* vol. iii. tab. i. fig. 1.); and some fragments of a species that appears to be distinct from any figured or described.

ECHINI.

Of this order of mollusca, numerous species occur both in a recent and fossil state. They are marine animals, having a body more or less round, covered with a crustaceous shell, and furnished with moveable spines; the mouth being placed beneath. The crust or covering is composed of an immense number of plates, varying in form in different families, and in some species amounting to nearly a thousand in one individual. It has numerous perforations, through which the tentacula of the enclosed animal are protruded. These pores form bands (*ambulacra*) that divide the shell into segments (*areæ*), the latter being more or less covered with tubercles, to which the spines are attached by strong ligaments. Upon the death of the animal, these ligaments undergo decomposition, and the spines almost constantly fall off,—a circumstance that explains the cause of their being so seldom found in connection with the shell, in a fossil state. The mouth is armed with five or six triangular teeth.

These animals feed upon crabs and the lesser kinds of shell-fish, which they seize and convey to the mouth by means of the tentacula, the spines being the instruments of motion.*

The remains of the numerous family of echinidæ occur in the chalk abundantly, and those of the genera Galerites, Ananchytes, Spatangus, and Cidaris (the latter more rarely), are among its most characteristic productions. Their cavities are commonly filled up with silex, which presents a perfect cast of the interior, after the crustaceous covering has been removed by chemical or mechanical agency; fossils of this kind are very frequently found among the beds of loose flints on the surface

* Rees's Cyclopædia, art. *Echinus.*

of the Downs, and, under the names of shepherds' helmets and crowns, are preserved by the peasantry as ornaments for the rustic sideboard of the cottage. This class of organic remains is so familiar to every one, that it is quite unnecessary to give a detailed description of the various species that occur in the chalk of the south-east of England.

CRUSTACEA.

The fossil remains of those species of Cancer, in which the crustaceous covering is hard and compact, are not unfrequent in the London clay at Highgate, Sheppey, &c.; and a few have been found in the Galt of Sussex: but the lobster, cray-fish, and other kind, whose structure is more delicate and fragile, but seldom occur in a mineralized state, and rank among the most rare and interesting objects in the cabinet of the oryctologist.

In the chalk near Lewes the claws of a delicate species of Astacus are often found, and rarely, portions of the crustaceous covering of the body. These remains consist of a delicate friable crust, and when first collected are of a dark chocolate colour, inclining to black; but they become pale, and lose much of their beauty, by exposure to the air and light. The inner surface only is seen in the specimens discovered by breaking the chalk; it is glossy, and covered with minute circular depressions formed by the bases of the spines. The external surface is armed with short spines and papillæ, and is invariably concealed by the chalk until the latter be carefully removed: a process which, from the delicacy of the fossil, and the hardness of the surrounding matrix, is exceedingly difficult and tedious, and can scarcely be accomplished by an inexperienced hand. Some of the specimens in my cabinet exhibit the claws, others the thorax, and a few the abdomen and tail. From these detached parts a restored outline of the original was delineated, and its relation to the recent species by this means ascertained. I have named the first species

Astacus Leachii, in honour of Dr. Leach of the British Museum. The genuine and specific characters are as follow: —

Gen. Char. Antennæ pedunculated, unequal, the exterior ones long and setaceous; the inner pair divided at the extremities; body elongated; legs commonly ten; tail foliaceous.

Spec. Char. Thorax scabrous, convex, six-lobed, marginate: head semicircular in front; hands chelate, muricated, twice the length of the thorax; pincers very long, armed with spines.

The thorax is longitudinally oblong, convex, covered with small tubercles and papillæ; it is divided into six lobes by a rounded dorsal ridge, and two lateral sulci; the margin is entire: the head appears to have been semicircular, or rounded in front, and is not distinct from the thorax.

The external antennæ are long, filiform, and setaceous, and are placed on squamous peduncles; the inner pair have not been discovered. The two chelate hand-claws are equal, and have their surface muricated, or beset with short erect spines. The pincers are very long, not muricated, but marked with three or four longitudinal, punctated furrows; each finger is armed with a row of obtuse, cylindrical spines, which are mutually received and inserted, when the claws are shut. The claws, including the pincers, are equal to twice the length of the thorax.

There are five legs on each side; the anterior pair is didactyle; the others appear to terminate in swimmers or paddles, but this circumstance cannot be accurately determined. The abdomen is composed of six granulated arcuate segments. The tail is foliaceous, marginate, granulated, and has a few longitudinal ridges; but the only known specimen does not exhibit the entire form.

This species appears to be distinguished from the recent animals of the genus, by the dorsal ridge and lateral sulci of the thorax; the great length and straightness of the pincers; and the peculiar form of their spinous processes: a claw is figured below: (*fig* 1.)

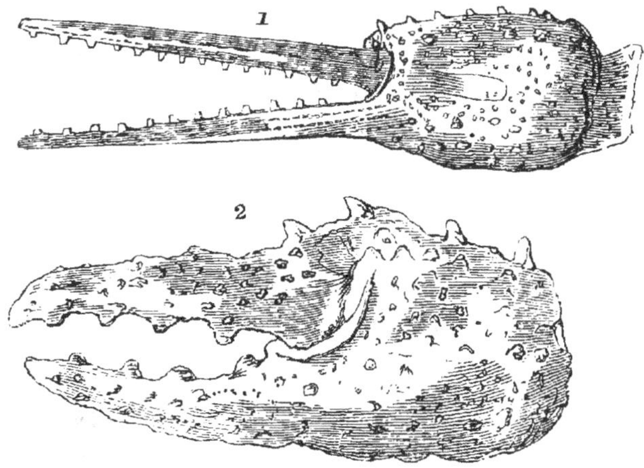

1. Claw of Astacus Leachii.
2. Claw of Astacus Sussexiensis.

Another species occurs in the Sussex chalk, and appears to be more generally distributed than the former. The chelate hand-claws are shorter, and the surface both of the claws and pincers is spirous; the latter are slightly curved, and armed with a row of obtuse tubercles. I have named it *Astacus Sussexiensis.* On the remains of this species M. Dasmarest has the following observations:— "Le crustace auquel appartenaient ces pinces, avoit la forme ordinaire des macroures, et ne presentait sur les pieces que nous avons vues, d'autres caracteres exterieurs que ceux qui consistaient dans la presence de trois fort tubercules sur chaque côté de la carapace qui etait d'ailleurs tres-rugueuse. Il était un peu plus grand que l'ecrevisse fluviatile." *

A chelate hand-claw is represented above (*fig.* 2.)

ANNELIDES, CONCHIFERA, MOLLUSCA, ETC.

The remains of the testaceous coverings and instruments of locomotion of the mollusca of the chalk are so very numerous, that we can notice but a few, and must refer to the catalogue for an enumeration of the various species. Among the most remarkable of the simple spiral shells are the two species of Cirrus, figured *a*, *Cirrus depressus*, *b*, *c*, *Cirrus perspectivus*: they are diminished five-sixths.

* Crustaces Fossiles, p.137. The learned authors describe, under the name of *Scyllarus Mantelli*, the remains of a crustaceous animal found on the coast of England, but do not mention its geological site.

The cast of a *Dolium* was found by Mr. Weekes, and is figured under the name of *D. nodosum* in Sowerby's Mineral Conchology. The Ammonites are not so abundant in the upper divisions of the chalk, as in the inferior strata hereafter to be noticed. Several large species, however, occur, viz. *A. peramplus*, *Lewesiensis*, and *catinus* ; and some that are peculiar, *A. Woollgari*, and *navicularis*.

Among the Conchifera, the Terebratulæ, Inocerami, and Plagiostoma, are the most numerous ; of the latter, P. spinosum is a very remarkable and peculiar species. It is readily distinguished from the other shells of the chalk, by the long slender spines attached to the upper valve.

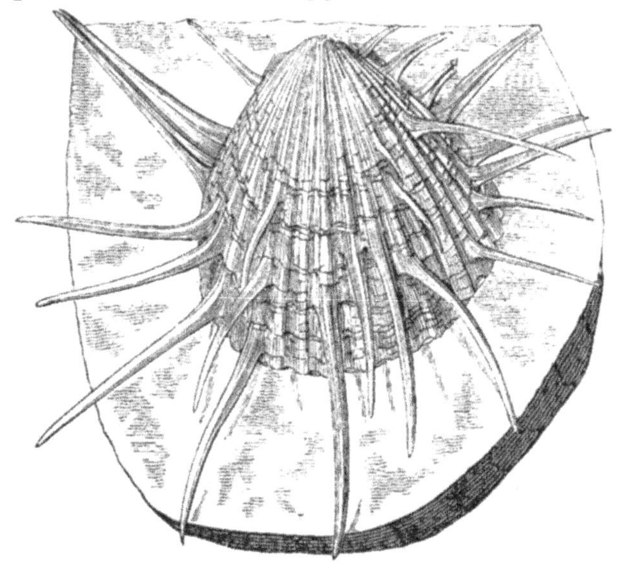

Each valve has from twenty-five to thirty rounded costæ, formed by intervening furrows, that radiate from the beaks to the margin; these are decussated by fine transverse striæ, and the lines of increase; the inner surface of the shell is also marked with corresponding impressions. The lower valve is most convex, and has the line of the hinge straight; the upper valve is spinous, rather depressed, and contains the angular sinus, by which the shells of this genus are characterised.

The spines arise from the ribs, but without any regularity, except that they are more numerous at the sides, than in the centre. They vary from fifteen to twenty in number, and are from half an inch, to two inches and a half in length; each spine has a groove on the under, and a corresponding ridge on the upper surface. They generally project from the shell, but in some instances lie close on the surface.

The beaks are convex, and incurved; the ears small, and even; the margin neatly denticulated. A specimen cleared from the chalk, exhibited no muscular impression.

There are several varieties of this species, of which the following are the most remarkable: —

Var. a. With both valves gibbous, and but few spines.

b. Valves depressed, spines numerous.

c. Valves gibbous, ribs regularly convex and even.

d. ————————, ribs channelled near the front.

This shell is one of the most common productions

of the upper chalk, but is less frequent in the lower beds; the hardness of the Sussex chalk renders it exceedingly difficult to clear the specimens, without destroying the spines.

The Terebratulæ are very abundant; a few of the common species are figured below.

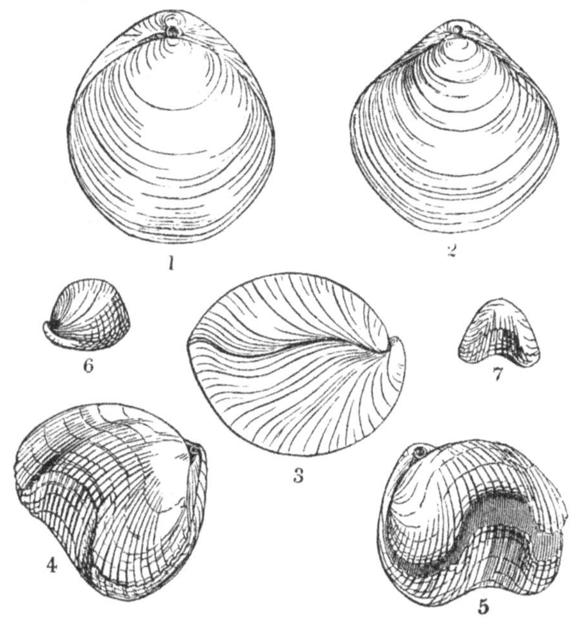

1. 3. Terebratula semiglobosa.
2. Terebratula carnea.
4. Terebratula plicatilis.
5. Terebratula octoplicata.
6, 7. Terebratula subplicata.

INOCERAMUS.

The shells of the genus Inoceramus are very remarkable, and ten species occur in the chalk of the South of England. These shells are more or less gibbous, and are commonly marked with transverse concentric ridges, and striæ; their constituent substance is invariably composed of crystallized carbonate of lime, of a radiated or fibrous structure. The hinge is a longitudinal furrow, transversely crenulated, extending on one side of the beaks only; its direction, as it regards the transverse diameter of the shell, being generally oblique. One species, the Inoceramus Cuvieri attains a large size: a fragment in my cabinet indicates a length of three feet, by two feet in width. An example of *Inoceramus Lamarckii*, which shows the hinge, is figured below, and will serve to convey a correct idea of the character of the genus.

INOCERAMUS LAMARCKII.

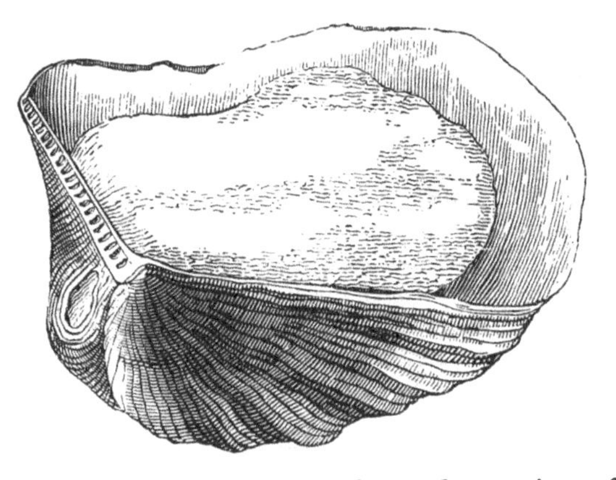

We cannot omit to notice here the curious fossil bodies that are occasionally observed on our flints, and which have been considered by some naturalists as silicified zoophytes. Their nature was first pointed out by the Rev. W. Conybeare, in a very interesting memoir, published in the Geological Transactions, vol. ii.

The shells of the larger Inocerami, appear to have been subject to the ravages of a peculiar parasitical animal, which destroyed the intermediate substance, leaving the outer and inner plates entire, and supported only by thin partitions. The specimens exhibiting these appearances are full of small oblong cells, connected by linear perforations; and these are either empty, or filled with chalk or flint; in the latter case, they give rise to a curious class of fossils, the nature of which Mr. Conybeare has very ingeniously explained.

SILICEOUS CASTS FORMED IN CELLS PRODUCED IN THE SHELLS OF THE INOCERAMI BY PARASITES.

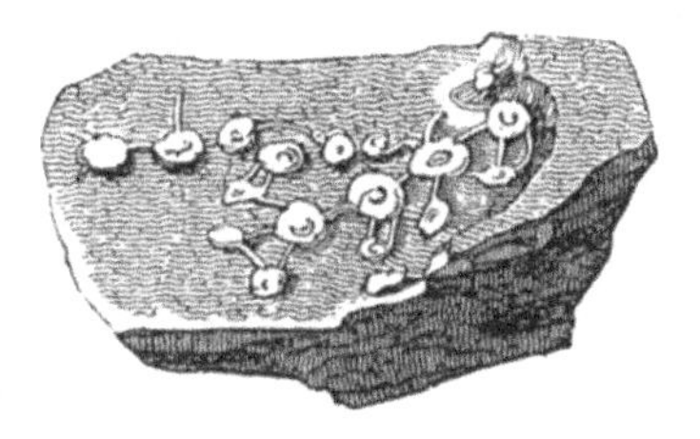

A specimen of this kind is here represented; it is part of a flint, moulded in the interior of an Inoceramus, having on its surface numerous irregular oblong bodies, more or less compressed, and united to each other by slender lateral filaments.

The investigations of Mr. Conybeare have clearly shown "that these are silicious casts, formed in little cells, excavated in the substance of certain marine shells, the work of animalculæ preying on those shells, and on the vermes inhabiting them. These casts, like the screw-stones of Derbyshire, must have been formed by the infiltration of siliceous matter, while in a fluid state, into the cavities of the shells, and which have been laid open and denuded by subsequent exposure to some agent capable of dissolving and removing the calcareous matter of the shell forming the matrix, while the siliceous casts remained unaltered."

Hippurites. — Specimens of the upper chamber of shells of this extraordinary genus, have lately been found in the chalk, near Lewes; it is the first instance of their occurrence in Great Britain. I have named the species *Hippurites Mortoni*, as a tribute of respect due to S. G. Morton, Esq. M.D. Secretary to the Academy of Natural Sciences of Philadelphia, whose important services to Geology I have already had occasion to notice.

ICHTHYOLITES, OR THE FOSSIL REMAINS OF FISHES.

The remains of fishes are of less frequent occurrence in a fossil state than those of some other animals; nor will this circumstance appear extraordinary, when it is considered that the softness of their structure, renders them liable to undergo putrefaction with great rapidity; and that such as die a natural death, rise to the surface of the water, and immediately become the prey of a multitude of assailants. A concurrence of circumstances, by which the death and envelopement of these animals may be almost simultaneously effected, seems therefore necessary to the preservation of their remains in the mineral kingdom. Hence some naturalists have supposed, that wherever petrified fishes occur in considerable numbers, it may be inferred that they perished by some sudden catastrophe which destroyed and overwhelmed them in shoals, in the very spots where they are now found entombed: this, however, has, probably, but seldom been the case.

Fossil fishes have been found in all the formations of England, from the old red sandstone to the tertiary deposits inclusive. They occur sparingly in the chalk of other parts of this island, but in Sussex are far from rare, and in the immediate vicinity of Lewes have been discovered in a more perfect state than in any other locality.

The specimens are generally distorted; and but few examples have been found, in which the number and situation of the fins, and other parts essential to the determination of the genus, or species, are distinctly exhibited; yet their general characters are sufficiently defined, to prove their want of identity with any known existing species.

The teeth and palates are remarkably beautiful, their original substance being heightened by an impregnation with iron, and their natural polish and sharpness remaining uninjured. The vertebræ and other bones are of a reddish brown colour, and very friable.

The fins and scales possess a glossy surface, are exceedingly brittle, and both in colour, and in the mode of their preservation, resemble the ichthyolites of Monte Bolca.

Of the cartilaginous fishes, the teeth of several species of *Squalus*, or shark, are most frequent; fins and vertebræ are also occasionally met with, but no decided examples of any other parts of these animals have hitherto been discovered.

The usual varieties are figured below. The fossil

FOSSIL TEETH OF SHARKS.

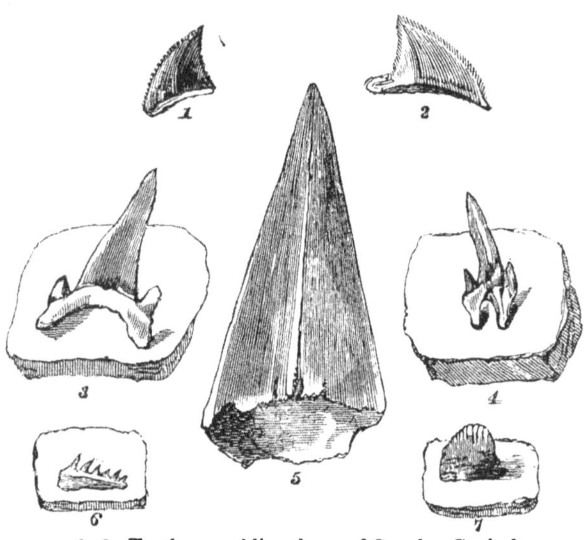

1, 2. Teeth resembling those of Squalus Cuvieri.
3, 4. Squalus Mustelus.
5. Squalus Zygæna.
6. Squalus, species unknown.
7. Tooth resembling that of Squalus Perlon.

palate teeth have hitherto been considered as belonging to a fish related to the Diodon; they are more or less of a quadrangular shape, having the outer surface convex, and composed of an exceedingly hard enamel, which in the centre is formed into sharp and slightly curved ridges; these are surrounded by a border of obtuse papillæ. The *Diodon histrix* has one tooth of this kind affixed to the os hyoides, and another to the palate or roof of the mouth. But the fossil teeth are sometimes found in considerable numbers, and of various sizes, forming a tesselated surface of several square inches; and so regularly disposed, the smaller palates being adapted to the intervals between the larger ones, that no doubt can exist of this having been the mode in which they were placed in the original. Hence, instead of each specimen being a distinct palate, like the corresponding teeth of the Diodon, they appear to have constituted the covering of the entire roof and base of the mouth.

PALATAL TOOTH OF AN EXTINCT SPECIES OF SHARK, OR DIODON.

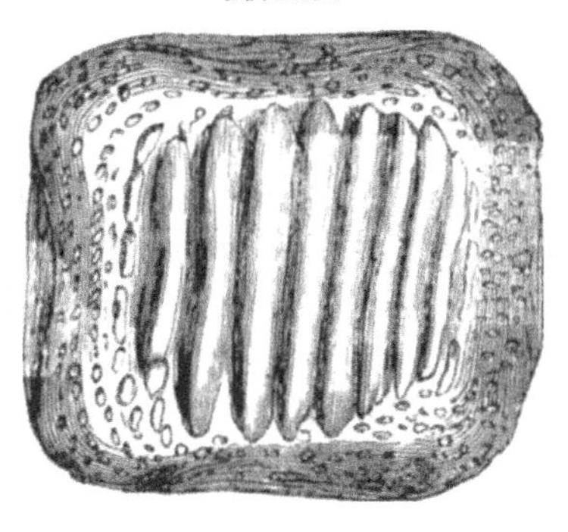

M. Agassiz (whose beautiful and truly scientific work on the Freshwater Fishes of Central Europe, affords an earnest of what may be expected from him when the history of the fossil fishes shall come under his examination) informed me, through Mr. Lyell, that he believed the teeth in question will be found to belong to a species of Squalus; a supposition by no means improbable. It is not unlikely that the large thick, radiated, fin-like processes, hitherto referred to the Balistes, are also referable to that universally prevailing genus, the Shark.*

Balistes.—The fishes of this curious genus have the head compressed, and close to the body, appearing as if it were a continuation of the trunk. The mouth is narrow, the teeth in each jaw are eight in number, of which the two anterior ones are the longest; there are also three interior ones on each side, opposite the intervals between the external row. The aperture of the gills is narrow, destitute of opercula, and placed above the pectoral fins; the branchiostegous membrane has two rays. The body is compressed, and carinated on each side; the scales are coriaceous, joined together, and rough, with sharp minute prickles. *They have two dorsal fins, of which the anterior one is armed with a strong spinous ray, concealed in a deep groove in the back, and can be erected or depressed by the animal at pleasure.*† Some species, as the B. *monoceros* (Unicorn file-fish), are furnished with a spine between the eyes.

The specimen delineated is evidently the defence of a fish, and so strikingly resembles the spine fixed between the eyes of certain species of Balistes, that there can be no hesitation in considering it to belong to a fish of that genus. It is of a dark chocolate colour, and possesses a fine

* Some years since, a block of chalk, containing upwards of a hundred of these bodies, was discovered by the workmen in Offham pit; it was sold to a stranger.

† Nouveau Dict. d'Hist. Nat. tom. xi p. 515.

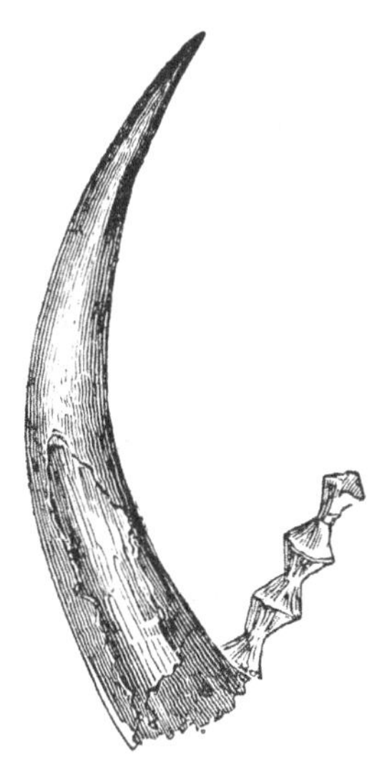

polish: several vertebræ are imbedded in the chalk near its base.

A magnificent specimen of a dorsal fin, or *radius*, of a fish allied to the *Balistes*, is one of the most interesting productions of the Upper chalk in my collection. It was unfortunately broken by the quarry-men, and the intermediate portion destroyed.

It consists of thirteen narrow parallel rays, divided by fine sulci, that gradually diminish in size as they approach the apex, which is broken off. The rays are anchylosed, or united to each other, the grooves or furrows penetrating but a short distance into the substance of the fin. The upper edge is serrated, having fifteen obtuse projections, with corresponding depressions. The inferior margin is entire, and near the base of the fin is furnished with numerous slender processes, or cirri, that occupy a space of three inches in length and an inch and a half in breadth; these are probably the remains of the tendinous expansion of the muscle by which the fin was erected and depressed.

Of the more perfect fishes, some resemble a species of Muræna; one is decidedly a species of Zeus, and is named Z. Lewesiensis in the catalogue. This ichthyolite is related to the genera *Stromateus*, *Chætodon*, and *Zeus*; but in its general form more closely resembles the recent individuals of the latter.

The fishes of the genus Zeus have the head compressed, and sloping, the upper lip arched, the tongue subulated, the body compressed, thin, and shining, and the rays of the first dorsal fin ending in filaments; in every essential particular of this description, the fossil alluded to will be found to correspond.

It is from six to eight inches long; and its width is nearly equal to the length of the body, exclusive of the head. It is covered with large, ovate, striated scales; the back and abdomen are ridged, and gently arched; and the body is thin, and compressed. The head is somewhat obtuse, and large in proportion to the body; the orbits project, and are placed high in the head. The lower jaw is straight, the upper one slightly arched; and both are destitute of teeth. The *opercula branchialia* are large, and there are six branchiostegous rays. The dorsal and anal fins are placed nearly opposite to each other, and extend over two thirds of the posterior part of the body, but do not unite with the tail; the rays of the dorsal fin appear to pass into long filaments, as in the recent Dory. The pectoral fins have not been observed. The caudal fin, or tail, is composed of numerous strong rays. The vertebræ are about twenty in number: in most instances the ribs still remain attached.

The above description is taken from a most beautiful specimen, which is figured below.

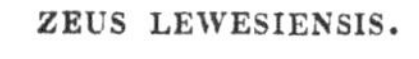

ZEUS LEWESIENSIS.

Abdominales.—The abdominal fishes are more frequent in the mineral kingdom, than those of any other order; they are distinguished by the ventral fins being placed behind the pectoral, or upon the abdomen. The remains of three species, belonging to as many genera, have been discovered in the Sussex chalk, all of which appear to differ from any previously noticed, either in a recent or fossil state.

The first appears to be somewhat related to the

genera *Salmo* and *Clupea*, but does not conform to the characters of either; it may, however, be convenient to affix some name as a temporary distinction, and, for reasons hereafter mentioned, the following has been chosen.

Salmo? *Lewesiensis*.

The body of this ichthyolite is of an elongated oval form, and covered with smooth, delicate, semicircular scales. The trunk is subcylindrical, the back slightly ridged, and the abdomen rounded. The head, so far as can be ascertained from the specimens in my collection, appears to have been of an obtuse form. The eyes are placed high on the head; the mouth and jaws resemble those of the *Salmo odoe*, but no vestiges of teeth are perceptible; the lips are rounded as in the Perch (Perca *fluviatilis*). The *opercula branchialia* consist of three or four plates, and in one example ten or eleven of the branchiostegous rays remain. The pectoral fins lie close to the gill-covers, and are composed of seven or more rays. The ventral fins are attached to the abdomen, and each has six or seven rays. The caudal fin is unknown; but the small adipose fin or process, so constantly observable between the dorsal fin and tail, in the recent fishes of the salmon tribe, is distinctly shown in one specimen.

The ventral fins being situated behind the pectoral, places this fossil fish in the order abdominales; while the relative situation of these parts, the adipose dorsal appendage, the structure of the opercula, and the rounded form of the abdomen, point out some affinity to the salmo. The absence of teeth, and the obtuse form of the head, separate it, however, from all the recent species.

SALMO ? LEWESIENSIS.

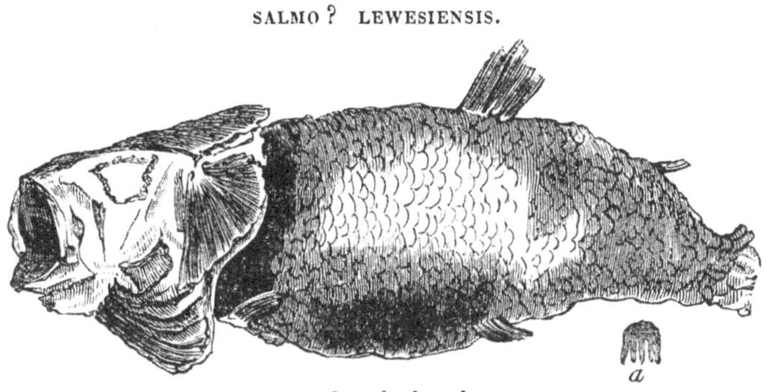

a, a detached scale.

In a remarkable specimen, the back of the animal is imbedded in the chalk, but the abdomen, head, &c. are distinctly exposed. This fish lies four inches in relief, is nine inches long, 2·5 inches wide between the pectoral fins, and one inch between the ventral; the latter being placed three inches below the former. The relative situation of these parts may probably have been altered, but the specimen is so little distorted, that the difference cannot be material. The head is considerably mutilated; it exhibits portions of the jaws, temporal bones, the plates of the opercula, and ten or eleven branchiostegous rays on each side; the latter are spread out from beneath the opercula, and meet under the lower jaw. Both the pectoral fins are preserved; the right one remains in its natural situation; the other is displaced and partly covered by the gills; each is composed of seven or eight rays. The ventral fins consist of six or seven rays, and are partially separated from the body of the fish. A dorsal fin of seven or eight rays remains. The tail is altogether wanting.

An extraordinary fact relating to this ichthyolite remains to be noticed: it is, that the body is very generally *uncompressed;* being almost cylindrical, and evidently as perfect in form as when recent. It would seem as if the fish had been suddenly surrounded by a soft cretaceous mass, which consolidated before the form of the original had been changed by decomposition.

The maxillæ of a fish which closely resemble those of the Esox or Pike, have established the species to which we have given the name of Esox *Lewesiensis.*

JAWS OF ESOX LEWESIENSIS.

The specimen here represented is evidently the jaws of a fish, whose recent prototype is unknown. The dentature of the maxillæ in certain species of

Esox or *Pike*, is very analogous, and in all probability, the relic before us will be found to belong to an extinct or unknown species of that genus.

The lower jaw is nearly perfect, and is attached to the chalk; of the upper, a part only remains. In the lower there are nine teeth *in situ;* these are not fixed in sockets, but united to the jaw by anchylosis. They have a glossy surface, and are exceedingly brittle; differing most essentially in this respect from those of the shark, and other fishes previously noticed. The two anterior teeth are nearly an inch in length, and possess a very peculiar form; they are broad at the base, and suddenly contracting, terminate in a point; they are convex behind, and rather channelled in front. The teeth are of various sizes; some being very short, and not attached to the edge of the jaw, but to a longitudinal depression on its inner surface. They are very irregularly disposed, and appear to have suffered some degree of displacement.

The ichthyolite we have next to describe, is in all probability abdominal, but the situation of the fins is so imperfectly known, that even this point is not positively ascertained. The determination of its generic characters is involved in still greater obscurity, and there does not appear to be any recent genus to which it can be correctly appropriated. It bears some affinity to the *Antherina Mugil*, and *Polymnemus*, but possesses characters obviously distinct from either of those genera. In the elongated form of the body, the number and situation of the fins, and in the dentature of the jaws, it resembles an ichthyolite figured by Cuvier*; and which is considered by that illustrious naturalist as approaching to the *Amia calva*† of Linné.

Both the fossils in question differ, however, from each other, and from the recent species, in many important particulars; and it is probable that they will hereafter be found to be but very remotely related, yet, in the present infancy of oryctological science, it may be excusable to retain them under the same genus, until their characters shall be accurately determined by the discovery of more illustrative specimens.

Amia Lewesiensis. — The length of this ichthyolite generally exceeds eighteen inches, the head being equal to one third of the whole; the width is about 4·5 inches. The body is of an elongated form, slightly compressed, scaly, and reticulated. We have attempted a restored outline.

AMIA LEWESIENSIS.

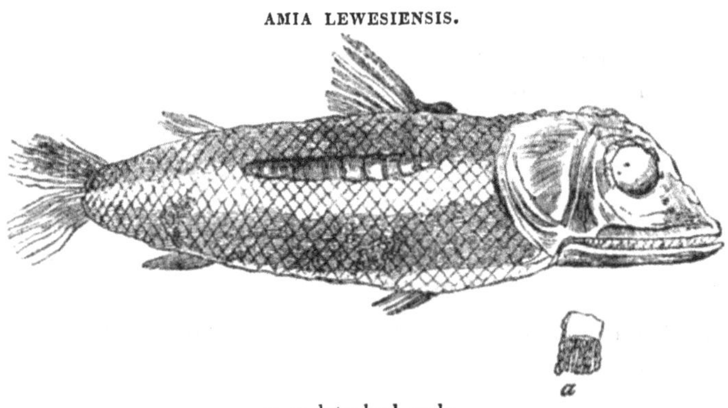

a, a detached scale.

The scales are of a rhomboidal shape, and beset with numerous, small, adpressed spines, producing

* Fossiles de Paris; Reptiles et Poissons, fig. 13.

† The *Amia calva* is a freshwater fish, inhabiting the rivers of Carolina.

a scabrous reticulated appearance, not unlike the surface of some kinds of Balistes. The head is angulated; the orbits large; the opercula smooth, and rounded; the jaws dentated, and nearly straight. The teeth in the upper maxilla are conical, pointed, and rather flat; there are about forty on each side, of which the eight or nine anterior ones are the largest. Those of the lower jaw are exceedingly small, and very numerous. The dorsal fins are two in number; the anterior one is placed in a sulcus, or groove, in the back, and appears to have been capable of erection or depression; it consists of eight strong rays, the two first being garnished with spines. The posterior dorsal fin is remote from the other, and composed of numerous delicate rays. The pectoral fins are placed on the thorax, near the lower margin of the opercula. The ventral fins are attached to the abdomen, opposite to the anterior dorsal fin; and the anal fin to the posterior one. The tail appears to have been rounded, but no perfect specimen of this part has been obtained. The *tongue* is occasionally preserved; it is of a triangular form, and its surface is covered with numerous papillæ. The air bladder is of an elongated oval shape, and lies in the abdomen, immediately beneath the spine.*

From the preceding description, which comprehends all that is at present known concerning this

* It may seem scarcely credible, that a part of such delicate structure should be preserved in a mineralized state, yet the fact is unquestionable; I have several specimens in my collection, in which it is clearly shown; and also the coprolites or fossil fæcal contents of the intestinal canal.

curious ichthyolite, the original appears to have borne some resemblance to the *Mugil;* but its dentated maxillæ, not to mention other obvious differences, distinguish it from the recent individuals of that genus.

The structure and situation of the anterior dorsal fin, and the reticulated scabrous surface of the body, is similar to what is observed in some species of *Balistes;* but the fossil before us does not present the slightest analogy, in any other respect, to that tribe of fishes.

The specimen figured by Cuvier, and described by Blainville, under the name of *Amia ignota**, possesses many characters in common with the fossil before us. It consists of the skeleton of a fish, attached to a block of gypsum. It is twelve inches long, and four inches high; the head being equal to one third of the length. It has two dorsal fins occupying the same relative situation with those of the Sussex fossil; the ventral fins also correspond; the lower jaw is furnished with many small pointed teeth, and the tail is rounded. But the angular form of the head in A.? *Lewesiensis:* the spinous rays of the anterior dorsal fin, and the scabrous structure of the scales, separate it most decidedly from the A. *ignota* of the French naturalists.

It would be uninteresting to particularize the detached portions of fishes which have been found in the Sussex chalk, unless the descriptions were accompanied by numerous figures. We shall therefore only notice those substances which, under the name of "supposed juli of the Larch," were de-

* Nouveau Dict. d'Hist. Nat. tom. xxviii. art. *Ichthyolites*, p. 69.

scribed in the Fossils of the South Downs, and which the sagacity of Dr. Buckland (assisted by the analysis of the substance in question, by Dr. Prout) has proved to be the fæcal remains of certain fishes, the peculiar manner in which they are coiled up, depending on the structure of the intestinal tube through which they have passed; their animal origin, as stated in my former work *, had long since been suggested by Mr. König. I have two specimens of the ichthyolite described under the name of *Amia*, in which coprolites occur in the body of the fish, and in both instances are lying near the air bladder: and it is remarkable that these are not spirally twisted, as in the generality of specimens, but appear to have been situated in the upper part of the intestinal canal, and, not having passed through the tortuous part of the tube, do not possess the spiral structure. Three coprolites of the usual character are here figured.

COPROLITES FROM THE CHALK.

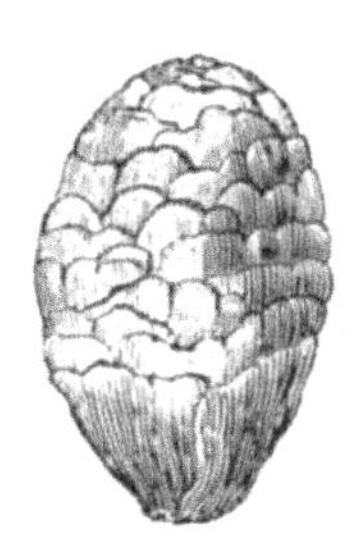
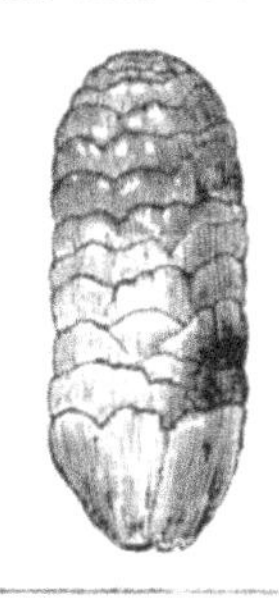
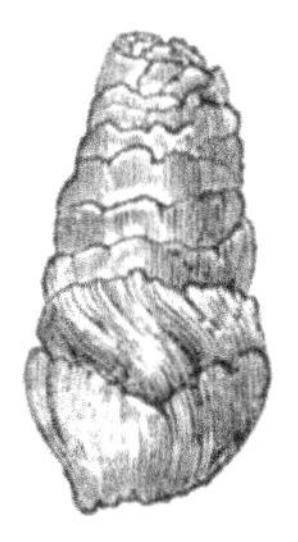

REPTILES.

The bones and teeth of several species, belonging to various genera of oviparous quadrupeds, have been found in the chalk near Lewes.

* Geology of Sussex, p. 103.

The most important of these are, unquestionably, the vertebræ of the Mososaurus, or Fossil Monitor of Maestricht. The quarries of St. Peter's mountain near Maestricht, have long been celebrated for the remains of one of the most extraordinary oviparous quadrupeds, hitherto discovered in a fossil state: several magnificent portions of the skele-

VERTEBRÆ OF THE MOSOSAURUS, FROM LEWES.

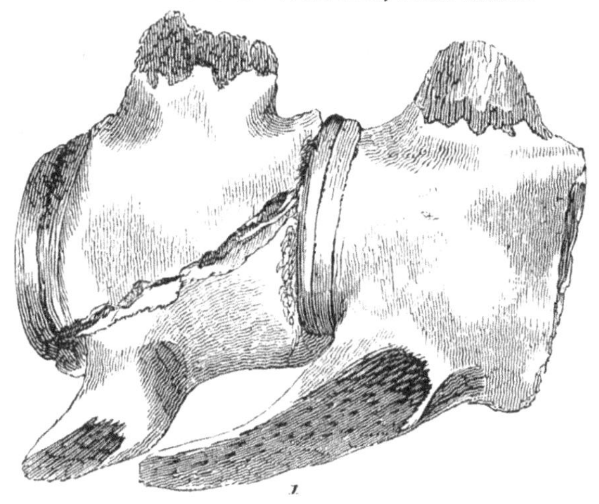

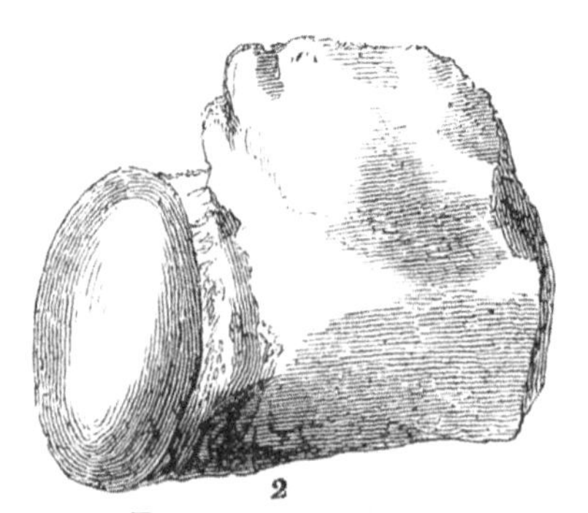

1. Two caudal vertebræ.
2. A dorsal vertebra.

ton of this animal, are figured and described in the spendid work on the fossils of that mountain, by Faujas St. Fond; and the nature of the original, has since been ably elucidated by M. le Baron Cuvier.

These remains have not previously been noticed in England; and have been found in Europe in the immediate vicinity of Maestricht only, where they occur in a soft, yellowish, calcareous freestone. This limestone reposes upon the flinty chalk, and contains beds of flints perfectly resembling those of the chalk formation.

The vertebræ here represented are from the Upper chalk near Lewes; and, being found in the same quarry, and at a short distance from each other, may probably have belonged to the same individual. Like those of the *crocodile*, *monitor*, *iguanas*, and the greater part of the saurian animals, they have the body convex posteriorly, and concave anteriorly: a structure that distinguishes them from those of the cetacea and fishes.

Fig. 2. appears to correspond most completely with the posterior dorsal vertebræ in the spinal column of the Maestricht monitor, figured by Faujas, Pl. 52.; particularly with the third and fourth vertebræ, reckoning from the left hand of the specimen. The body of the vertebra is rather compressed, about two inches long, and 1·4 inch high; the face is slightly elliptical. The convexity of the posterior extremity is but slight, and the concavity of the opposite side of a corresponding depth, the surface being perfectly plain and smooth. The spinous process, of which a fragment only remains, is compressed, and occupies the anterior four fifths of the body of the vertebra.

The specimen fig. 1. contains two vertebræ articulated to each other. They are shorter than the one above described, and each has an inferior apophysis. In their general characters, they resemble the vertebræ delineated in Pl. 7. and 8. of Faujas: their bodies are compressed, and their length and height nearly equal; their faces are elliptical in a vertical direction, the transverse diameter being 1·1 inch, and the longitudinal 1·5 inch. The dorsal apophyses are narrower than in the preceding example. The inferior apophysis is strong, and rounded at the base, and suddenly contracts into a spinous process, which, when entire, was probably several inches in length. As this appendage is placed rather laterally, it was suspected that another might exist on the opposite side, and that the union of the two would form a triangular bone, corresponding to the *os en chevron* of the crocodile, and other animals of the lizard tribe. To ascertain this point, the chalk was removed so far as was practicable, but not the slightest trace of another process could be discovered. This circumstance was extremely embarrassing; and the difficulty of explaining it was increased, upon perceiving that the apophyses in question were perfectly anchylosed to the bodies of the vertebræ, and not united by suture, as in the recent lacertæ. A careful perusal of Cuvier's observations on the osteological characters of the monitor of Maestricht enabled me, however, to explain this apparent want of agreement, in a very satisfactory manner; the researches of that philosopher having shown that the *posterior caudal vertebræ* possess the structure here described, "*l'os en chevron n'y est plus articulé, mais soudé, et fait corps avec elles.*" The situation of this inferior process, presents also another striking proof of the identity of the vertebræ before us, with those of the Maestricht monitor. In the lizard tribe in general, the chevron bone is placed at the *junction* of the vertebræ; and in the monitor, at the *posterior* part; but in the animal of St. Peter's mountain, it is attached to the *middle* of the vertebræ, as in the specimens before us.

That the reader may form his own opinion upon this interesting subject, Cuvier's anatomical description of the vertebral column of the Maestricht animal is here subjoined. The extract is rather long, but it will not be deemed irrelevant, when the importance of extreme accuracy in these researches is duly considered.

"Toutes ces vertèbres, comme celles des *crocodiles*, des *monitors*, des *iguanes*, et en général de la plupart des sauriens et des ophidiens, ont leur corps concave en avant et convexe en arrière, ce qui les distingue déjà notablement de celles des cétacés, qui l'ont à-peu-près *plane*, et bien plus encore de celles des poissons, où il est creusé des deux côtés en cône concave.

"Les antérieures ont cette convexité et cette concavité beaucoup plus prononcées que les postérieures. Quant aux apophyses, leur nombre établit cinq sortes de ces vertèbres.

"Les premières ont une apophyse épineuse

supérieure, longue et comprimée; une inférieure terminée par une concavité; quatre articulaires dont les postérieures sont plus courtes et regardent de dehors, et deux transverses, grosses et courtes; ce sont les dernières vertèbres du cou et les premières du dos. Leur corps est plus long que large, et plus large que haut; les faces sont en ovale transverse, ou en figure de rein. D'autres ont l'apophyse inférieure de moins, mais ressemblent aux précédentes pour le reste; ce sont les moyennes du dos.

"Il en est ensuite, qui n'ont plus d'apophyses articulaires; ce sont les dernières du dos, celles des lombes, et les premières de la queue; et leur place particulière se reconnoît à leurs apophyses transverses qui s'allongent et s'aplatissent. Les faces articulaires de leur corps sont presque triangulaires dans les postérieures.

"Les suivantes ont outre leur apophyse épineuse supérieure et les deux transverses, à leur face inférieure deux petites facettes pour porter l'os en chevron; les faces articulaires de leur corps sont pentagonales.

"Puis il en vient qui ne différent des précédentes que parcequ'elles manquent d'apophyses transverses. Elles forment une grand partie de la queue, et les faces de leur corps sont en ellipses, d'abord transverses, et ensuite de plus en plus comprimées par les côtés. L'os en chevron n'y est plus articulé, mais soudé, et fait corps avec elles.

"Enfin, les dernières de la queue finissent par n'avoir plus d'apophyses du tout.

"A mesure qu'on approche de la fin de la queue,

les corps des vertebres se raccoursissent, et presque, dès son commencement, ils sont mois longs que larges et que hauts. Leur longueur finit par être moitié moindre que leur hauteur."*

From this investigation, I think we may, without hesitation, refer the vertebræ before us to the fossil animal of Maestricht.† The specimen fig. 2. is evidently one of the posterior dorsal vertebræ; those represented fig. 1. are two of the posterior vertebræof the tail.

In conclusion, it may be observed, that Cuvier has ascertained that the original animal formed an intermediate genus between the lizards with a long and forked tongue, including the *monitors* and *common lizards;* and those with a short tongue and dentated palates, comprising the *iguanas*, *marbres*, and *anolis*. This genus, he thinks, would only have been allied to the crocodile, by the general characters of the lizards. The length of the entire skeleton appears to have been nearly twenty-four feet; the head being equal to a sixth of the whole length. The tail must have been very strong, and the width of its extremity so considerable, as to have rendered it a powerful oar, by which the animal could stem the most agitated waters.

From this peculiar structure, and from the character of the organic remains with which those of

* Animaux Fossiles, tome iv. Animal de Maestricht, p. 20, 21.

† This opinion is confirmed by the observations of Mr. König, who obliged me by comparing the drawings of the Sussex specimens with the vertebræ of the Maestricht monitor, in the British Museum, and expressed himself perfectly convinced of their identity.

the Maestricht animal are associated, M. Cuvier concluded that the original was an inhabitant of the ocean ; a circumstance very remarkable, since none of the existing lacertæ are known to live in salt water.*

Crocodile. — There are three teeth in my cabinet which possess the external characters of those of the crocodile, but do not contain within them traces of the supplementary tooth. Baron Cuvier, to whom I showed them in 1831, decided that they must have belonged to some species of this genus.

Undetermined Reptiles. — A lower jaw, with twelve smooth, pointed, slightly convex teeth, was figured in the Fossils of the South Downs, as the jaw of a fish. There can scarcely admit of a doubt that it belongs to a saurian animal: a figure is annexed. The original is five inches long; the fangs of the anterior teeth, like those of the crocodile, are hollow, fixed in sockets, and not attached to the jaw ; but their smooth polished surface, and flattened form, separate them most decidedly from the animals of that family. The posterior teeth are affixed to the edge of the jaw, a mode of dentature

* Vide Cuvier's interesting description of the remains of the animal of Maestricht, *Oss. Foss.* tome iv.

In the course of this volume it will be seen that the Mososaurus had an *os tympani*, approaching more nearly to that of the monitor than to the crocodile. As the bones of the extremities have not been discovered, it is by no means certain that the original was exclusively aquatic, much less that it was confined to the ocean.

Teeth of the Mososaurus have been found in America by Dr. Morton: they so perfectly approximate to those of Maestricht, that they must have belonged to the same species.

JAW OF A REPTILE.

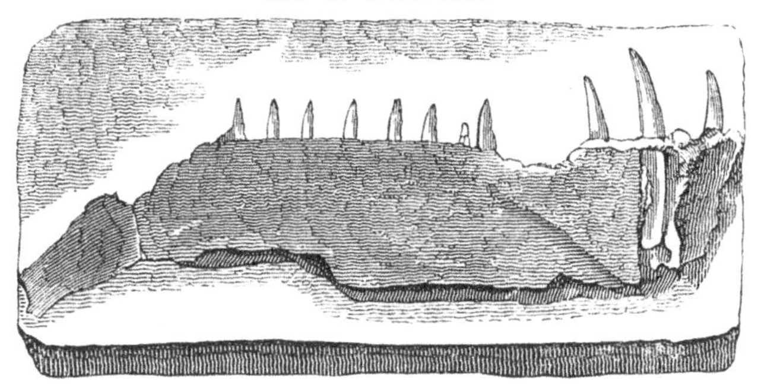

observable in many kinds of fishes. The structure of a vertebra found with the jaw is decidedly that of a fish, the conical cavities being very deep; and it possesses the annular markings so constantly observable in the vertebræ of fishes. A cylindrical bone was also found, but was too much injured to allow of any correct inference being drawn from it. The posterior extremities of a lower jaw of a reptile were found in the same block of chalk, with portions of the upper and lower maxillæ bearing many teeth, and corresponding with those of the last mentioned fossil; and fragments of other maxillæ have since been discovered: the materials at present in my possession are, however, too imperfect to admit of the zoological relations of these remains being accurately determined.

CHAP. VII.

3. CHALK MARL.

This deposit constitutes the foundation of the chalk hills, its outcrop forming a fillet or zone round their base, and connecting the detached parts of the range with each other.

The marl is commonly soft and friable, but indurated blocks occur which possess the hardness of limestone. It is of a light grey colour, inclining to brown, and frequently possesses a ferruginous tinge derived from oxide of iron. It consists principally of carbonate of lime and alumine, with an intermixture of silica, a very small proportion of iron, and perhaps of oxide of manganese.

Where denuded, the surface of this deposit composes a fertile tract of arable land, including some of the best farms in the country.

In the range of low cliffs near Southbourn, the grey marl is seen rising from beneath the chalk, and reposing on the firestone, with which it is intermingled at the line of junction. Its separation from the superincumbent bed of chalk without flints is well defined, and may be traced with but little difficulty. From this spot it extends, with scarcely any interruption, to Shoreham river, its outcrop being interposed between the foot of the Downs and the basseting edge of the galt.

In western Sussex it occupies the same relative position; the lower chalk passing insensibly into the grey marl, and the latter into the malm rock.

In its course through this tract of country, it forms a few hillocks or mounds of low elevation, which are remarkable only for the abundance and variety of their fossil remains. I shall proceed to notice a few of the more interesting localities.

A low bank at Middleham, in the parish of Ringmer, near the seat of the Rev. J. Constable, contains *hamites*, *turrilites*, *nautilites*, *ammonites*, and *inocerami*.

Stoneham, near Lewes. From a marl bank in a field adjoining the turnpike-gate, I have collected the same kinds of fossils as at Middleham; also *rostellariæ*, *auriculæ*, *scaphites*, &c.

Hamsey Marl Pits. The hillock, of which these pits present a vertical section, is insulated by the river Ouse. The quarries are situated to the north of the church, and are about 25 feet high. The strata are slightly inclined, and vary from a few inches to a foot or more in thickness; the indurated layers are separated by intervening seams of a soft loose marl, of a dark colour. The face of the rock is traversed by innumerable crevices, which, in some instances, are parallel with the stratification, and in others assume a vertical or transverse direction.

The lowermost strata are of a bluish grey colour, indicating a transition to the galt, into which the grey marl passes, at the depth of a few yards. These quarries contain sulphuret of iron, and spicular crystals of carbonate of lime; the former often composes the constituent substance of the fossils, the calcareous spar occurs in the fissures and cavities in the marl. The organic remains found in these pits are very numerous; they consist of several species of *ammonites*, *nautili*, *turrilites*, *scaphites*, *hamites*, the teeth and vertebræ of sharks, &c.

Offham Pit. This excavation lies on the road-side, between Offham and Cooksbridge, to the east of the road near the seat of Thomas Partington, Esq.; it produces *ammonites*, *nautili*, *turrilites*, *scaphites*, &c.

Clayton, near Hurstperpoint. A marl pit at this place has afforded to the researches of Mr. Weekes, *turrilites*, *hamites*, *ammonites*, *scaphites*, &c.

In other localities of the marl, the fossils are less abundant than in those above enumerated, and the *turrilites*, *hamites*, and *scaphites* but rarely occur.

On the surface, a narrow belt of this deposit appears to encircle Lewes Levels, separating the latter from the edge of the chalk hills: this want of continuity, however, does not extend beneath the surface: the marl is invariably found upon sinking through the alluvial clay, of which the Levels are composed. Protrusions of the marl through the clay occur in some situations, and these form islands when the levels are inundated; a circumstance that, previously to the improved state of the navigation of the Ouse, was of very frequent occurrence.

MINERALS.

The mineralogical productions of the grey marl are few, and offer but little variety: they consist of various modifications of sulphuret of iron, and crystallized carbonate of lime.

1. Crystallized carbonate of lime.

This mineral is frequently semi-diaphanous, varying in colour from a lightish grey to a gall-stone yellow. It occurs in inconsiderable veins, and occasionally in groups of crystals, lining the cavities of the marl: the usual form of the crystal is that of an acute rhomboid; of this kind some interesting specimens have lately been discovered at Hamsey.

2. Sulphuret of iron, or iron pyrites.

This substance, from the decomposition of its surface, is generally of a yellowish rusty brown colour externally. It occurs in a variety of irregular fantastic shapes, and oftentimes bears the impression of organic bodies, forming casts of *terebratulæ*, *pectenites*, *madreporites*, and the inner volutions of *scaphites*. Small spherical masses with an elongated stem, their surface beset with obscure pyramidal crystals, and exposing a brilliant radiated structure internally, are not uncommon. One specimen in my possession contains, within a cavity, small crystals of *sulphate of lime*.

Crystals of pyrites terminating in the quadrangular pyramid of an octohedron, and disposed in irregular groups, are often imbedded in the casts of ammonites and other fossil remains; and the marl pits at Hamsey contain masses of this mineral, bearing the form of a species of *Eschara*, somewhat resembling *E. foliacea*.

3. Oxide of iron, in the state of a reddish brown powder, is frequent in cavities of the marl, and has probably been produced by the decomposition of

iron pyrites; the greater part of the marl fossils have acquired a ferruginous colour from this mineral.

4. Clay slate. The occurrence of this substance in the marl is clearly accidental, having been derived from some regular bed of argillaceous slate of anterior formation to the chalk marl. The only examples hitherto discovered were imbedded in the marl at Southerham, near Lewes; the largest is about two inches square, and nearly half an inch thick: the edges are sharp, and the specimen appears to have suffered but little from attrition.

ORGANIC REMAINS.

The grey Chalk marl, in its course through Sussex, abounds in organic remains, which differ both in their nature, and in the mode of their preservation, from those of the superincumbent bed of lower chalk, and of the galt beneath.

Ammonitæ, *nautili*, *pectenitæ*, and *inocerami* are the most common productions of the pits near Lewes: which also contain *turrilitæ*, *scaphitæ*, *hamitæ*, &c. These remains of testacea very rarely exhibit any vestige of their original shelly covering, but consist of casts of indurated argillaceous limestone, of an ochraceous or a ferruginous colour, more or less distorted by compression. Fish and crustacea are rare: zoophytes related to the alcyonia and spongiæ are not of unfrequent occurrence. Echini are not abundant. Wood, in the state of that described as occurring in the chalk, is occasionally met with: and very rarely, traces of confervæ.

Of the simple univalves, the shells of the genera voluta, buccinum, rostellaria, auricula, and trochus are the most remarkable.

In this division of the chalk the multilocular genera nautilus, hamites, ammonites, scaphites, and turrilites, appear in great numbers, and in many localities are by far the most abundant of the organic remains. We subjoin figures of two species of turrilites.

TURRILITES FROM HAMSEY.

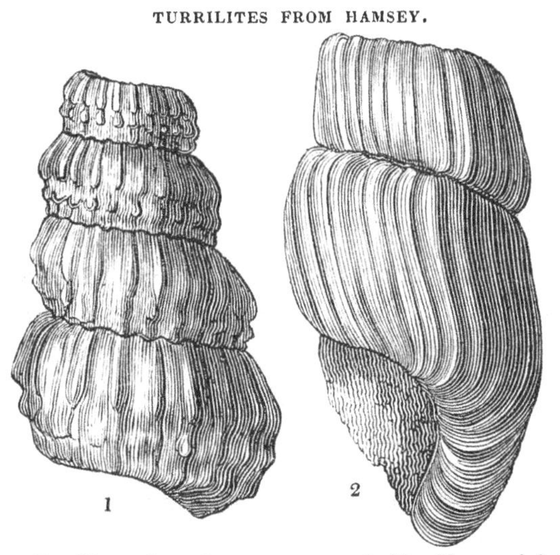

1. Turrilites tuberculata. 2. Turrilites undulata.

There are, probably, no localities in England so rich in the various species of *turrilites*, as the marl pits in the vicinity of Lewes ; and the first British specimens of the genus, as well as of *scaphites*, were found by the writer in Hamsey marl banks. A specimen of *turrilites tuberculata*, nearly two feet long, the only instance in which traces of the siphunculus are visible, was discovered in a marl bank on the estate of the Rev. J. Constable, at Middleham, near Lewes, and is figured by Mr. Sowerby, Min. Conch. tab. 74.

FOSSILS FROM HAMSEY.

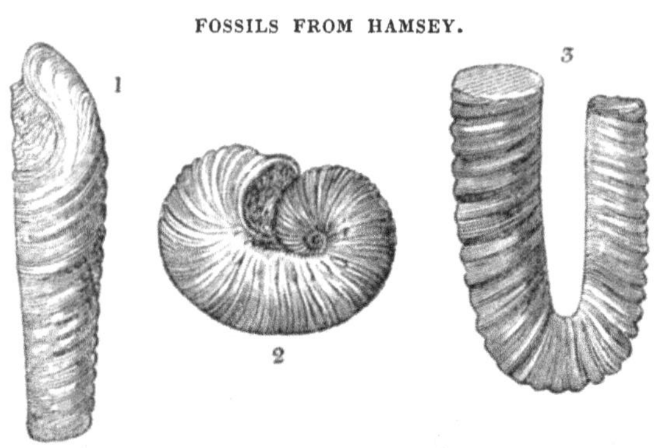

1. Baculites obliquatus. 2. Scaphites striatus. 3. Hamites intermedius.

The bivalves of the chalk marl need not detain us long: a very characteristic shell is a large pecten, *P. Beaveri*. There are also species of *plicatula* and *cardita*, which do not occur in the other divisions of the chalk.

---

4. FIRESTONE, OR MERSTHAM BEDS.

The chalk marl rests upon an arenaceo-argillaceous deposit of a greyish green colour, composed of marl and grains of silicate of iron* ; in some

* The green particles of the firestone (*Glauconie crayeuse*) of Havre have been analysed by M. Berthier, who found them to consist of

Silica - - -	50
Protoxide of iron -	21
Alumine - - -	7
Potash - - -	10
Water - - -	11

M. Brongniart names these grains, a granular chlorite of iron.—*Géog. Min.* p. 249.

places, in the state of sand; in others, forming a stone sufficiently hard for building. The transition from the marl to the firestone is in many localities so gradual, and the sandy particles are so sparingly distributed, that the chalk marl may be said to repose immediately on the galt; in others, however, the characters of the firestone are very peculiar, and some geologists have deemed them sufficiently important to rank this deposit as an independent formation.

The low cliffs, near Southbourn, expose a section of the arenaceous variety of the firestone. Eastward of Beachy Head, and to the west of Holywell quarries, the chalk marl is seen under the chalk without flints, dipping about 5° to the S.W.*; proceeding along the beach, a bed of greyish sand emerges from beneath the marl, to the east of the first martello tower; this quickly rises, till it constitutes one half of the cliff, and beneath it is seen a stratum of friable sandstone of a deep green colour; at the distance of about forty or fifty yards, the cliff, which is twenty feet high, is entirely composed of these strata, with the exception of a covering of alluvial loam on the summit. The alluvial tract called "the wish" obscures the beds near this spot, but they re-appear at about a hundred yards to the eastward, where the firestone forms the base of the cliff, and is

* The lowermost bed of marl, and which is in contact with the firestone, is almost wholly composed of the remains of ramose milleporites, madreporites, &c., so as to form a ridge or reef of corals: in this bed we found a long cylindrical zoophyte of the same kind as those which occur in the vale of Pewsey, in Wiltshire; it is partly composed of chert.

covered with rubbly marl of a greenish-yellow colour; the latter being twelve, and the former six feet in thickness. Approaching the sea-houses, the firestone occupies the middle of the cliff, resting on grey marl six feet thick, with scarcely any intermixture of sand: chalk marl, regularly stratified, lies above it; and were we to judge of the geological character of the firestone from this locality only, we should certainly consider it to be a subordinate bed of the chalk marl.* At the sea-houses, the firestone gradually descends and forms the base of the cliffs, which are there of an inconsiderable height; the buildings along the sea-shore obscure the strata to the eastward, and prevent the junction of the firestone and galt from being seen; specimens of the latter are observed, however, on the shore at low water, and it rises to the surface not far from the library. The fossils

SECTION OF THE CLIFFS NEAR SOUTHBOURN.

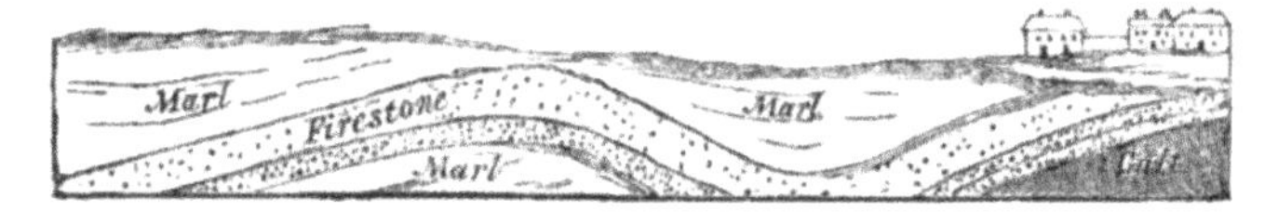

discovered in the firestone of Southbourn are, with but few exceptions, similar to those which are common in the chalk marl: viz. ammonites varians, mantelli, &c., turrilites, scaphites, &c.

In attempting to trace the firestone through the interior of the county, we find its course, in many localities, but obscurely indicated, and in some, the prevalence of a few green particles in the lower

* See the section of the cliffs at Southbourn in the plate.

beds of chalk marl is the only proof of its presence. It is not until we approach the Adur that its characters are sufficiently developed to merit attention. It may be observed at Edburton and Poynings; and at Steyning, which stands near the northern escarpment of the Downs, to the west of the Adur, a bluish grey marlstone emerges from under the chalk marl, and forms a terrace of inconsiderable breadth; this is the first appearance of that variety of the firestone which, in the west of Sussex, is called "*malm rock*," and which, interposed between the chalk marl and the galt, must be regarded as the equivalent of the more arenaceous strata of Southbourn. This variety of the firestone was first pointed out to me by John Hawkins, Esq. F.R.S., of Bignor Park, a gentleman to whom I am indebted for much interesting information relating to the geology of Sussex, and whose contributions to geological science, form so important a feature in the Transactions of the Geological Society of Cornwall. Mr. Martin of Pulborough, in his Geological Memoir of a Part of Western Sussex, observes that "the transition from the pure chalk to the malm is through an intermediate chalk marl; its outcrop is obscure, and does not form a country distinct from the other chalk beds, nor is it quarried for domestic purposes; but when accidentally opened, is found to abound in a great variety of marine fossils. This marl passes into the *malm*, of which there is a thin stratum, succeeded by a bed of green sand, and that by the more indurated malm rock, which again gradually resolves itself into the galt beneath. The broadest exposure of the great body of malm rock is between Sutton and Bury, under shelter of the projecting hills of Duncton and Sutton. The wells are from 30 to 100 feet deep, and the average depth of the whole stratum may be about 70 or 80 feet." From Steyning, the firestone, gradually acquiring greater breadth and thickness, may be traced along the northern edge of the Downs to Sullington, Amberley, Bury, Barlavington, Sutton, Elstead, Nursted, &c.; it forms a terrace of considerable breadth along the eastern edge of the Alton chalk hills, and on it are situated Selbourne, Binsted, Bentley, &c.* On the southern margin of the Hog's Back, in Surrey (the commencement of the North Downs), it is scarcely seen, but appears at Reigate; and may be traced by Merstham, Godstone, and through Kent, to its termination on the coast, near Folkstone, forming a boundary line between the chalk hills and the galt which constitutes the vale of Holmesdale. In some localities in Sussex and Hampshire a bluish chert occurs in this deposit.

Organic Remains. — The firestone contains the same fossils as the grey marl; and a few species

* Mr. Murchison, in his excellent Memoir on the Geology of the North-western Part of Sussex, &c. observes: "These terraces are covered by a tenacious greyish white soil, celebrated for its wheat crops. From the rapid decomposition of this rock the roads are worn into deep hollows, and in many places present sections twenty or thirty feet deep. The harder beds are used for building. The deep and woody glens which intersect the escarpment of the firestone, offer the most picturesque varieties of landscape."

that have not, I believe, been found in any other bed. One of these is the *Ostrea carinata*, figured in White's Natural History of Selborne, and which has been found at Southbourn; and a species of ammonite, *A. planulatus*.

AMMONITES PLANULATUS, FROM THE FIRESTONE, SOUTHBOURN.

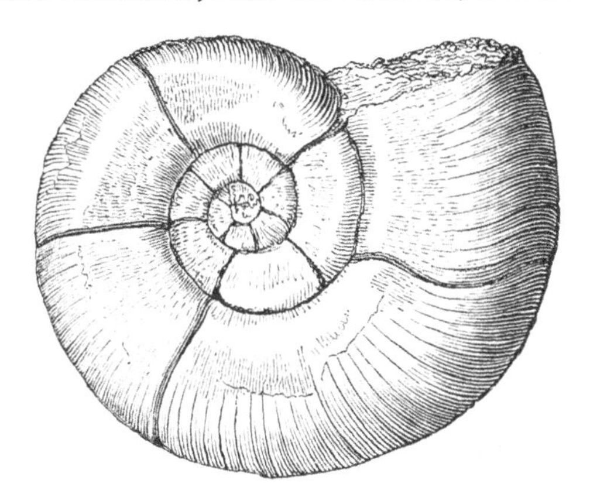

The *Gryphæa vesiculosa* occurs at Nursted and Amberley; and an echinus, named by Mr. König, in honour of Mrs. Murchison, the highly talented lady of the late president of the Geological Society of London. The most interesting fossil is, however, a species of fucus, which I observed in the stone on the road-side near Bury, on my first visit to the Roman villa at Bignor, soon after the discovery of that splendid remain of antiquity, and with beautiful specimens of which, I have been supplied through the kindness of Mr. Hawkins. This fucus is figured by M. Adolphe Brongniart, in his Végétaux Fossiles, under the name of *Fucoides Targionii*: wood, and the claws of a species of astacus, also occur.

FUCOIDES TARGIONII, FROM NEAR BIGNOR.

5. GALT*, OR FOLKSTONE MARL. BLUE CHALK MARL.

A bed of stiff marl, varying in colour from a light grey to a dark blue, and abounding in ammonites, nautili, and other marine shells, succeeds the firestone, emerging from beneath the northern edge of that deposit where the latter is visible, and forming the base of the chalk marl where the firestone is absent. It generally constitutes a valley within the central edge of the chalk of Sussex, Hampshire, Surrey, and Kent, and may be traced, with but little interruption, from Southbourn, through Laughton, Ringmer, Plumpton, New-Timber, Steyning, Bignor, &c. in Sussex, to Folkstone in Kent, near which town it forms a cliff, ce-

* *Galt*, or *gault*, a provincial term used in Cambridgeshire: it is employed by geologists from its not being so likely to mislead, as a name derived from the colour, or chemical and mineralogical character, of a deposit; since those characters may vary in different localities, or be equally applicable to some other formation.

lebrated, in the geology of England, for the beauty and variety of its organic remains.

The denuded surface of this bed forms a soil remarkable for its tenacity, and which, in many parts of Sussex and Surrey, is distinguished by the provincial term "black land:" it is thus described by Mr. Young: — "At the northern extremity of the Downs, and usually extending the same length, is a slip of very rich and stiff arable land, but of inconsiderable breadth; it runs for some distance into the vale before it meets the clay. The soil of this narrow slip is an excessively stiff calcareous loam, on a clay bottom; it adheres so much to the share, and is so very difficult to plough, that it is not an unusual sight to observe ten or a dozen stout oxen, and sometimes more, at work upon it. It is a soil that must rank amongst the finest in this or any other country, being pure clay and calcareous earth."* It generally occupies low situations, and where uncultivated, is covered by rushes and other plants, that affect a moist and clayey soil.

This deposit seldom exceeds 100 feet in thickness. It may be traced, with but little difficulty, from near Laughton Place, six miles N.E. from Lewes, through Ringmer, Hamsey, Offham, Plumpton, near Ditchling, Clayton, New Timber, &c. to Beeding, and along the northern edge of the Downs to Bignor Park, &c.

The identity of this bed with the blue marl of Folkstone†, the galt of Cambridgeshire, and the

* Young's Agricultural Survey of Sussex.

† The blue marl of Folkstone has been well described by Mr. W. Phillips, and more recently by Dr. Fitton. Folkstone is built upon the

malm of Surrey *, cannot for a moment be doubted; not only is there a perfect agreement in their physical characters, but also in their geological position, and organic remains. The marl of Folkstone is said, by Mr. Phillips, to contain 30 per cent. of carbonate of lime; and that of Ringmer, upon being submitted to the action of acids, indicates a like proportion.

FOSSILS OF THE GALT.

Organic Remains. — The fossils of the galt, like those of other argillaceous strata, are remarkable for their beauty, the pearly covering of the shells being in most instances preserved. They consist of several species of *ammonites* and *hamites*; *nautili*; a delicate pellucid *belemnite*; many species of *nuculæ*, *inocerami*, &c. and of *turbinolia*, or *caryophillea*. Two of the most characteristic fossils are here delineated: viz. *Inoceramus sulcatus*, and *concentricus*.

---

green sand, and the cliffs on the east of the town are from eighty to ninety feet high, the upper part of which, for a considerable distance from their termination at Copt Point, consists of the blue marl. Crystallized sulphate of lime occurs in this bed, and numerous remains of shells with their pearly lustre still preserved. There can be no doubt that this deposit is altogether analogous to that underlying the chalk at Malling in Kent, in Cambridgeshire and Oxfordshire, and which, in the latter counties, is provincially termed Galt. — *Geological Transactions*, vol. v. p. 37.

* At the foot of the chalk hills near Godstone, and Bletchingley, the galt rises from beneath the grey marl; and I have collected from these localities precisely the same species of ammonites, belemnites, nuculæ, &c., as those which occur at Ringmer and Laughton.

INOCERAMI FROM RINGMER.

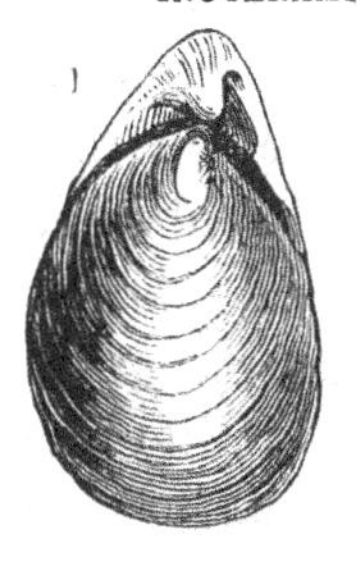

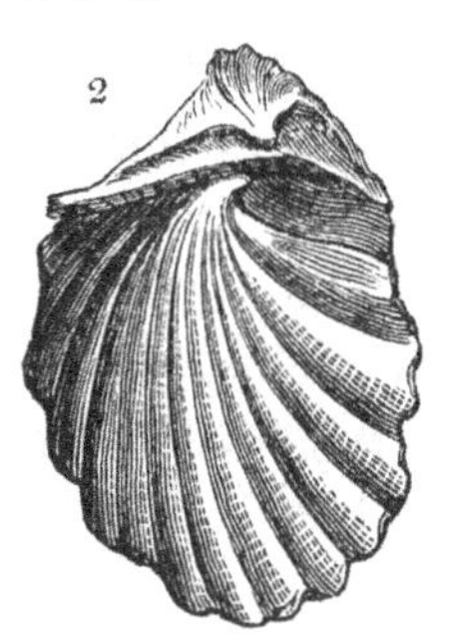

1. Inoceramus concentricus. 2. Inoceramus sulcatus.

The crustacea of the galt are but few, and are so remarkable that we must notice them in detail.

CRUSTACEA OF THE GALT.

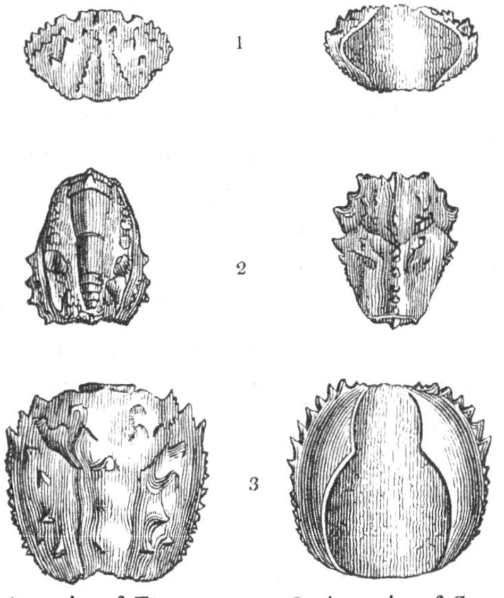

1. A species of *Etyus*. 2. A species of *Corystes*. 3. A species of *Arcania*.

To the kindness of William Elford Leach, M. D. of the British Museum, I am indebted for the following identification of their affinities.

1. " A species of a new genus of the family *Leucosiadæ* *, nearly related to the genus *Arcania*."

In these specimens the shell or crust of the thorax alone remains. It is of a suborbicular form, rather inflated, obscurely trilobate, with twelve or thirteen aculeated tubercles; the margin is dentated.

2. " A species of a genus of the family *Corystidæ* †, allied to a new Indian genus in the cabinet of Dr. Leach."

The shell is oblong, ovate, depressed; the surface covered with minute granulæ, the margin bidentated near the front. No vestiges of the legs, antennæ, or claws, remain.

3. " A species of the genus *Etyus*, of the family *Canceridæ*."

Transversely obovate, obscurely trilobate; the surface covered with irregular papillæ.

4. " These belong to a genus of the family *Corystidæ*, intimately related to *Corystes*."

This species is longitudinally obovate, convex, with a tuberculated dorsal ridge, having a row of three tubercles on each side. The shell is truncated posteriorly, and the margin laterally tridentated. The abdomen is composed of six or seven arcuate segments, and there are three or four legs on each side.

* The recent *Leucosiadæ* have two or four small quadriarticulate antennæ inserted between the eyes. The tail is naked; they have eight legs, all furnished with claws; and two chelate hand claws. — *Rees's Cycloped.* art. *Cancer.*

† The *Corystidæ* have four antennæ; the external pair approximate, setaceous, ciliated, and very long. The eyes remote and pedunculated. The shell is oval, and longer than wide; the tail folded under the body when the animal is in a state of repose. They have ten legs; the anterior pair chelate, the others terminating in an acute elongated nail or claw. Vide *Lamarck*, *Animaux sans Vertèbres*, tome v. 233.

5. Fragments of the abdomen of two species of Astacidæ.

---

6. SHANKLIN, OR LOWER GREEN SAND.

Silicious sands and sandstone of various shades of green, grey, red, brown, yellow, ferruginous, and white, with subordinate beds of chert, limestone, and fuller's earth, constitute the group of deposits to which the name of *Shanklin-sand* formation is now appropriated. These beds are but obscurely seen in the east of Sussex, but as we proceed towards the west they gradually rise into hills of considerable altitude, and form a striking feature in the physical geography of the country. They admit of a triple division, as was first shown by Dr. Fitton in his admirable memoir on the beds below the chalk.*

"The first or uppermost consists of sand, with irregular concretions of limestone and chert, sometimes disposed in courses oblique to the general direction of the strata. The top of this sand in the vicinity of Folkstone and Hythe forms an extensive plateau, resembling that of the Blackdown range of hills in Devonshire.

"The second consists chiefly of sand, but in some places is so mixed with clay, or with oxide of iron, as to retain water: it is remarkable for the great variation in its colour and consistency.

"The third and lowest group abounds much more in stone; the concretional beds being closer together and more nearly continuous."

* Annals of Philosophy, 1827.

In the south-eastern part of Sussex, this formation occupies but an inconsiderable extent on the surface; and, in many instances, a few insulated hillocks are the only indications of its presence. At about three miles N.W. of Eastbourne, the green sand is covered with a thin layer of galt, which occurs immediately beneath the surface; the sand abounds with rounded fragments of coniferous wood, that occur in a bank on the road-side, near the Folkington road. The specimens are incrusted with a covering of grey sand containing small pebbles of quartz, and internally are of a light reddish brown, clouded with darker shades of the same colour. The wood is calcareous, and bears a good polish, the transverse sections displaying, in a distinct and beautiful manner, the radial insertions and annular markings, which denote the annual growth of the tree.

In some instances, the wood is studded with the remains of a small species of Fistulana, of a pyriform shape, about 0 3 inch long, bearing some resemblance to F. *lagenula*, or F. *ampullaria*, of Lamarck; the bivalve part of the shell has not been detected, but is in all probability enveloped in the indurated sandstone with which the tubes are filled. This species of Fistulana appears to be new, and may be distinguished by the name of F. *pyriformis*.

In a bank on the south side of the road, leading from Selmeston Fair Place to Chilver Bridge, fossil wood of the same kind as that of Willingdon, has been noticed by Mr. Wm. Figg, jun.

At Chilley, near Pevensey, a bed of sandstone very strongly impregnated with bitumen, occurs beneath a thick layer of marsh land, or silt. It was discovered a few years since, by the late Mr. Cater Rand, of Lewes, while superintending the execution of some improvements in the drainage of Pevensey Levels.*

This substance is of a dark chocolate colour, is easily scraped with a knife, and emits a strong bituminous odour. Exposed to the action of the hydro-oxygen blow-pipe it burns with a bright flame, and fuses into a steel grey enamel.†

SECTION FROM NEAR SOUTHBOURN TO PEVENSEY.

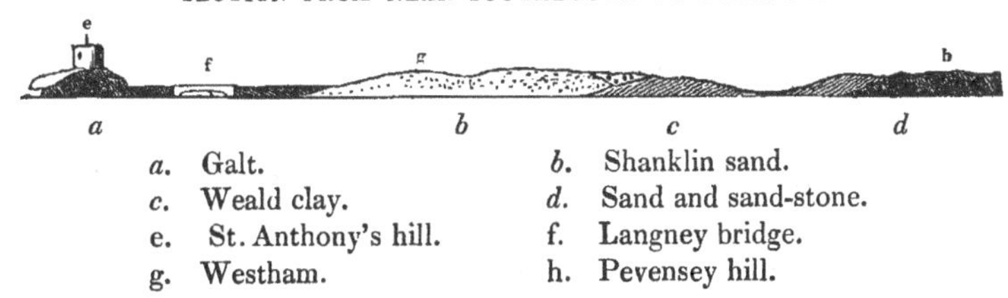

a. Galt.
b. Shanklin sand.
c. Weald clay.
d. Sand and sand-stone.
e. St. Anthony's hill.
f. Langney bridge.
g. Westham.
h. Pevensey hill.

At Langney point, near Southbourn, and at Pevensey, the sand may be observed, although much covered by alluvium; and may be traced westward, in a line parallel with the outcrop of the galt, through Arlington, Selmeston, Laughton, Norlington, &c. to Ditchling, where sand and sandstone of a red colour, suddenly rise from beneath the galt, near the foot of the Downs, and form the mound on which that town is situated.

To the west of Ditchling, the sand gradually as-

* This bed was worked by the Romans, who employed it in the construction of part of Pevensey Castle. In the alluvial clay near Chilley, Mr. Rand discovered the remains of a Balista, and a considerable number of enormous balls of bituminous sandstone; the latter were in all probability intended to supply the engine which (as is well known) was employed for hurling large stones.

† A specimen analysed by my brother contained 15·4 per cent of bitumen.

sumes a bolder aspect, and constitutes an elevated tract, which runs parallel with the northern escarpment of the South Downs. This range passes into Surrey, and finally into Kent, terminating in the cliffs of Folkstone and Hythe.

An instructive section of the sands, which much resembles that of Dunnose-point in the Isle of Wight, is seen at Stone-pound-gate, near Hurstperpoint, on the Brighton road. The bank is nearly 30 feet high, and has on the top a layer of loam and clay, beneath which is a bed of bluish clay, with an intermixture of sand; this reposes on the greenish ferruginous sand, that forms the base of the hill. Sand of a deep reddish brown colour mottled with pure white, appears on the south side of the turnpike-gate; and even here, although on so small a scale, the triple division of the beds is visible. The weald clay comes out from beneath the sand, to the north of this place.

The pretty village of Hurstperpoint, well known from the select and interesting collection of fossils and antiquities of Richard Weekes, Esq. F.S.A. &c., stands on this deposit: and in the plantations at Danny, many interesting sections of the red and ferruginous sands are exposed.

At Henfield Hill a very interesting section is formed by the road leading from Steyning over Broadmere Common, as Mr. Martin * has ob-

* Geol. Mem. p. 29. Farther west, Mr. Martin remarks, that the road from Sutton to Petworth affords a good display of the whole series: 1. The arid upper sand at Crouch Common; 2. The argillaceous, slaty, and moist strata between Shophambridge and Strond; 3. The green sand and blue limestones from Strond on through Byworth.

served. The mill stands on the ferruginous sand with quartzose ironstone; the argillaceous beds constitute the middle portion of the hill: the green sand with limestone is seen at the base, forming a low escarpment towards the common, which is composed of weald clay.

SECTION OF HENFIELD HILL.

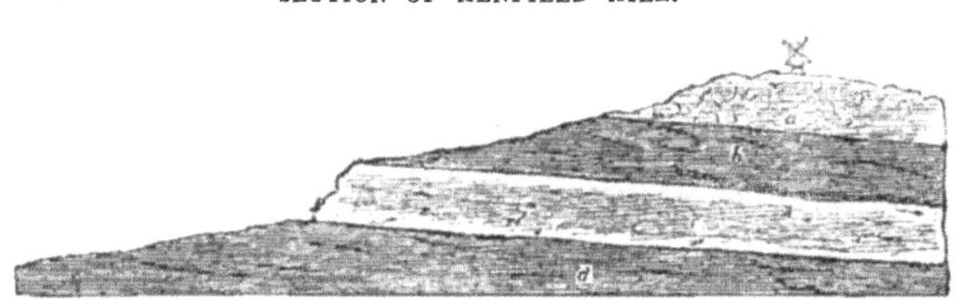

a. Upper ferruginous beds of the Shanklin sand.
b. Argillaceous or middle division.
c. Lowermost beds, with arenaceous limestone.
d. Weald clay. The strata incline to the north.

In the western part of Sussex, layers of chert or hornstone, provincially termed "*whinstone*," prevail in the sand near Petworth, at Bexley-heath, Black-down Hill, &c. The chert above mentioned is a variety of hornstone; it occurs massive, is either of a greyish yellow, or greenish colour, has a conchoidal fracture, and a glimmering lustre. To my excellent friend, John Hawkins, Esq. of Bignor Park, I am indebted for the following account of its characters and position.

"This stone is a compact mass of quartz, but not homogeneous, for it contains iron, and perhaps some other substance. It occurs in great abundance in the beds of our building stone, a ragged sandstone, which constitutes a chain of hills running E. and W. on the north side of the Arun and Rother: the strata there have a regular dip to the south, and basset out on the north. At Petworth, and on the top of the high hill to the eastward, this hard stone is dug in great quantities for repairing the roads *, and I know not in any country a better material. It is usually disposed in irregular beds in the sandstone, but occasionally forms veins, which intersect the strata; I call the beds irregular, because they vary much in breadth and appear not to continue far.

"The Whinstone† shows a transition into the sandstone, and they are certainly of contemporaneous formation. In some situations near Petworth, great lenticular masses of this substance are imbedded in the friable sandstone, and these follow the same sedimentary line as the beds of sandstone, although separated from each other by very wide intervals: they are therefore unquestionably in situ. These masses frequently measure eight or ten feet by two or three, and are invariably surrounded by a more friable and ochraceous sandstone than the rest of the strata."

The whinstone is sometimes traversed by veins of chalcedony, and it also contains cavities lined with mammillated concretions of the same substance.

At Parham, near the village of Storrington, and in the adjoining parish of Rackham, the sand is highly ferruginous, and contains irregular concre-

* It was also noticed on Bexley Heath, and on the sides of Blackdown Hill, by Mr. Lyell.

† *Whinstone*, the name by which the chert is distinguished in Western Sussex, is probably of Saxon origin; it is unknown in the south-eastern part of the county.

tions of ironstone. These beds are peculiarly interesting, from the abundance and variety of their organic remains; which consist of casts of shells of the genera Venus, Trigonia, Gervillia, Thetis, Modiola, &c.

A bed of calcareous grit, which sometimes passes into a blue limestone, occurs at Dean Farm, near "Petworth, and may be traced to Stedham, Rogate, "Lyss, Headley, and Godalming: in the last-mentioned localities, the grit is a conglomerate of "quartz grains and pebbles, held together by a "calcareous cement, and is called *Bargate stone*."* This bed extends eastward through Surrey and Kent. Mr. Murchison observes, that the lowest beds of the sand are marked by a great change of appearance; green particles prevail in the mass, and all the hills immediately adjoining the valley of the Weald clay afford a yellowish sandstone, filled with green particles, which is used for building, and is the only bed of the Shanklin sand formation, west of the river Arun, that has been observed to contain shells, or rather casts,—probably of ammonitæ and terebratulæ.

Tilvester Hill, on the south of Godstone, in Surrey, presents a most interesting section of this formation; and as the main road from London to Lewes passes directly over it, and a considerable fault or displacement of the strata is exposed on the road-side near the top, this locality generally attracts the notice of travellers. For this reason, as well as

* Mr. Murchison, in Geolog. Trans. new series, vol. ii. p. 101.

from the section being an excellent example of the beds of chert, &c., a sketch of the hill is given in the plate of sections, which the following description will explain. From Fellbridge waters, a few miles to the north of East Grinstead, the Weald clay prevails, and its peculiar limestone, with *paludinæ*, is dug up in a farm at the foot of Tilvester Hill. Ascending the hill, sand resembling that of Red-hill, near Reigate, appears, containing veins and large concretionary masses of ironstone: this is succeeded by grey sand and layers of chert, alternating with marl, and sand of various colours. From their first emergence, to near the summit of the hill, the strata are highly inclined, dipping at an angle of about 40° to the north: they then reappear, with but a slight degree of inclination, the red sand forming a bank on the side of the road, nearly 30 feet high. On the northern slope of the hill, the chert is seen immediately beneath the turf, and is quarried for the roads. Godstone is situated on the sand, but the windmill near it stands on the galt, which forms a plain or terrace extending to the foot of the neighbouring chalk hills, where the firestone, marl, and chalk, appear in their usual order of superposition. The following is a more particular enumeration of the strata at Tilvester Hill :—

1. First and lowermost deposit, Weald clay with Sussex marble.
2. Red and fawn-coloured sand, with ironstone.
3. Grey sand, with veins of a reddish colour.
4. Greyish green sand.
5. Olive green, and ferruginous sand.

6. Grey sand, with veins, and lenticular masses of chert: thickness five feet.*

7. Sandstone, with veins of ironstone: thickness four feet.

8. Chert in layers, alternating with sand and marl: thickness twelve feet.

9. Grey, green, and red mottled sand: thickness two feet.

10. Pale yellow sand.

11. Diluvial loam and rubble.

The fullers'-earth pits of Nutfield, near Bletchingly, so celebrated for magnificent specimens of sulphate of barytes, are situated in the range of Shanklin sand hills, of which Tilvester Hill is a part.

Organic Remains.—In some parts of England, the Shanklin sand contains a numerous and highly interesting assemblage of organic remains; but in Sussex, fossils rarely occur, and Parham Park, and its vicinity, are the only productive localities at present known. Casts of univalves and bivalves, particularly of trigoniæ, gervilliæ, and rostellariæ, are there found in profusion, in the indurated blocks of ferruginous sand.†

Many new species have been discovered by Mr. Martin, in the middle and lower group of sands in the vicinity of Pulborough: among those for which I am indebted to his liberality are *Mya mandibula*, *Trigonia spinosa*, *Nucula impressa*, *Mytilus edentulus*, *Pholadomya*, *Lenia*, *Lucina*, *Modiola*, &c.

* The chert in this bed, like that of Western Sussex, occurs in large lenticular blocks in the sand, and not in regular horizontal layers.

† Those genera and species which have been determined are mentioned in the catalogue: ammonitæ and terebratulæ are said to occur in the sandstone west of the river Arun.

CHAP. X.

THE WEALDEN FORMATION.

THE very appropriate term of *Wealden*, was suggested by Mr. Martin, to designate the strata which in the South-east of England, are interposed between the lower arenaceous beds of the chalk formation and the Portland oolite; the Purbeck limestones and shales being considered as the lowest members of the group. The fluviatile origin of these various deposits is now universally admitted: and we may be permitted to express our gratification, that our humble labours have, in some measure, assisted in the establishment of a fact, of so much interest and importance to British Geology.

Taken in a general view, this formation may be described as a series of clays and sands, with subordinate beds of limestone, grit, and shale, containing freshwater shells, terrestrial plants, and the teeth and bones of reptiles and fishes; univalve shells prevailing in the upper, bivalves in the lower, and saurian remains in the intermediate beds; the state in which the organic remains occur manifesting that they have been subject to the action of river currents, but not to attrition from the waves of the ocean.

The entire area comprised between the escarpments of the lower arenaceous beds of the chalk of Kent, Sussex, Surrey, and Hampshire is occupied by this formation, which forms an anticlinal axis, of considerable elevation, the direction of which is from nearly east to west; the strata diverging E. N. E. and W. S. W., and dipping beneath the inferior members of the chalk formation. The district of the Wealden may be described as an irregular triangle, the base extending from near Pevensey in Sussex, to Seabrook in Kent; and the apex being situated in western Sussex, near Harting Combe. An inspection of the map will render this description more intelligible. At the same time, the reader should bear in mind, that the above statement must be taken in the most general sense; for the strata have suffered so much displacement, that instances may be found of the dip being towards every point of the compass.

For the convenience of study, the Wealden strata may be separated into three principal divisions; namely, the Weald clay; the Hastings beds, including the strata of Tilgate Forest; and the Ashburnham or inferior limestones and shales, which are presumed to be the lowermost strata in Sussex. The formation, therefore, will admit of the following synoptical arrangement: it must, however, be remarked, that these subdivisions are arbitrary; for the strata are, in truth, exceedingly variable, both in their nature and extent.

WEALDEN FORMATION.

1. WEALD CLAY.

Characters.	*Oganic Remains.*	*Localities.*
Stiff clay, of various shades of blue and brown; with subordinate beds of limestone and sand; septaria.	Paludinæ, Cypris faba, cyclades, bones of reptiles rarely; scales and bones of fishes.	The Wealds of Sussex, Surrey, and Kent; forming the vale between the Downs and Forest Ridge.

2. HASTINGS BEDS.

Horsted Sand.

Grey, white, ferruginous, and fawn-coloured sand, and friable sandstone, with abundance of small portions of lignite.	Traces of carbonized vegetables.	Little Horsted, Uckfield, Framfield, Bexhill, Chailey, Fletching, Eridge Park, Tunbridge Wells, &c.

Strata of Tilgate Forest.

Sand and friable sandstone, of various shades of green, yellow, and ferruginous; surface oftentimes deeply furrowed. *Tilgate stone*, very fine compact bluish or greenish grey grit, in lenticular masses; surface oftentimes covered with mammillary concretions; the lower beds frequently conglomeritic, and containing large quartz pebbles.	Ferns, stems of vegetables: bones of saurian animals, birds, turtles, fishes, &c. Shells of the genera *unio, cyclas, cyrena, paludina*, &c., lignite, wood.	Loxwood, Horsham, Tilgate and St. Leonard's forests; Chailey, Ore near Battel, Hastings, &c., Rye, Winchelsea.
Clay or marl: of a bluish grey colour; alternating with sand, sandstone, and shale.	Bones, and shells, but rarely. Ferns; and stems of vegetables.	Tunbridge Wells; near Reigate.

Worth Sandstone.

White and yellow friable sandstone and sand.	Ferns, and arundinaceous plants. Lignite, &c.	Worth, near Crawley; St. Clement's Caves, Hastings, &c.

3. ASHBURNHAM BEDS.

A series of highly ferruginous sands, alternating with clay and shale, containing ironstone and lignite.	Ferns, and carbonized vegetables.	Lower part of Hastings Cliffs; near Buxted; West Hothly, Crawley, &c.
Shelly limestone, alternating with sandstone, shale, and marl; and concretional masses of grit.	Cypris. Shells of the genera cyclas and cyrena; lignite, carbonized vegetables.	Archer's Wood, near Battel; Brightling, Pounceford, Burwash, Hurst Green, Eason's Green, Etchingham.

---

1. WEALD CLAY.

A tenacious blue clay, containing subordinate beds of sandstone and shelly limestone, with layers of septaria of argillaceous ironstone, forms the subsoil of the Wealds of Sussex and Kent, and separates the Shanklin sand from the central mass of the Hastings beds. It constitutes a low tract, from five to seven miles in breadth, which forms a zone round the Hastings sands, and extends from the Sussex coast, near Pevensey, to Petworth and Harting Combe in the west of the county, and from thence passes to near Tunbridge and the Isle of Oxney in Kent. At its western angle at Harting Combe, it forms a narrow gorge for several miles, which is flanked on the north by the lofty hills of Blackdown, and on the south by the corresponding ridge, which extends from Holder and Bexley hills to near Petworth; the valley then suddenly expands to three times its former breadth, by the retirement at a right angle of the sand escarpment of Blackdown and Haslemere.*

Presenting no remarkable characters on the surface, except that the soil it produces is favourable to the growth of the oak, its range and extent scarcely require further observation. Its outcrop forms a valley between the hills of the Shanklin sand on the one hand, and the Forest Ridge on the other, throughout the northern and north-western division of the "southern denudation of the chalk;" but in the south-eastern part of Sussex, where the Shanklin sand is scarcely seen on the surface, it constitutes a valley at the foot of the northern escarpment of the Downs, its beds of limestone and sandstone forming longitudinal ridges.

The Sussex marble, so strikingly characteristic

* Mr. Murchison's Memoir. Geol. Trans. vol. ii. p. 104.

of the Weald clay in England, occurs in layers that vary from a few inches to a foot or more in thickness, and are separated from each other by seams of clay, or of coarse friable limestone.

POLISHED SLAB OF SUSSEX MARBLE.

This limestone is of various shades of bluish grey, mottled with green, and ochraceous yellow, and is composed of the remains of freshwater univalves, formed by a calcareous cement into a beautiful compact marble. It bears a high polish, and is elegantly marked by the sections of the shells which it contains; their constituent substance is a white crystallized carbonate of lime, and their cavities are commonly filled with the same substance, presenting a striking contrast to the dark ground of the marble. In other varieties the substance of the shells is black, and their sections appear on the surface in the form of numerous lines and spiral figures. Occasionally a few bivalves (*Cyclas*) occur, and the remains of the minute crustaceous coverings of the *Cypris faba* very constantly.

The marble is frequently found in blocks or slabs, sufficiently large for sideboards, columns, or chimney-pieces, and but few of the ancient residences of the Sussex gentry are without them. There is historical proof of its having been known to the Romans, "and in the early Norman centuries it was much sought after, and applied, as the Purbeck marble was, when cut into small insulated shafts of pillars, which were placed in the *triforia*, or upper arcades, of cathedral churches, as at Canterbury and Chichester. At the first-mentioned, the archiepiscopal chair is composed of it. Another more general use was for the slabs of sepulchral monuments, into which portraits and inscriptions of brass were inserted. In the chancel at Trotton, there is a single stone, the superficial measure of which is nine feet six inches, by four feet six inches; and another, in the pavement of the cathedral of Chichester, measures more than seven feet by three and a half." York Cathedral, Westminster Abbey, Temple Church, Salisbury Cathedral, and most of the principal Gothic edifices in the kingdom, contain pillars or slabs of this marble. It is singular that, in Woodward's time, an opinion prevailed, that these pillars, &c. were artificial, and formed of a cement cast in moulds; but, as that author remarks, " any one who shall confer the grain of the marble of those pillars, the spar, and the shells in it, with those of this marble got in Sussex, will soon discern how little ground there is for that opinion, and yet it has prevailed very generally. I met with several instances of it as I travelled through England, and had frequent opportunities of showing those who asserted these pillars to be factitious, stone of the very same sort with that they were composed of, in the neighbouring quarries."

Numerous examples of the durability of this limestone have been noticed above; yet, from long exposure in damp situations, it undergoes decomposition, and the petrified testacea may then be extricated almost entire. The specimens hereafter figured, are examples of this kind; and the slab delineated, p. 184. is of the most compact and beautiful variety that occurs in the south-eastern division of the county.

Mr. Young observes, that this limestone affords a very valuable manure, equal to chalk.

The shells belong to the genus *Paludina*, the recent species of which inhabit fresh water; and they are associated with the shelly remains of a minute crustaceous animal of the genus Cypris (*C. faba*, Min. Conchology, pl. 485.) that occur also in a freshwater limestone in France.* The Sussex marble has been found in almost every part of the Weald clay; from Laughton, near Lewes, to near Petworth, Kirdford, Newdigate, Charlwood, south of Tilvester Hill in Surrey, and at Bethersden in Kent.

In some localities the clay, alternating with the marble, contains the remains of cyclades, and other shells too imperfect to be determined. At Shipley, near West-Grinstead Park, in sinking a well in the Weald clay, masses of broken spiral univalves (apparently *potamides*) and *cyclades* were found at the depth of thirty feet; and layers of the same

* Description Géologique des Environs de Paris.

kinds of fossils were passed through, at the depth of 100 feet*: and in a well near Cowfold similar remains were observed. From observations made in many places in Sussex, the marble would appear to occupy, chiefly, the middle beds of the Weald clay, &c.: there are, probably, five or six alternations of it; and in the upper divisions of the clay, septaria, composed of a deep red argillaceous ironstone, occur, in layers of two or three feet in thickness, which dip in conformity to the general inclination of this formation.†

At Resting-oak Hill, on the Chailey road, about four miles to the N. N. W. of Lewes, a fine section of these septaria was lately exposed, by the removal of the summit of the hill, for the improvement of the road. The indurated marl, which formed the nuclei of the septaria, was of a pale yellowish grey colour, and so hard and fine as to be employed by the peasantry and carpenters to set their tools‡;

* To the politeness of Sir Charles Merrick Burrel, Bart., M.P., I owe the knowledge of this interesting fact.

† In the west of Sussex, the following, according to Mr. Martin, is the order in which the strata of the Wealden occur, — viz.

1. Stiff clay, with concretional ironstone. Bones of vertebrated animals; paludinæ, cyclades, cyprides.
2. Yellow and fawn-coloured sand. *Endogenites erosa.*
3. Clay with Sussex marble. Bones, paludinæ.
4. Micaceous fawn-coloured sand, with masses of calcareous grit.
5. Clay with Sussex marble.
6. Fawn-coloured micaceous sand alternating with beds of red clay, friable sandstone, and calcareous grit. Bones of vertebrated animals.
7. Soil, a clay country of uniformity of aspect, and of moderate fertility.

‡ At Wapping-thorn Hill, near Steyning, septaria of argillaceous ironstone were lately found, and applied to the same purposes. I am indebted for this information to the polite attention of the Earl of Egremont.

the ironstone surrounding it contained the remains of shells of the genera cyclas and paludina in great abundance; the scales and bones of small fishes; and myriads of the shelly coverings of the Cypris faba

---

ORGANIC REMAINS OF THE WEALD CLAY.

Scales and bones of fishes, apparently of a very small species - -	Resting-oak Hill.
Bones of saurians.	
Tooth of Crocodile -	Swanage Bay, Isle of Wight.*
Cypris faba - -	in the marble, and septaria.
Paludina fluviorum -	in the marble.
——— extensa - -	———
——— elongata - -	Resting-oak Hill.
Cardium turgidum -	Swanage Bay.
Melania attenuata -	———
——— tricarinata -	———
Cyclas membranacea, Min. Conch. tab. 527.	Resting-oak Hill.
Pinna - - -	Swanage Bay.
Venus - - -	———
Potamides? - -	Shipley, near Cowfold.

* It is to the discrimination of Dr. Fitton that we are indebted for the first correct list of the characteristic fossils of the Weald clay. (Annals of Philosophy, November, 1824.) Those from the Isle of Wight, enumerated in the text, are on the authority of his masterly paper on the geological relations of the beds between the chalk and Purbeck limestone.

"A line of blue shale extends along the upper part of the clay in the west of Sussex, abounding in some of the fossils above mentioned, and in ferruginous septaria, which are coated with a conglomerate of casts of paludinæ, cyclades, cyprides, and bones of fishes." *

---

2. HASTINGS BEDS.

The alternating sands, sandstone, and shale, &c., which form the central group of the Wealden, are distinguished by the name of "Hastings Beds" by Dr. Fitton; and we consider the term as most appropriate, since the cliffs in the vicinity of Hastings present the most instructive and extensive section that can be obtained, and must be examined by every one who is desirous of investigating the geological characters of this important series of the British strata.†

The Hastings beds consist of numerous strata of sand, sandstone, grit, shale, &c., some of which enclose a great variety of organic remains. On the coast they extend from Bexhill, in Sussex, to Ham Street, near Aldington, in Kent, forming a line of irregular cliffs, thirty or forty miles in length, and

* Mr. Martin.

† The vicinity of Hastings will indeed become classic ground to the British geologist; for the admirable sketch of the geological phenomena which the cliffs exhibit, and of the history of the Wealden formation in general, from the pen of Dr. Fitton, has added to their natural attractions all the interest which extensive observation, profound scientific research, and sound philosophy can bestow.

from twenty to upwards of 600 feet in height. Crowborough Hill, which is situated in the interior, near Tunbridge Wells, and is the highest point of the range, is 804 feet above the level of the sea. The anticlinal axis of the strata may be placed near Winchelsea, from whence the beds diverge towards the North Downs on the one side, and towards the South Downs on the other.*

Horsted Sand. — The Hastings beds, on their first emergence from beneath the Weald clay, consist of sand and friable sandstone of various shades of grey, yellow, and ferruginous colour, with occasional interspersions of ironstone, and a great intermixture of *small linear* portions of lignite. The sandstone is composed of siliceous sand, with particles of mica, held together by a ferruginous cement, sometimes having a considerable proportion of calcareous matter. These beds alternate with a stiff grey loam or marl. The lignite is disseminated in minute portions through the grey sands, a character by which these upper strata may be identified: it appears to have originated from the carbonization of plants of the fern tribe.

These beds are well displayed at *Little Horsted*, five miles N. N. E. of Lewes, and form the hill on which Horsted church, and the seat of E. Law, Esquire, are situated. On the east side of the road

* In the Isle of Wight, the Hastings sands consist of an alternating series of beds of sand, more or less abundant in ferruginous matter, and containing courses of *calcareous grit*, with clay mixed with sand; subordinate beds containing fullers' earth; wood, more or less changed; wood-coal, and ironstone: the proportion of the clay to the sands is very great. — *Dr. Fitton, Annals of Philosophy.*

the strata appear in the following succession, dipping towards the south-west; —

1. Sand and sandstone, ferruginous.
2. Grey sand, with minute particles of lignite.
3. Sandy marl.
4. Sand and sandstone, highly ferruginous.
5. White sand and sandstone.

The Horsted sand and loam occur along the northern border of the Weald of Sussex, and the southern of that of Kent.

Tilgate Beds. — Beneath the Horsted beds irregular alternations of sand and sandstone succeed; the lowermost stratum containing large concretional or lenticular masses of a compact calciferous grit or sandstone, which was formerly extensively quarried in the neighbourhood of Tilgate and St. Leonard's forests, near Horsham. * The lower portions of this bed in some localities form a conglomerate, and contain pebbles of quartz and jasper, some of which may be of chemical origin, but others have evidently been transported from a distance. There are three or four layers of the Tilgate stone, varying in thickness from 2 to 3 inches, or 1½ to 2 feet each.

These beds rest on blue clay and shale, which separate them from the next subdivision. They are the principal repository of the saurian remains of the waters which deposited the Hastings sands, and are therefore by far the most interesting strata of the formation. They extend from its western

* These strata were first described by the author in a paper read before the Geological Society, in the beginning of 1822; and further noticed in the "Fossils of the South Downs;" and also in a memoir read before the Geological Society of London, in June, 1822.

extremity at Loxwood, to Hastings, where they occupy the upper part of the cliffs, as will hereafter be particularly mentioned.

Worth Sands, and Sandstone.—The blue clay, &c. which support the Tilgate beds, are succeeded by, a series of arenaceous strata, some of which afford a fine, soft, building-stone, which is extensively dug at Worth, near Crawley. This sandstone is, for the most part, of a white or pale fawn or yellow colour, and occasionally contains leaves and stems of ferns, and other plants. The Worth sands occur in great force at Hastings, occupying the middle part of the cliffs.

---

3. ASHBURNHAM BEDS.

Alternations of sand, friable sandstone, shale, and clay, occur beneath the Worth stone: they are, for the most part, highly ferruginous, and enclose rich argillaceous iron ore, and large masses of lignite. They occupy the base of the cliffs at Hastings, and appear in many places in the interior. They are succeeded by beds of shelly limestone, alternating with shale, and including layers of a fine grit, precisely similar to the Tilgate stone; so much so, that specimens from the respective strata could not be distinguished from each other. This bed abounds, like that of Hastings, with carbonized vegetables (ferns) casts of bivalves, and other organic remains. The discovery of this bed is important, and the correctness of placing it so low in the series might have been doubted, had not a bed of the shelly *bivalve* limestone of Ashburnham, been found lying upon it, as will be particularly described in our account of the strata at Pounceford.

STRATA AROUND HASTINGS.

We have already remarked, that the cliffs extending along the Sussex coast, from Bexhill to Winchelsea, present the most illustrative section of the Hastings strata, in England. At Bexhill, the Horsted sand, sandstone, clay, &c., rise into a cliff which forms the hill on which that watering-place is situated. Near Bulverhithe, sand and sandstone, in thin laminæ, alternating with beds more or less argillaceous, form an irregular line of cliffs.

Saint Leonard's near Hastings.—The immense excavations, and removal of the cliffs, occasioned by the erection of the buildings and terraces of this new town, have exposed many interesting sections of the strata. Immediately behind the causeway, near the western entrance of the town, the blue calciferous grit of Hollington, is quarried half-way up the cliff. The grit lies beneath beds of fawn-coloured sandstone; the latter, though very friable, is in some parts sufficiently compact for building. In certain places a coarse grit occurs in layers of from one to a few inches in thickness, between the blue limestone; it contains immense numbers of small teeth, and scales of fishes, minute pebbles of quartz, &c.; it appears to have been formed of the detritus of the rock on which it reposes, and corresponds to the finer varieties of the Tilgate conglomerate, hereafter to be noticed. The blue grit contains *paludinæ* in abundance, and the *Cyclas parva*. In the sands and soft sandstone, vegetable remains are most abundant, but they are too imperfect to admit of accurate determination; they belong to the genera *Sphænopteris*, *Lonchopteris*, and, perhaps, *locopodites* and *calamites*. Some bones of the large saurians, which have rendered the strata of Tilgate forest so celebrated, have been discovered in the sand near Bexhill; and a portion of a femur and leg-bone, *os tympani*, and vertebra of the *Iguanodon*, were shown to me by Dr. Johnston, on my visit to Hastings in 1832.

Proceeding eastward, we find, to the west of Hastings, the first indication of the Tilgate beds, in the form of large masses of grey calciferous sandstone, which lie scattered along the sea-shore. The blocks of this substance, that have been long exposed to the action of the waves, are almost white, and hence the name of "*White Rock*," given to a large portion of the cliff which has slipped from its original situation, and is now exposed on the beach, near one of the stations of the blockade service. The largest mass of the strata, in this place, is about 30 feet high, and consists of—

1. Loam and vegetable mould; the summit of the cliff.*

2. Sand, and friable sandstone, of a fawn

* See the section of "White Rock," in the plate.

colour, and laminated structure, from 3 to 5 feet thick.

3. *Tilgate stone*, bluish grey, with numerous casts of bivalves, 2 feet.

4. White and fawn-coloured sand, 15 feet.

5. *Tilgate stone*, with bivalves, 1½ foot.

6. Thin layers of a coarse friable aggregate, with remains of fishes, vegetables, and paludinæ, 2 to 6 inches.

7. Laminated ferruginous sandstone, with layers of blue clay and shale; containing innumerable traces of carbonized vegetables, 3 to 4 feet.

The *Tilgate stone* resembles, in every respect, the sandstone of the forest from which its name is derived; it abounds in casts of a small species of cyclas, which is comparatively rare in the stone of the interior of the county. The thickness of the masses that are *in situ*, does not exceed two feet; but there are detached blocks on the beach which are upwards of four feet. The alternations of the sand and Tilgate stone, and the concretionary form of the latter, are strikingly displayed at this interesting and picturesque spot.* The sandstone occurs, not in continuous layers, but in irregular lenticular masses imbedded in the sand, yet preserving the same sedimentary line; as if its consolidation were owing to the infiltration of a fluid holding carbonate of lime in solution, which produced a subcrystalline structure in certain portions of the bed, subsequently to their mecha-

* See the section in the plate.

nical deposition.* This opinion is corroborated, by the fact of many of the blocks being covered with obtuse, mammillary projections, of various sizes, from two to five or six inches in diameter; and which, when the lines of stratification are washed away by the action of the waves, appear like clusters of depressed spherical bodies. But in the quarries in the forest, where a similar structure oftentimes prevails, these projections have an external coating of sand, and bear decided proofs of original stratification, in the numerous annular sedimentary lines with which they are encircled. The thin layers of coarse aggregate consist of a grey sand, loosely held together by a calcareous cement, and are remarkable only for the innumerable remains of fishes, shells, &c., which they contain; this bed is, probably, the equivalent of the more friable varieties of the conglomerate of the forest. It does not, in any instance, possess the compactness of the indurated masses of the latter, but contains the same kinds of fossils; it is from this bed that the specimens of *Endogenites erosa*, so frequently thrown upon the shore by the waves, are derived.

The ferruginous sand, No. 7. rests on a sandy shale, which occurs in thin laminæ, and presents traces of ferns and other vegetables; it may probably be identical with those beds of Tilgate forest, that contain the *sphenopterites* and *lonchopterites*. Beneath this is a coarse, friable, yellowish sand-

* The whin-stone of the Shanklin sand, in Western Sussex, occurs also in lenticular masses; this remark will apply to the compact portions of almost all arenaceous deposits.

stone, of a laminated structure, having the surface of the laminæ deeply furrowed; it reposes on a dark-coloured shale, which forms the base of the cliffs.*

A few yards behind White Rock, a bank of ferruginous sand and clay is cut through by the road; the strata are highly inclined, dipping to the west, at an angle of 25°. This mass probably belongs to the uppermost portion of the series.

A diluvial valley intervenes between the White Rock and the west cliff, where the ruins of the castle are situated. The highest point is about 260 feet high; and in 1827, a fine section of the cliff was formed by the labourers employed in carrying on the improvements at Pelham Place. It consisted of,—

1. (Uppermost bed).† White sand and friable sandstone, with veins of ferruginous sand — the Worth beds — about 100 feet thick.

2. Loam, shale, clay, and sand, 30 feet.

3. Clay, approaching to fuller's earth, enclosing undulating veins of lignite, 10 feet.

4. Soft sandstone in horizontal layers, alternating with clay and shale; contains traces of lignite, 10 feet.

The lignite in this place corresponds in every

* The *White rock*, which formed so interesting and picturesque an object, was being broken up at my last visit to Hastings. It was evidently but a very small part of a mass of strata, which at some remote period had been separated (probably by the subsidence of the soft inferior beds) from the inland cliffs. At low water, reefs of rocks may be seen to the south-west, which partake of the same inclination as the strata of white rock.

† Traces of clay and shale, apparently situated *above* the *Worth sands*, were observed, in the late excavations near the Castle; but the relative position of the beds could not be ascertained.

particular with that which has been noticed at Bexhill, Newick, Waldron, and Tilgate Forest. It occurs in nearly horizontal layers, which become extenuated into mere lines. It is very brittle, possesses the lustre of jet, and contains nuclei of a ligneous appearance, somewhat approaching to that of Bovey coal. The largest masses do not exceed two inches in thickness.

Pursuing our course eastward, another deep valley occurs, through which is the romantic entrance to Hastings by the London road; at the east cliff, the strata rise into a majestic range, from 400 to 500 feet high *, which extends with but little interruption to Haddock Point, a few miles from Winchelsea. The upper part of these cliffs consists of yellow and ferruginous sand, in which two or more layers of the *Tilgate stone* are imbedded; the middle portion is composed of the Worth sand, and sandstone; and the lowermost of ferruginous sand, and dark shale, with carbonized vegetable remains, lignite, and rich argillaceous iron ore. The following section presented by the cliffs near Eaglesbourn, (a spot well known to visitors, from the romantic beauties of the neighbouring fish-ponds,) will convey an idea of the whole.

1. Uppermost beds; fawn-coloured sand, and friable sandstone, about 10 feet.

2. *Tilgate stone*, from 2 to 6 feet.

3. Clay, loam, &c., alternating with sand and sandstone, containing lignite, &c., 20 feet.

4. White and fawn-coloured sandstone (*Worth sandstone*), 100 feet.

* Fairlight Down, the highest point, is 540 feet above the level of the sea.

5. Ferruginous sand and sandstone, alternating with dark blue shale, and reddish clay; lignite, ironstone, vegetable remains; these beds form nearly the lower third of the cliff. The strata are slightly inclined towards the east.

The *Tilgate stone*, presenting the same characters as the beds at White Rock, may be traced along the upper part of the cliffs, to within a short distance of the town, and its ruins are scattered along the shore, particularly near Fairlight Point and Eaglesbourn; at which places large portions of the stems of *Endogenites erosa*, are frequently thrown up by the waves.

At Hollington, near Hastings, the Tilgate stone also occurs; and at Ore, between Battel and Hastings, it is quarried for repairing the roads. In a quarry on the road-side, near Ore, the following section is presented: —

1. Loam, 6 to 8 feet.
2. White and fawn-coloured sandstone, 10 feet.
3. Grit, 3 to 4 feet.
4. Blue clay, depth unknown.

The reader will perceive hereafter, that this section corresponds with that of many quarries in Tilgate Forest.

Having thus briefly noticed the geological phenomena which the cliffs in the vicinity of Hastings present to our observation, before quitting this part of our subject, we would remark, that, notwithstanding the irregularity of the strata, and the ruinous state in which they occur, certain well-marked characters are exhibited.

1. A series of ferruginous and fawn-coloured

sands, sandstones, and clays (*the Horsted beds*), forming the uppermost part of the series; these appear at Bexhill, behind St. Leonard's and the White Rock, and on the top of West and East Cliffs.

2. The *Tilgate beds*, with their peculiar calciferous grit, and accompanying sand, clay, and shale; these occur at White Rock, are wanting on the West Cliff, but reappear on the East Cliff in a very striking manner.

3. A series of white and yellow sands, and sandstone, (the *Worth sandstone*) containing ferns, and other vegetable remains.

4. Shale, ferruginous sand, and sandstone, with ironstone, lignite, and other carbonized remains of vegetables, in vast quantities; forming a part of the Ashburnham beds.

The organic remains of the cliffs at Hastings, though comparatively presenting but little variety, are nevertheless important, from their identity with those of Tilgate Forest. The following were collected in a few hasty visits to Hastings; and if research were made under more favourable circumstances, there can be no doubt a far more extensive series might easily be obtained.

Horsted beds.

Lignite; carbonized remains of vegetables, too imperfect to admit of being determined.

Bones of saurians.

Tilgate grit.

Casts of *cyclades, cyrenæ?* and *paludinæ* in abundance.

Ribs of saurian animals.
Bones of birds.
Bones of turtles.
Radii of fish allied to silurus.
Jaw of an unknown fish.

*Tilgate coarse grit, and shale, &c.**

Tricuspid teeth of fishes.
Molar teeth of fishes.
Lozenge-shaped scales.
Rolled fragments of bones.
Paludina carinifera. Min. Conch. Tab. 509.
Paludina lenta †; *and P. elongata.*
Avicula? very thin shells, apparently of this genus.
Cyclas media, Min. Conch. Tab. 527.
Cypris faba.†
Endogenites erosa.
Traces of carbonized vegetables.

Worth sandstone, &c.

Three or more species of ferns.
Stems of arundinaceous plants.

Ashburnham beds.

Lignite and carbonized vegetables.

At Rye, and Winchelsea, beds of the Tilgate stone occur in sand of the same character as that

* From the coarse grit and shale, exposed by the removal of the face of the cliffs, behind the new town of Saint Leonard's, I have obtained, through the liberality, of the Marquess of Northampton, numerous specimens of Paludinæ, Cyclades, teeth, and minute bones of several species of fish, &c.

† Found in calcareous grit, at Hollington, near Hastings, by Dr. Fitton.

of the Hastings cliffs, above described.* The rock on which Winchelsea is built, is surrounded by an alluvial marsh, and presents a cliff more or less abrupt on every side. In a quarry, near the ancient gateway leading towards Rye, a layer of the *Tilgate stone*, nearly five feet thick, occurs in ferruginous sand; and in a cliff considerably lower than the base of this quarry, another bed is seen *in situ*. The mound on which Rye is situated presents a similar arrangement of the strata. The Tilgate stone, in these places, contains casts of the small cyclas, which occur in such abundance at White Rock.

HASTINGS BEDS IN THE INTERIOR OF THE COUNTRY.

From the coast to the westernmost point of this formation, the *Tilgate grit* is but sparingly distributed near the surface, except in the immediate vicinity of the forests. The arenaceous strata, and the lower ferruginous sands and clays, occupy the whole country, and in some localities form romantic groups of rocks, as at Maresfield, Uckfield, Buxted, and Fletching, where large masses lie bare, as when left by the waters of the ocean.

* The occurrence of the Tilgate stone, in these localities, was first described by Mr. Lyell; vide " Notice on the Iron-sand Formation of Sussex," by the author; and it may be remarked, that although this Notice appeared in the Geological Transactions for 1826, it was read before the society so long since as June 14th 1822. The same volume contains a memoir on the strata at Hastings; with a drawing of the coast, by Mr. Webster.

UCKFIELD ROCKS.

Near "the Rocks," the seat of —— Streatfield, Esquire, about half a mile west of Uckfield, a group of sandstone rocks occurs, under circumstances of considerable beauty and picturesque effect. The path that leads to this interesting spot lies to the right of the road, and by a circuitous route, conducts the spectator to the centre of a wood, where a beautiful lake, nearly surrounded by rocks, suddenly opens to the view. The cliffs, overhanging the water, are from twenty to thirty feet high, and are surmounted by forest trees and underwood. In some places the rocks are nearly perpendicular; in others, they descend with a gentle slope to the water's edge, the declivity being covered by a luxuriant vegetation. On the northern margin, a projecting point of high rock is perforated by a natural archway, that has been enlarged by art, and this leads to a recess in the sandstone on a level with the bosom of the lake. From this point the picturesque beauty of the scene is exhibited to peculiar advantage. On the opposite shore, the base of a rock that juts into the water is in like manner excavated into an arch, beneath which a little shallop was moored at the time of my visit. In one of the vertical cliffs, some fine young birch trees had taken root between the thin layers that separate the strata, and in almost every fissure of the rocks numerous plants had insinuated themselves, and by the beauty and variety of their foliage, relieved the monotonous and sombre appearance of the smooth grey sandstone. On the less elevated masses, lichens, mosses, and heaths were growing in great profusion and luxuriance. The strata are nearly horizontal, and partake of the characters of those already described.

A fine lake, overhung with sandstone rocks, and crested with a noble wood, near the seat of the Earl of Sheffield, in the parish of Fletching, may also be mentioned as affording another example of the picturesque scenery, to which the irregular surface of the sandstone gives rise in certain situations. Here, as in other parts of its course, the soil is in general sterile; but some spots near Fletching are remarkable for flourishing oaks: these contain six parts of sand, one part of clay, and a considerable proportion of finely divided vegetable matter.

In the parishes of Maresfield and Buxted there are also fine escarpments of the sand rocks.

Near West Hothly, a deep and extensive valley is flanked by a line of perpendicular cliffs of sandstone, of a very imposing character. From the circumstance of an enormous mass of rock being supported by a smaller one, it is called "*Big-upon-little.*" This group of rocks, like those above mentioned, clearly owe their present appearance to those disrupting and denuding operations, whose effects are every where so strikingly manifest.

Before proceeding to an examination of the course and extent of the Tilgate beds, it may be convenient in this place to take a brief notice of two mineral substances; one of which was formerly of great importance to the inhabitants of the south-east of England in an economical point of view; and the other led to expensive and useless undertakings, of which it is presumed the present state of geological science will prevent the recurrence. These substances are *ironstone,* and *lignite,* or, as it is commonly called, *coal.*

IRONSTONE.

Ironstone. This substance was formerly extracted from the ferruginous sandstone strata; it is internally of a dark steel grey, and generally very hard and compact; occasionally it is laminated, and separates into thin flakes upon exposure to the air. It occurs either in irregular concretions in the sand, or is stratified and alternates with beds of sandstone. The globular masses often contain nodules of argillaceous earth, round which the ironstone is disposed in concentric layers.

In some parts of the county the ironstone is of excellent quality, and extensive founderies were anciently established in different parts of its course; "the almost inexhaustible quantity of wood, with which the country was covered in the early centuries, and the numerous lakes and morasses, which the total neglect of drainage had occasioned, being circumstances peculiarly favourable for the conversion of the iron ore into bars. For this purpose the lords of the several manors which lay within the woodland district, collected the rivulets into large ponds, and erected mills and furnaces. The iron, so procured, was at first principally used for agricultural implements;" but Fuller also observes, in his Worthies, that "it is almost incredible how many great guns were made of the iron of this county. The total decline of the manufacture in Sussex is to be attributed to the establishments in Scotland and Wales, in which pit-coal is used, the superior cheapness of fuel having enabled them to monopolize the trade." There is now but a single foundery in the eastern division, and which belongs to the Earl of Ashburnham. According to the present practice, it requires fifty loads of charcoal, and fifty loads of ironstone (twelve bushels to each load) to make thirteen tons of pig iron.*

Lignite occurs in many of the shales and clays; and at Bexhill, the south-eastern extremity of the Forest-ridge, towards the sea, indications of this substance induced some enterprising individuals to sink a shaft; and at the depth of 160 feet a layer of lignite, resembling Bovey coal, was found: the undertaking was abandoned after an outlay of some hundred pounds sterling!

At Waldron, and Newick, seams of fibrous lignite have also been noticed. The strata pierced in Newick-Old-Park, about one mile from the banks of the Lewes canal, were: 1. Loam. 2. Sandstone and clay. 3. Marl and sandstone. 4. Laminated sandstone. 5. Indurated clay rubble. 6. Alternations of clay and sandstone. 7. Shale. 8. Coal of the Bovey kind, examined to the depth of eleven inches. It bassets out on the side of a

* Dallaway's History of the Western Division of the County of Sussex, vol. i. p. 161. folio, 1815.

rivulet, and may be observed in the bank of the adjoining hop-grounds. This lignite has much the appearance of jet: it is of a velvet black; does not soil the fingers; has a resinous lustre, and conchoidal fracture; is very brittle, and burns with a bright flame.

TILGATE FOREST.

In attempting to trace the Tilgate grit from Hastings to the west of Sussex, with the exception of Ore, previously mentioned, the first quarry that has been opened is on Chailey North Common, about seven miles north of Lewes.* A rivulet, which runs along the valley on the north of Chailey, marks the termination of the Weald clay, and the emergence of the Horsted sands with lignite; the latter beds continue on to the valley, beyond the little public house on the road side, near the eighth milestone, where the Tilgate beds basset out, and form a bold ridge or hill. The strata in this place are disposed in the following order: —

1. Ferruginous and fawn-coloured sands, and sandstone.
2. *Tilgate stone*, from two to four feet thick.

* That it approaches the surface, in many places, between Ore and Chailey, cannot however be doubted, but its course is hid by the deep bed of diluvial clay and loam that constitutes the subsoil of this district; the sinking of a well, or some other artificial opening, will one day or other detect it: and as it is decidedly one of the best road materials in the county, a knowledge of its probable range and extent is of great importance in many parts of Sussex.

3. Clay and shale.

4. White sandstone, and sand.

Proceeding towards the west, we arrive at Tilgate Forest, and within an area of eight or ten miles in length, and three or four in breadth, the compact varieties of the sandstone have been extensively quarried, the rapid increase of Brighton, rendering the repair of the roads that lead to that town, from the metropolis, a subject of the first importance. Before we enter into a detailed account of the more interesting localities in this tract, we refer to the section (in the plate) from Brighton to near Tilgate Forest, where the relative position of the strata is distinctly shown.

1. The Crag: the cliffs of Brighton.

2. Chalk, forming the range of Downs.

3. Firestone,
4. Galt, } but obscurely seen on the surface.

5. Shanklin sand; a fine section at Stone-pound gate.

6. Weald clay, and Sussex marble; at St. John's common.

7. Emergence of the upper beds of the Hastings sands, at Taylor's bridge.

8. Tilgate beds, near Cuckfield, &c.

If this section were continued still further north, we should find the white sands and sandstone, with fern-leaves, at Worth, near Crawley: then the ferruginous sand, with ironstone, &c.

Although, as we have elsewhere remarked, the subdivisions of the strata under examination are so variable, that scarcely any two quarries present precisely the same characters, yet a certain order of superposition prevails in the distribution of the principal masses, which order is never inverted; this is well illustrated in the quarry, represented in the frontispiece to the present volume.*

The upper part of the pit consists of, —

1. Loam, from five to seven feet.

2. Sand, and soft friable sandstone, of various shades of yellow, green, grey, and ferruginous; eight feet.

3. A bed of fine, compact, bluish-grey, calciferous sandstone (*Tilgate grit*), imbedded in sand; varying in thickness, from one to nine feet; the lower portion a conglomerate.

4. Blue clay or marl — the base of the quarry; thickness unknown: wells are sunk in it, to the depth of seventeen feet.

Of these strata, the calciferous grit alone requires notice. It differs in a few particulars from that of "White Rock:" — it is less homogeneous; is generally of a darker shade of bluish or greenish grey; occurs in masses that are more tabular, and the lower portions form a very compact conglomerate, enveloping large rolled pebbles of variously-coloured quartz, jasper, and small ones of pure white quartz, and flinty slate. It also contains organic remains, in far greater quantities than the stone at Hastings; bones, more or less rolled; teeth, &c., are to be seen in almost every fragment; and the car-

* This ancient quarry is now filled up, and the spot it occupied converted into arable land; a method invariably adopted by the proprietors of the estates in the environs of the forests, &c. after the harder sandstones have been extracted.

bonized remains of vegetables are equally abundant. It is exceedingly compact, and offers great resistance to the hammer; it scintillates with steel, effervesces strongly with acids, and varies considerably in purity: it contains about 25 per cent. of carbonate of lime. Lenticular and tabular crystals of carbonate of lime frequently occur in the hollows of the stone, and in the medullary cavities of the bones, fissures of the lignite, &c. This bed is separated into irregular layers, of from three to twelve inches in thickness, by blue marl or clay; and the compact masses, when removed by the quarrymen, and cleared of the surrounding sand or clay, present the same concretional or lenticular forms as those on the shore, at Hastings, that have been washed out by the breakers; the same kind of mammillary projections on the surface of the stone, are also observable.* The *conglomerate* is one of the most remarkable features of these strata; it appears to be composed of the grosser materials of the bed, such as pebbles, rolled masses of sandstone, bones, &c., and even the finer parts of it are made up of grains of sand, and comminuted bones, teeth, &c.; the whole has evidently been the sediment of a current of water, subsequently consolidated by a subcrystalline, calcareous cement: almost all the bones found in it are more or less rolled, the

* In many of the quarries, near Tilgate Forest, the concretional formation of the stone is clearly shown; although, from its being incrusted with sand, it appears like blocks of the friable sandstone; and as it maintains the same sedimentary line as the sand in which it is imbedded, the layers appear like tabular strata, till examined with attention.

greater part being mere fragments, rounded by attrition. This conglomerate, varying in texture and composition, occurs in greater or less force, in almost every quarry, in this part of the Tilgate strata. In some instances layers of a coarse friable grit, with calcareous concretions, occupy the upper part of the beds, as is the case in many places in the town of Cuckfield, where it rises so near the surface as to be exposed in digging the foundations of houses, &c. In a quarry to the east of Cuckfield, this bed of coarse grit is seen *in situ**, the section offering the following series:—1. loam; 2. friable sandstone; 3. *coarse grit;* 4. sand and friable sandstone; 5. *calciferous grit*, the *lower part conglomeritic;* 6. *blue clay* or *marl.*

The sections above described present the most remarkable variations of the strata around Tilgate Forest; and it is unnecessary to multiply examples. In all these localities, organic remains are more or less abundant. The bones of enormous saurians; of turtles, fishes, and birds; remains of large arborescent ferns; shells of the genera cyrena, paludina, unio, &c. are found in all the beds, but more particularly in the calciferous sandstone, and the sand which covers it; in the ferruginous sands above these, they are seldom, if ever, discovered; and in the blue clay beneath, they are almost equally rare; as the latter bed, however, contains nothing worthy the attention of the proprietors, it has been but little explored, and its contents are, consequently, but imperfectly known.

* Vide the section in the map and sections.

On Wivelsfield common, about 5 miles to the south-east of Cuckfield, a thin bed of the Tilgate stone, containing carbonized vegetables, occurs beneath a ferruginous friable sandstone; and near Lindfield a calciferous sandstone, more calcareous than that of the forest, is quarried on the estate of Mr. Crawford: in both localities bones of reptiles have been noticed.

Approaching Horsham, we find numerous quarries of sandstone along the southern edge of the forest, and on many farms and villages in the vicinity of the town, as Stammerham, Coltstaple, Southwater, Slinfold, Broadbridge, Strood, Nuthurst, Oakhurst, Warnham, &c.; and at Wisborough Green, Skiff Common, Headfold Wood, and Loxwood, ten miles farther westward, and the extreme point of the Wealden formation in Sussex. The most considerable quarry in the vicinity of Horsham, is that of Stammerham, two and a half miles to the south-west of that town, and of which the following is a section: —

	Feet.	In.
1. Clay and loam - - -	10	0
2. *Calciferous sandstone;* surface deeply undulated (*prov.* rough causeway) -	0	4
3. The same stone, but more indurated, in two layers, used for repairing the roads, (*prov.* scrubstone) - - -	1	4
4. Ferruginous sandstone; pulverized for bricks - - - -	1	0
5. Blue soapy marl - - -	1	6
6. Hard calciferous grit, used for roads,		

	Feet	In.
and rough paving (*prov.* ground pinning-stone) - - - -	1	0
7. Compact *calciferous sandstone* of finer texture than any of the above. It occurs in large slabs, and forms excellent paving-stone for kitchens, &c.; it is slightly marked with undulating furrows on the upper surface - - -	2	0
8. Marl, sunk through, but not worked	4	0

9. Stone in slabs, reached by boring, depth unknown.

Organic remains are comparatively rare in these beds; bones of reptiles have been discovered, and vegetables, in a carbonized state, are not uncommon. The stone is not so uniform in colour as that of Cuckfield, and is more micaceous; the conglomerate has not been observed in any of the localities around Horsham.

The furrowed state of the surface of the sandstone, above mentioned, is of frequent occurrence in the arenaceous strata of Sussex, and has evidently been produced by the rippling of water.

I shall offer a few remarks in this place on the appearance alluded to. The deep undulations with which the slabs are covered, must have attracted the notice of all who have observed the pavements in the towns, and villages, where the Horsham stone is employed. In many instances, the slabs are so rough, as to be made use of in stable-yards, and other situations where an uneven surface is required to prevent the feet of animals from sliding in passing over. It is scarcely possible that any one who examines the markings produced by the undulations of water on the sand and mud of the shallows and margins of lakes, rivers, estuaries, and sea-shores, can hesitate to conclude, that characters so perfectly analogous as those observable on the Horsham stone, have been produced by a similar operation. When a large surface of the stone is cleared from the superficial soil which covers it, a most interesting appearance is presented, and the spectator is struck with the conviction that he is standing on the dried shallows of an ancient river. In some places the furrows are deep, affording evidence of the water having been much agitated, and the ripple strong; in other instances, the undulations are gentle, and are frequently intersected by cross ripples, from a change in the direction of the waves. On some parts of the surface, there are slightly elevated longitudinal ridges of sand, made up of gentle risings, disposed in a crescent-like manner, and these closely resemble the sand ridges which are produced by the little rills which flow back into the sea or river at low water. Some of the slabs are covered with thin angular ridges, irregularly crossing each other, like the fissures in septaria, and which have obviously been caused by deposition into crevices made in the sand or mud by drying. A considerable portion of the stone, the flat as well as the furrowed surfaces, is covered with small cylindrical bodies, which have been moulded in the hollows occasioned by some species of vermes. Similar forms of a large size also occur; these resemble the trails left by *myæ*, and have probably been produced by some freshwater bivalve. Since the foot-marks of reptiles, and crustacea, have been discovered on the surface of sandstone in some parts of England and Scotland, I examined the slabs in the quarries of Horsham with considerable attention, in the expectation that similar indications might be detected, but without success; yet, as reptiles are known to have existed in vast numbers on the land and in the water, at the period of the deposition of these strata; and as the surface of the stone exhibits proofs that it was deposited in shallow water, and was occasionally left dry, it is probable that, sooner or later, such impressions will be discovered. These observations were made during a visit to Stammerham quarry, near Horsham, in company with Mr. Lyell, in 1831.

At Sedgwick, near Stammerham, large quantities of the thinner layers of sandstone, or grit, are quarried for the purposes of paving, roofing, &c.; this variety is sent to many parts of the county, and is distinguished by the name of *Horsham-stone*. The dip of the strata around Horsham and Tilgate Forest, is exceedingly variable; the general inclination is towards the S. E.; but on the forest, according to the information of an intelligent old quarry man, the beds formerly worked there were found to be nearly horizontal over a considerable area, so as to render the drainage of the quarries very difficult. *

In the westernmost portion of the Hastings beds,

* In 1829, in a quarry near *Rusper*, bones of the extremities and several vertebræ of the Iguanodon were discovered; and the latter are now in my collection, through the kindness and liberality of —— Tredcroft, Esq. of Horsham.

the strata rise in gentle undulations immediately to the east of Kirdford, in the parish of Wisborough Green, where at Headfold Wood Common, near Loxwood, their outline makes an angle corresponding to that formed by the Weald clay and superior strata.* These beds, at their first emergence, consist chiefly of clay, but enclose small slabs of a slightly *calcareous sandstone*; and, beneath these, large tabular masses of a calcareous grit, which is very similar to certain beds of the Stammerham and Slinfold quarries; unlike, however, the latter, the calcareous grit of Wisborough Green is based upon a deep mass of red ferruginous marl, and is only found in detached portions. The strata are arranged in the following manner: —

1. Calcareous sandstone (containing about 30 per cent. of carbonate of lime), in clay, with casts of viviparæ, &c.
2. Tabular calcareous grit, with bones of large saurian animals.
3. Deep red clay, with argillaceous ironstone.
4. Blue clay, with selenites, the lowermost bed.

In these strata, Mr. Murchison discovered vertebræ, and a femur of the Iguanodon of enormous magnitude (3 feet 7 inches long, and 34 inches in circumference at the largest extremity), and portions of other bones of some of the saurian animals of Tilgate Forest. The greater part of

* The western extremity of the Hastings sands was first noticed by Mr. Murchison, the excellent President of the Geological Society, and described by him in a highly interesting memoir on the Geology of the north-western extremity of Sussex, and the adjoining parts of Hants and Surrey, published in the Geological Transactions, vol. ii. part i. New Series.

these fossils were found at Headfold Wood Common, in ferruginous marly clay, about five feet below the surface, and beneath a stratum of calcareous sandstone; at Loxwood, bones have also been found in clay.

The Tilgate grit is quarried in many places south of Riegate; and at Heaver's Wood, in the parish of Horley, contains *cyclades* and bones of reptiles; and near Flanchford, *paludinæ*, and *uniones*, occur in abundance.*

Organic Remains of Tilgate Forest.—These are exceedingly numerous, and present as interesting an assemblage as any hitherto recorded. In this division of the work, we propose to give a general idea only of the fossils contained in these strata; their characters will be fully detailed in the chapter expressly devoted to the subject.

Vegetable Remains.—Stems of large plants, allied to the genera Dracæna, Cycas, &c.

Clatharia Lyellii.

Endogenites erosa.

Leaves, &c. of plants of the fern tribe.

Cycadites.

Equisetum Lyellii; at Pounceford.

Lonchopteris Mantelli; in the Worth sandstone.

Sphenopteris Mantelli; in the Tilgate stone.

Carpolithus Mantelli.

Carbonized vegetables; too imperfect to determine their characters.

Testacea.—The testacea which are found in these beds, but seldom have the shells remaining,

* Communicated by W. Constable, Esq. of Dover's Green, near Riegate.

and generally occur in the state of casts and impressions; their generic characters are, therefore, seldom sufficiently manifest to afford any positive conclusions. Those we have noticed appear to belong to the following genera: — Neretina, paludina, melanopsis, unio*, psammobia, cyclas, &c.

Remains of fishes. — Teeth, scales, and bones of some species allied to lepisosteus, siluri, &c.

Turtles. — Three kinds: the first resembling a freshwater, the second approaching to a terrestrial, and the third to a marine species. The bones, and detached parts of the sternum and buckler, are very abundant in the *calciferous grit*, as well as in the sand, and clay.

Saurian animals. — The bones, teeth, scales, &c., of four (if not more) gigantic animals, of the saurian order, viz.: —

Iguanodon. — The teeth, and many enormous bones: these remains have not been found in any other deposit, nor in any other country.

Megalosaurus. — The teeth, vertebræ, ribs, and some of the bones of the extremities: the only other known locality is the slate of Stonesfield.

Crocodile. — The teeth, vertebræ, ribs, and other bones of a species allied to the Gavial, or Gangetic crocodile; and of three or more species.

Plesiosaurus: — teeth; caudal, lumbar, dorsal, and cervical vertebræ, of the long-necked species.

Birds: — bones of some species of Ardea.

* Casts of shells of these genera sometimes form the entire mass of the layers of sandstone; and it is worthy of remark, that they occur also in the shale and grit of the coal measures, and in the ironstone of Derbyshire.

In this list of the fossils of Tilgate Forest, the reader cannot but remark on the almost entire absence of the remains of the inhabitants of the sea. The vegetables are either terrestrial, or belong to those tribes which affect a marshy soil; the shells are lacustral; and the reptiles, with but one or two exceptions, are those which bear a resemblance to the living species that now inhabit the banks of lakes and rivers, of tropical regions; the fish, also, appear to be of freshwater origin; but on this point, some degree of uncertainty may exist. The *absence of ammonites, belemnites, and other multilocular* shells of our ancient seas, and of the echinodermata and zoophytes, whose remains are so frequent in the sands and chalk *above* the Hastings sands, and Weald clay, and in the Portland stone and other deposits *beneath*, is also another remarkable feature in the oryctological characters of the strata of Tilgate Forest.*

LOWER GROUP OF THE WEALDEN STRATA, INCLUDING THE ASHBURNHAM BEDS.

In this division of the Hastings sands formation, we include the beds of shale, forming the lowermost strata in the cliffs at Hastings; and the inferior limestone, shale, and blue clays, which appear in various places in the interior of the country.

* The *ammonites*, *nautili*, &c. mentioned in Messrs. Conybeare and Phillips's Geology of England as occurring in the iron-sand, as well as those figured in Sowerby's Mineral Conchology, which are referred to that formation, must not be admitted as exceptions, until their several localities have been carefully examined; the confusion that formerly existed in the geological nomenclature of the sands below the chalk, having given rise to many erroneous geological habitats.

But the strata under consideration have suffered such extensive displacement; and, even where accessible to observation, present so many obstacles to an accurate knowledge of their position and relations, that we can only attempt an outline of their general characters, for future observers to fill up and correct.

They may be defined as a series of shelly limestone and shale, alternating with blue clay, and containing subordinate beds of grit, ironstone, limestone, and sandstone.

Limestone, of a dark, bluish-grey colour, containing immense quantities of bivalve shells, which are more or less changed into a spathose calcareous spar, and whose characters and forms are but occasionally preserved, is the most characteristic deposit of this group. In texture, appearance, and fracture, it bears considerable resemblance to the Sussex marble, but is readily distinguished by the semilunar markings which the sections of the bivalves present. The shale, by which it is accompanied, is also of a dark blue colour, very stiff, sometimes slaty, and laminated; and abounds in the same kinds of shells as the limestone, in a white friable state.

At Archer's Wood, near Battel, on the estate of the Earl of Ashburnham, extensive limeworks have been carried on for the last century; shafts are sunk through the shale and clay, to the depth of 100 or 120 feet, to the limestone, which lies beneath, and is dug up and converted into lime; the thinner slabs are used for paving. These strata occur in the following order: —

1. Loam, of an ochreous, or grey colour, with thin layers of friable shelly limestone.

2. Compact blue limestone with bivalves (provincially termed *greys*), alternating with shale.

3. Compact blue limestone and shale, alternating in layers of from 14 to 20 inches each in thickness : these are almost destitute of organic remains. There are not less than fourteen or fifteen alternations of the limestone and shale, making a total thickness of upwards of 100 feet. The workmen informed us, that the beds at Archer's Wood generally dip towards the south, or south-west: the inclination of the strata, however, from the displacement they have suffered, is so variable in different localities, as to render it impossible to obtain an approach to accuracy. The faults are very numerous : one of them is so considerable, that it is called the "Sixty Fathom Fault." Strata of a similar kind are worked in Darvel's Wood, about four miles north-west of Battel; at Brightling, on the estate of John Fuller, of Rose Hill, Esq.; near Burwash; at West Down, Willingford, and Hurst Green.*

One of the most interesting localities of these beds is at *Pounceford*, on a farm belonging to the Earl of Ashburnham, situated in a deep glen, about a mile to the right of the turnpike road, leading from Cross-in-hand to Burwash.

* At Tetham, about a quarter of a mile from Crowhurst Place, near Battel, the workmen employed in digging calciferous grit, discovered vertebræ, and bones of the extremities of an enormous reptile, in soft ferruginous sandstone; traces of ferns were also observed. Some of the bones were preserved through the discrimination of Nathaniel Kell, Esq. of Battel, who kindly added them to the collection of the author.

The Forest Ridge, along which the main road passes, consists of sand and friable sandstone, more or less ferruginous. Descending the narrow winding way that leads to Pounceford, occasional openings occur in the white varieties of the sandstone, along the sides of the deep valleys; shale and blue clay appear on the surface; and springs issue from the defiles of the glens, the water being thrown out by the argillaceous partings of the strata. Before reaching the bottom of the valley, to which the road conducts, limeworks are seen, on a rising ground on the left hand; and, to the right, a pathway leads, by a farm-house, to a deep glen, where a quarry has been opened, and from whence an incrusting spring has its source. This quarry, moreover, presents a most interesting section to the geologist; for there, *in situ*, is seen a bed of the Tilgate stone, beneath a layer of the Ashburnham bivalve limestone: it is not a little extraordinary, that the occurrence of calciferous grit, in this division of the Hastings beds, should have so long eluded observation. The quarry, of which a sketch, by my friend Warren Lee, Esq., is annexed, is composed of—

	Feet	In.
1. Loam - - -	8	0
2. Bed of laminated shale, clay, &c., containing thin septaria of ironstone, with bivalves, and a layer of *very hard, dark, blue limestone*, 2½ inches thick; similar to the most compact masses of Archer's Wood	2	5
3. Bluish grey loam, or soapy marl -	3	9
4. Friable sandstone, with traces of carbonized vegetables - - -	5	6

	Feet	In.
5. *Compact calciferous grit*, containing casts of Cyclas parva, and media; lignite, and traces of carbonized vegetables; and bones and teeth of reptiles* - -	1	10
6. Blue clay, and shale — depth unknown.		

The strata dip to the N. N. E. with an angle of about 5°. The incrusting spring is seen in the engraving, issuing from between the limestone beds above the Tilgate grit.†

QUARRY AT POUNCEFORD, 1827.

* This stone so entirely agrees in its mineralogical characters with that of Tilgate Forest, that specimens from the respective localities could not be distinguished from each other. The same bed is quarried at Etchingham, three miles north-east of Burwash.

† This spring has been diverted from its former course by the works carried on for the purpose of extracting the grit for a road material. It now empties itself into a stream higher up the valley, and its lapidescent powers are much diminished. A bank, over which it formerly ran, contains large blocks of calcareous tufa, which enclose mosses,

Proceeding to the opposite valley, on the sides of which the limeworks are situated, indications of the calciferous grit may be observed at the depth of about twenty feet; and it is a curious fact, that the perforation of this stratum (which could not be avoided in sinking the shafts for limestone) was formerly one of the most expensive processes in the whole works, and cost the Earl of Ashburnham a very large sum annually; the excellence of the stone, as a road material, and for which it is now quarried, having escaped observation, till the system of M'Adam became more generally known and appreciated.

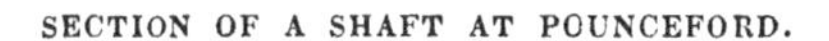

SECTION OF A SHAFT AT POUNCEFORD.

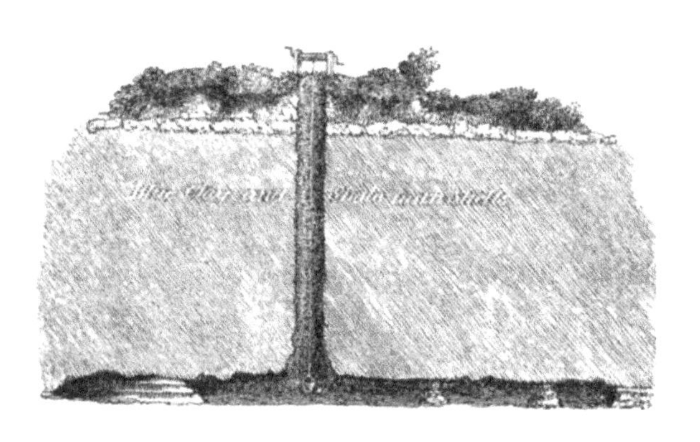

---

twigs and branches of trees, and stems of equiseta, with snail shells. This recent deposit is now covered by a layer of peat and bog-earth, on which brushwood and equiseta are growing in great luxuriance. No spot in Sussex is more interesting than this romantic glen, in which may be seen, at the same time, the effects of the ancient and modern operations of nature.

The limestone here, as in Archer's Wood, and at Brightling, lies at the depth of from 30 to upwards of 100 feet, and is extracted by means of shafts, sunk in the most favourable situations; the stone is dug from under the shale to a considerable distance, the excavations forming an area of many acres, occasional props of limestone being left to support the strata, as is shown in the annexed section of one of the principal shafts.

The sections formed by the shafts are hid by the frame-work employed; but, from the specimens of the beds pointed out to us by the workmen, it appears that loam, friable sandstone, and a thin layer of grit, with a succession of blue laminated shale, more or less shelly, are generally passed through, before the limestone is reached: the latter is irregularly divided into three layers, but the thickness of the whole seldom exceeds two feet. It is more friable than that of Archer's Wood; and the softer masses are composed of white brittle shells, held together by clay or shale. The lime it produces is in great estimation * among the agriculturists of that part of Sussex.

The dip of the strata around Pounceford, as in all the surrounding district, is as various as possible, the disruption of the beds being so great that faults, and *horses*, as the risings of the strata are called in Sussex, are observable in almost every quarry. The bottoms of the deep glens are

* A kiln, containing 600 bushels, is converted into lime in twenty-four hours; it sells for 6*d*. per bushel.

generally selected, for the purpose of extracting the stone, as the labour of digging through the beds of sand and soft sandstone is thereby avoided.

Near Swife's farm, in a deep valley, a branch of that of Pounceford, an interesting section was exposed in 1830, at the time of my visit, in company with my friend Mr. Lyell.

SECTION EXPOSED IN A QUARRY AT SWIFE'S FARM NEAR POUNCEFORD.

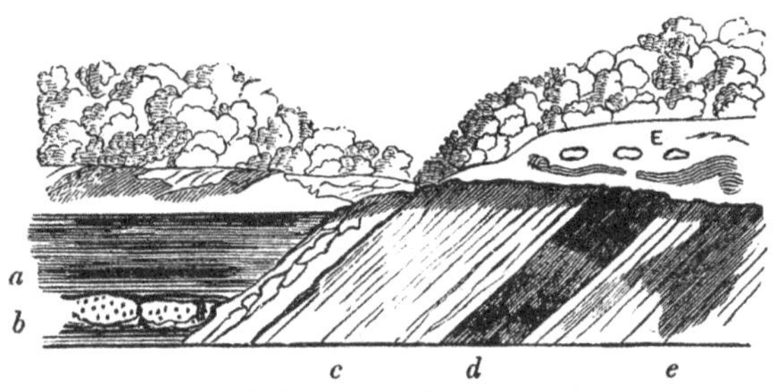

a. Clay, shale, and sandstone.
b. Calciferous grit.
c. Shale and clay highly inclined.
d. Beds of limestone like that of Ashburnham.
e. Shale and clay.
E. Openings of shafts.

1. Uppermost bed (beneath a superficial covering of loam) soft ferruginous sandstone.

2. Laminated clay and shale, with *cyclades*.

3. Clay and shale, with shells in immense quantities, chiefly *cyclades*; some of the layers are wholly composed of testaceous remains, in the state of a white friable mass, so perfectly resembling the shelly clay beds of certain parts of the

tertiary series, that they could not, by the inspection of hand specimens, be distinguished.

4. Calciferous grit, like that of Tilgate; with bluish grey limestone. Contains abundance of *Equisetum Lyellii.*

5. Shale and clay.

6. Limestone, with cyclades; like the "Greys" of Asburnham.

The sketch, p. 226., represents a "*horse*," or fault, near the above section.

In the strata at Swife's farm, were collected, vertebræ and ribs of saurians, and turtles; teeth of crocodile, scales of fishes, and a great number of *Equiseta Lyellii*, *sphenopterites*, and traces of other ferns. Shells of the genera *cyclas* (the most common), *psammobia*, *mytilus*, *melanopsis*, *paludina*, &c.

I shall now proceed to mention some other interesting localities of the harder varieties of grit, limestone, and ironstone, whose geological position is not at present satisfactorily determined; but which appear to belong to the inferior group of the Wealden, rather than to the upper strata.

Pit-wood Quarry near *Cross-in-hand.* — This quarry lies in a valley to the right of the road leading to Burwash, and exhibits the following section: —

1. Soft fawn-coloured sandstone, 22 feet.

2. Argillaceous beds, with an intermixture of sand, and thin laminæ of marl or clay.

3. Compact stone, exceedingly hard, in many parts highly ferruginous; a single layer, two feet thick: it is employed to repair the roads. This

bed abounds in organic remains, and particularly in ferns of a very large size, but exceedingly imperfect. Large *uniones* also occur. Some portions of this layer are entirely composed of casts of *paludinæ elongatæ*, the shells having disappeared: these masses are very ferruginous. Traces of lignite are disseminated throughout the beds.

The fossil plants of this quarry are highly interesting; the stems are generally carbonized, and the fronds wanting; but the cavities made by their impressions are filled up with hydrate of alumine, like the plants in the ironstone of Yorkshire and Staffordshire. I have observed the same kind of ferrugino-argillaceous sandstone near Forest Row, and Tilgate Manor-house near Crawley.

Lansdown Quarry near *Heathfield.*—No calciferous sandstone is here worked; and in the blue shale, which alternates with it, ribs and portions of the sternal plates of freshwater turtles, and bones of reptiles have been observed.

In *Kingsdown Quarry*, in the same parish, grit, with lignite and ferns, occur.

A shelly limestone that alternates with blue clay and shale, at Eason's Green, in the parish of Framfield, appears to belong to this series; and a very beautiful compact variety is dug up in Barnet's Wood, adjoining the ninth milestone, on the turnpike road, leading from Lewes to the Blackboy's public house; this limestone is strongly impregnated with iron, and traversed by veins of white calcareous spar; the bivalves are the *cyclas parva*. The same beds occur at Rotherfield.

We have not observed the Ashburnham limestone farther west than Framfield; nor, perhaps, should we expect to find it but in the eastern part of the county, when we consider the low situation in the series which it occupies, and the direction of the line of elevation which broke up the strata of the south-east of England.

In the deep valleys near Maresfield, Uckfield, and Buxted, beds of shale, with the *cyclas parva*, are found; and the wells sunk through the sandstone strata on the south of Uckfield, in some instances, reach the same deposits. At Streale, near Buxted, shale with similar organic remains has been collected.*

In the parish of Maresfield, adjoining the new road to Tunbridge Wells, a sub-crystalline limestone, containing *cyclas media*, and *parva*, has been dug up in the wells belonging to the cottages.

Near West Hothly, in the valley of rocks already mentioned, and in others of the deep glens, bivalve limestone and sandstone occur, but their relation to the other members of the formation has not been ascertained.

In Surrey, we find traces of the calciferous grit,

* The argillo-calcareous strata of Sandown Bay, in the Isle of Wight †, are evidently identical with those above described. "Near the termination of the cliffs, towards the middle of the bay, are several thin strata of a stone composed wholly of *bivalve shells*, in a calcareous matrix, much resembling the Purbeck stone, but the shells are larger. These strata are from one inch to three inches in thickness, and separated from each other by beds of shale and fibrous carbonate of lime; they have the same inclination as the strata lying above them. These strata are the lowest in the island."‡

† Conybeare and Phillips's Geology of England, p. 157.

‡ Specimens from these limestone beds could not be distinguished from those of Ashburnham.

ferruginous sandstone, and limestones, in various places to the south of Riegate; and the bones of saurians have been observed in these beds, as well as in the softer sandstones and sands. At Oakwood Hill, in the parish of Wootton, near Ockley, in Surrey, several vertebræ, and a fine carpal bone of an enormous reptile, were discovered in 1828, and presented to me by the late Walter Burrell, Esq., M. P. for Sussex. At Norwood Hill, near Charlwood, Mr. Constable found *cyclades*, and a shell resembling the *melania*.

In the Weald of Kent, these beds have not been examined with due attention. In the neighbourhood of Tunbridge Wells, there are many interesting localities, which diligent research would, I am persuaded, soon render as productive in organic remains as any of the strata in Sussex. Near the pretty hamlet of Southborough I collected from a quarry on the road-side, leaves of ferns, scales of fishes, and teeth of crocodiles; from the sandstone on Tunbridge Wells Common, *paludinæ*, teeth of crocodile, and *megalosaurus*, and minute teeth and scales of fishes; and, in short, in almost all the quarries around the Wells, the peculiar organic remains of the Wealden were found more or less abundantly.

In the sandstone pit on Langton Green, masses of ferruginous sandstone occur, which abound in casts of a small *paludina*, and *cyclas*, and a remarkable bivalve, which has not been named; with myriads of the casts of *cypris faba*. The sandstone of the High Rocks, near Eridge, contains traces of ferns and of arundinaceous plants. The quarries between Langton and Groombridge, afforded to our researches *paludinæ*, vertebræ, teeth of crocodiles, and scales of fishes.

It would be tedious to enumerate all the localities from which the fossils of the Wealden have been collected, for although the strata in certain parts of the country are richer than in others, yet there is no spot so barren, from which industry and well directed research may not succeed in extracting objects of interest.

Organic Remains.—The fossils of the inferior group of strata of the Wealden agree in their general characters, with those of the middle and upper divisions; the only essential difference appears to be, that in the Ashburnham limestone, and the shales and clay beneath, bivalves predominate; and that the remains of saurian reptiles are less abundant than in the superior deposits. There is no *paludina* limestone in this group that could be taken for the marble of the Weald clay; nor does the latter contain, in any of its calcareous beds, such an abundance of *cyclades* and other bivalves, as to form a limestone resembling that of Ashburnham.

We shall now proceed to describe the organic remains of the Wealden, and those of the strata of Tilgate Forest in particular; the observations on the recent discoveries will then follow, and we shall conclude the work with a review of the geological phenomena which have been presented to our notice.

CHAP. XI.

DESCRIPTION OF THE ORGANIC REMAINS OF THE WEALDEN, AND PARTICULARLY OF THOSE OF THE STRATA OF TILGATE FOREST.

Before we enter upon the description of the fossils of the Wealden, let us consider what would be the nature of a delta or an estuary, formed by a mighty river flowing in a tropical climate, over primary and secondary strata, through a country clothed with palms, arborescent ferns, and the usual vegetable productions of equinoctial regions, and inhabited by turtles, crocodiles, and other amphibious and terrestrial reptiles. In such a deposit we should expect to find sand more or less consolidated, with layers of clay and silt; containing water-worn fragments of the harder portions of the rocks, in the form of pebbles or gravel; bones, teeth, and scales, more or less rolled, of the amphibia that had lived and died on the borders of the river, and had been transported down its waters; the branches, and stems, and leaves of the vegetables that grew on its banks, intermingled with fresh water shells, and a small proportion of marine productions; a few bones of aquatic birds might also probably be observed. The strata of the Wealden present precisely such characters, and such an assemblage of animal and vegetable remains.

Vegetables. — These consist of the petrified trunks of large plants, belonging to that tribe of vegetables of the ancient world, which is so common in the carboniferous strata, and appears to have held an intermediate place between the Equiseta and the Palms; of the stems of a gigantic monocotyledonous vegetable; the foliage of ferns; and the stalks of arundinaceous plants. Of these, the most interesting is that which has been described as *Clathraria anomala*, in the Geological Transactions; but as we had previously named it in honour of Charles Lyell, Esq., F. R. S., &c., the author of the "Principles of Geology," a work which places its author among the most distinguished of modern geologists, we claim the privilege of original discoverers, and retain that specific designation.

Clathraria Lyellii. Plate I. figs. 1. 2. 6.

This vegetable appears to have possessed a thick epidermis, or *false bark*, formed by the union of the bases of the leaves, and covered externally with distinct rhomboidal scales, each scale being surrounded by an elevated ridge. These cicatrices, or scales, have evidently been formed by the attachments of the petioles of the leaves, or by the bases of the leaves themselves; the form of the latter is not positively known, although from some imperfect traces on the stone in a specimen bearing the impressions of the cicatrices of the bases of the leaves, there is reason to conclude that they were of a lineari-lanceolate form. The axis, or interior part of the trunk, originally enclosed by the bark, occurs in the state of solid subcylindrical blocks of sandstone, attenuated at their base, the surfaces of which are marked with longitudinal interrupted ridges, and, in some instances, are deeply imbricated; they are generally of a dark brown colour. A specimen surrounded by the cortical covering, is represented Plate I. fig. 2., and a branched example of the interior, in Plate I. fig. 1. A cellular interstitial substance seems to have existed between the internal part and the bark, and which is frequently found attached to the former, appearing like a fibrous carbonized integument, an eighth of an inch thick. Cicatrices are occasionally seen on the stems (*a*, *a*, fig. 1.), the nature of which has not been determined; they have been supposed to indicate the setting off of branches, but in the ramose specimens, the appearance is by no means similar; they are more like what is observable in the *Dracæna draco.* "This tree possesses a thin outer bark, marked by the cicatrices of the leaves; and within that, an internal, somewhat reticulated surface, in which there is a singular plexus of the vessels, formed where the dragon's blood was secreted, to which the cicatrix in the fossil vegetable bears a striking resemblance."* The casts of the internal axis of the fossil represent the true stem, and are composed of solid sandstone throughout, and show no traces of internal organisation, Plate I. fig. 1.; the original, therefore, probably possessed a structure too delicate to be preserved by mineralisation. The recent plants which present the greatest analogy to these

* Geolog. Trans. vol. i. p. 423. New Series.

fossils, are those of the genera *Cycas*, *Zamia*, *Yucca*, and *Dracæna*. The impressions of the petioles on the bark, bear a great resemblance to those on the stems of *Cycas revoluta* *, and *C. circinalis;* but, as M. Adolphe Brongniart observes, the dichotomous stem or branch brings it nearer to the *Yucca* and *Dracæna ;* the internal part of the caudex is, however, very different to that of the true monocotyledonous plants. In the *Dracæna*, there are not an internal axis and distinct bark, as in the fossil; and the bases of the leaves in the former are more transverse.

Among the numerous specimens of this plant in my cabinet, are two portions of the axis of the same stem, which fit into each other, the one extremity being slightly convex, and the other concave, to receive it; the first of these is figured (Plate I. fig. 6.), and is remarkable for the oblique transverse direction which its fibres assume. When these two portions are placed together, an elliptical opening is left on one side, indicating (as M. Adolphe Brongniart remarks, who obliged me by examining this specimen,) the origin of a lateral branch, or rather of a *floral axis*, the *axis of the panicle*, as in the *Dracæna ;* a proof that the *C. Lyellii* approximates very nearly to *Dracæna*, or rather to *Xanthorrea*, the stem of which has exactly the same structure as to its essential cha-

* A beautiful plant of this species exists in the unrivalled collection of living palms, of Messrs. Loddiges of Hackney. The scorings on the stem, formed by the shedding of the leaves, are very like those on C. Lyellii; the leaves that remain are sent off from the upper part of the plant.

racters, and is sometimes dichotomous, like the Clathraria, but the recent plant has not the bases of the leaves consolidated into a sort of bark, as in the fossil.* There are occasionally found in the Tilgate sandstone, irregular ramose fossils, having an obscurely fibrous structure, and these bear so much resemblance to the roots of the *Dracæna,* that there appears every reason to conclude they are the remains of those of the Clathraria Lyellii.

Endogenites erosa, Plate I. fig. 7. 4. and 5.

This is the only other vegetable, of any considerable size, that occurs in the strata of Tilgate Forest. A small specimen, exhibiting that peculiar eroded appearance of the surface denoted by the specific name, is so beautifully represented in the plate, as to convey a most correct idea of the original. This singular plant is of various forms and sizes: it is generally more or less flattened, attenuated at the base, and swells out at intervals, like some of the *Cacti,* and *Euphorbia.* Some examples are hatchet-shaped, and nearly flat, being from three to four feet long, twelve inches broad at the widest part, and not more than two or three inches in thickness: others are subcylindrical. All of them are covered with a dark carbonaceous matter,

* "Ce qui distingue cependant cette tige fossile de celles de ce genre de la Nouvelle Hollande, c'est que, dans la plante vivante, les bases des feuilles qui forment cette fause écorce, sont distinctes et seulement réunies par une matière résineuse, analogue au sang-dragon. Dans la Clathraria Lyellii au contraire, l'écorce paroit d'un seul morceau et formée par la soudure complète et intime des bases des feuilles: ces feuilles sont aussi beaucoup plus grosses dans la plante fossile et en moins grand nombre autour de la tige."—*Prodrome d'une Histoire des Végétaux Fossiles.*

which is removed by washing. The external surface is scored with small meandering grooves, and deep longitudinal furrows; a transverse section exhibits numerous pores, formed by the division of vessels, proving the monocotyledonous structure of the original (vide Pl. I. fig. 5.); and a magnified portion of the outer surface is shown Pl. I. fig. 4.

The constituent substance of these fossils is a dark grey sandstone; the cavities of the vessels are generally lined with crystallised quartz; and the fissures are frequently filled with white calcareous spar. We are not acquainted with any recent or fossil vegetables that are identical with this *endogenite*. M. Adolphe Brongniart informs us, that it bears some resemblance to the base, or the short and almost subterranean stems, of some recent species of ferns, that are not arborescent. Mr. Stokes observes, that a mass of monocotyledonous wood, from Upper Egypt, figured in the great work on that country, published by authority of the late Emperor Napoleon, has considerable resemblance to this fossil: and among some interesting specimens of the trunks of fossil palms, from Antigua, presented to us by the Honourable Mrs. Thomas, of Ratton, near Eastbourn, there are examples which expose a structure in many respects analogous. M. Brongniart is inclined to refer it to the arborescent ferns, rather than to the Palms.*

* Prodrome d'une Histoire des Végétaux Fossiles, par M. Adolphe Brongniart, Paris, 1828. 8vo.

CYCADITES BRONGNIARTI.

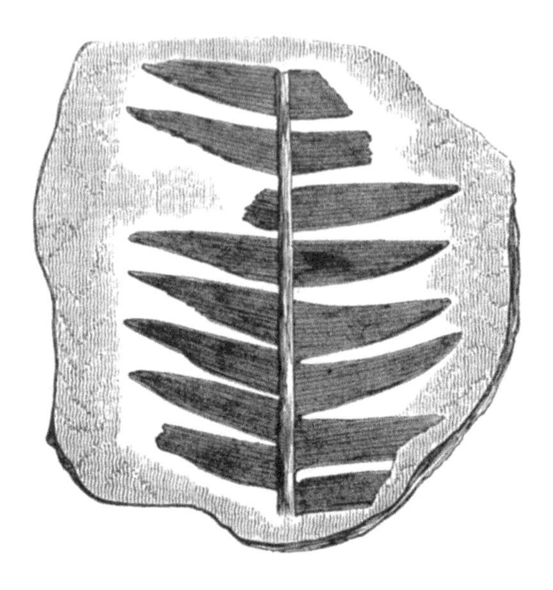

Cycadites Brongniarti.—In sandstone: Tilgate beds. This beautiful and rare plant, for the delineation of which I am indebted to Mr. William Constable, who discovered it in sandstone at Heaver's Wood Common, three miles south of Riegate, is more nearly related to the *C. tenuicaulis* of the *Cornbrash*, than to any other fossil with which I have compared it. It affords me the highest gratification to have this opportunity of offering my humble tribute of respect to that distinguished botanist, M. Adolphe Brongniart, whose philosophical and most interesting researches on the fossil flora of our planet, have already produced such important results; and the continuation of whose labours promises to effect in the botany, what the illustrious and lamented Cuvier has so successfully achieved in the zoology, of the ancient world.

FOSSIL FERNS FROM HEATHFIELD.

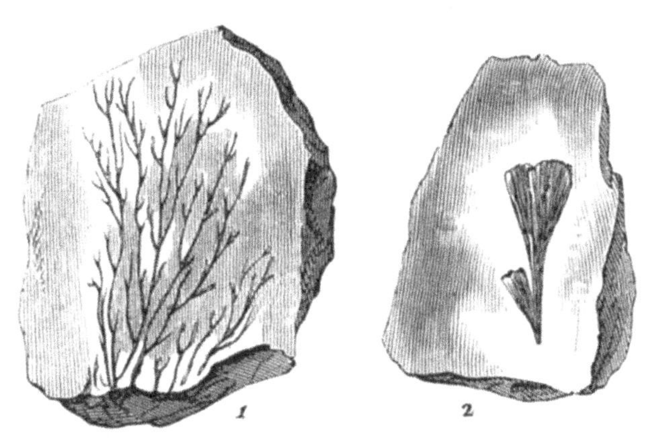

1. Sphenopteris Sillimani.
2. Sphenopteris Phillipsii.

Sphenopteris Sillimani. — In ferruginous calciferous grit. I have seen but the specimen here figured of this delicate fossil fern; it is from a quarry in the parish of Heathfield, where imperfect traces of vegetables are abundant. The specific name is in honour of that eminent philosopher and geologist, Dr. Silliman, Professor of Chemistry and Mineralogy in Yale College, Newhaven, Connecticut, Editor of the American Journal of Science; a man alike respected for his high scientific attainments, and beloved for his private virtues.

* Phillips's Yorkshire, Plate vii. fig. 18.

Sphenopteris Phillipsii. — Ironstone, Heathfield.

The delicate fern which is represented above, is from the ironstone of Pitwood quarry, Heathfield, mentioned in p. 227., where *paludinæ, uniones,* and stems, and imperfect remains of large ferns occur in abundance. It very much resembles *Sphenopteris latifolia* of the sandstone shale of Yorkshire*, figured and described in the excellent work of Mr. Phillips, to commemorate whose valuable services to Geology I have chosen the specific name.

The stems of large ferns distributed throughout the ironstone, in the quarry at Pitwood, are most remarkable; in no other part of the Wealden are fossil vegetable remains seen in such abundance, and yet, with but one exception no traces of the foliage remain. The example alluded to is a small frond in contact with a branch, which, like most other specimens in this locality, is converted into hydrate of alumine: and the frond, although in juxtaposition with the branch, is yet so placed, that the contact may be accidental; and the foliage differs so remarkably from that of all known ferns with similar stems, that Mr. Phillips and Mr. Stokes, who have examined the specimen, think it probable they belong to different genera; as a doubt remains on the subject, the specimen is not named. In the same quarry I observed numerous traces of Lycopodites.

* Phillips's Geology of Yorkshire, Plate VII. fig. 18.

Sphenopteris Mantelli.

Sphenopteris Mantelli. The form of this elegant fern is so remarkable, that it is easily recognised. Messrs. Stokes and Webb, from the shape of its frond, supposed that it resembled the plants of the genus *Psilotum;* but from the disposition of the vessels in the ultimate segments, more nearly approaching to that observable in those of the genera *Trichomanes* and *Hymenophyllum,* and from the circumstance of all the divisions of the frond being, in the fossil, bordered by a decurrent membrane, those gentlemen formed a new sub-genus for its reception, which they named *Hymenopteris.** M.

* Geolog. Trans. vol. i. new series.

Adolphe Brongniart does not, however, admit the propriety of this distinction, but refers the fossil to his sub-genus *Sphenopteris.** "The ramification of the fronds, indicated by the different planes in which the branches are disposed, appears to be very distinct from the mere distichous frond of the recent ferns," and is another instance of the difference existing between the vegetation of the ancient and modern condition of our planet, even in those genera that are in many respects analogous. The specimens present considerable variety both in the form, and disposition, of the fronds; perhaps, hereafter, these variations may be found sufficiently permanent and important, to constitute specific characters. This fern has been discovered in the hard calciferous sandstone, ironstone, shale, and clay; it is generally in a carbonized state, and occurs in small portions, so that a knowledge of the entire form of the original plant is still a desideratum. It appears to have been very abundant, for no considerable block of stone is without traces of it: but it generally occurs in confused masses.

* "Cette fougère appartient au sous-genre que j'ai nommé *Sphænopteris*, et non au s. g. hymenopteris: le nom de *psilotoides* ne me paraît pas très convenable en ce que cette plante diffère totalement du genre psilotum; si par cette raison vous me permettiez de lui donner le nom de *Sp. Mantelli*, on éviterait une comparaison qui ne me paraît pas exacte, et on consacrerait le nom de la personne qui a la première prouvé la présence de fougères dans des terrains aussi modernes."—Extract of a Letter from M. Adolphe Brongniart to the Author.

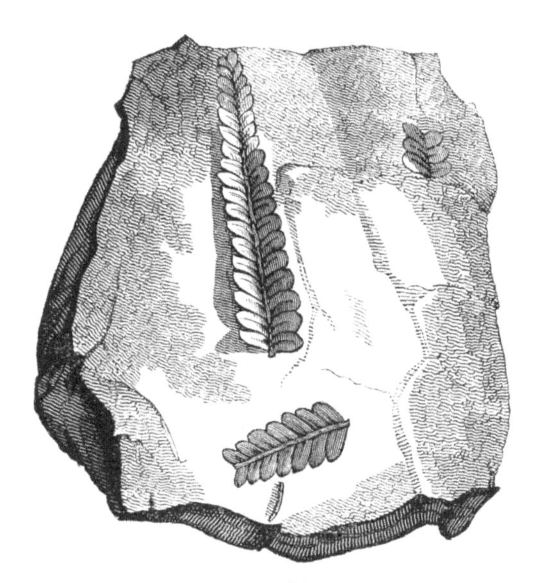

Lonchopteris Mantelli.

This beautiful filicite "approaches in habit to some of the tropical *Nephrodia ;* but the ramifications and interlacing of the veins, at the tips of the pinnæ, are peculiar, and serve to distinguish it from other species:"* a magnified view of this structure is shown Pl. I. fig. 3. There is a plant in the coal shales which has reticulated leaves, but the form and size of the pinnæ are altogether different. This fern is found in the Worth sands, both in the Forest, and in the cliffs at Hastings; and in the shales and clays of Tilgate Forest: it has but rarely been noticed in the harder strata; it always occurs in a carbonized state. It was described by Messrs. Stokes and Webb, under the

* Geolog. Trans. vol. i. new series, p. 423.

name of *Pecopteris reticulata;* but M. Adolphe Brongniart, who has carefully examined its structure, refers it to his genus *Lonchopteris*, with the above specific designation. One of the most curious and beautiful vegetable fossils in my collection is a leaf of this fern in hard grit, in which the parts of fructification are displayed in the most perfect manner.

Lycopodites? — Horsted Sand, near Chiddingly. Several delicate fossil plants have been collected in a marly sand, in a low bank on the road-side leading from the Dicker to Chiddingly: they resemble the vegetable remains referred by M. Brongniart to this genus, but they are so imperfect that their true characters have not been ascertained. In the same sand-bank, the *Lonchopteris Mantelli* is also met with; and one of the layers of sand is full of innumerable traces of a species of Sphenopteris, too obscure to make out its characters, but the stems and branches of which are disposed as if the plants were standing erect on the spot where they originally grew, and a stratum of sand had been gently deposited upon them. In a layer of reddish sandstone at Stammerham, I observed a similar appearance.

Calamites? — In the soft sandstone of Hastings, we perceived the remains of stems of plants, flat from compression, and in such a state of decay as to admit of no specification, yet they afforded certain evidence of the former existence of large arundinaceous plants in the Wealden. Traces of Cycadites and Sphenopterites are disseminated throughout these arenaceous strata.

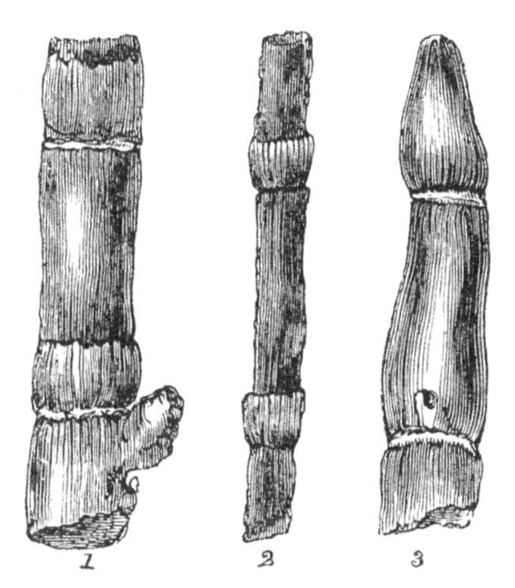

Equisetum Lyellii.

Equisetum Lyellii. — In grey and blue grit and limestone, Pounceford.

The occurrence of the Equiseta in the Wealden strata, is very local; but where these plants are found, they prevail in great numbers: in no locality are they more abundant than in the quarries at Pounceford, and the surrounding country. This species, which I have named in honour of Charles Lyell, Esq., Professor of Geology in King's College, London, &c., is very distinct from any previously noticed; when perfect, it probably attained a height of two feet or more: the termination of the cryptogamous head is seen in fig. 3: and, in fig. 1. and 2., traces of the striated sheath are displayed.

Carpolithus Mantelli.—Geol. Trans. vol i. new series, p. 423. "This fossil seed-vessel is occasionally found in the grit and sandstone of Tilgate

Forest. The surface bears marks of the ramifications of the veins, probably the impressions of the vessels of the integument, and not those of the nucleus itself.

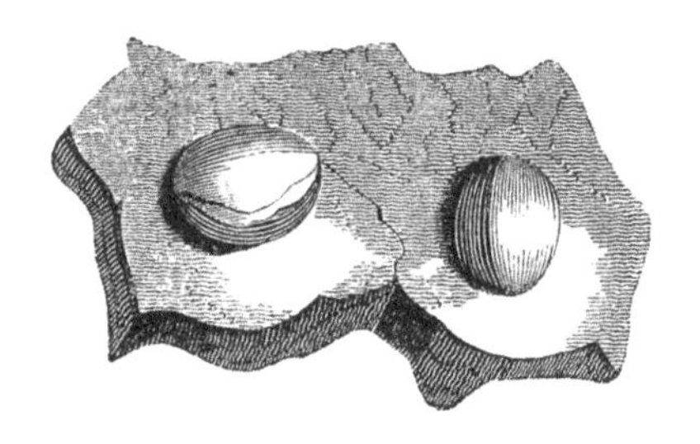

Carpolithus Mantelli.

This carpolithe resembles the grains or kernels of some kinds of palms, as the *Areca;* but the disposition of the nerves is not precisely the same:" such are the remarks of M. Adolphe Brongniart, who considers it very probable that this fossil is the petrified seed-vessel of the *Clathraria Lyellii.* Several specimens were found among the vegetable remains in the block of stone, in which the remains of the *Hylæosaurus* are imbedded.

The other vegetable remains that have been noticed in the strata of Tilgate Forest, are too imperfect to admit of discrimination. It may, however, be observed, that no unequivocal example of a dicotyledonous tree has been discovered, and also, that the fossils above described appear to be peculiar to the Wealden: they are not known to occur in the coal measures, nor in the chalk, or tertiary deposits. M. Adolphe Brongniart remarks, that the only fossil vegetables at all similar are some discovered in Denmark and Germany, in strata apparently of the same epoch.*

TESTACEA.

The shells of the Wealden, as already mentioned, are almost exclusively fluviatile: in vain might we seek for the ammonites, nautili, and other marine species, so abundant in the chalk. As in all fluviatile deposits, the number of genera and species is but few. In the harder grits and the limestones, the substance of the shell, if it remain, is commonly changed into calcareous spar; very generally it is altogether wanting; the casts of the interior being the only traces of the testacea that are left. The various species of shells of the

* "Observations sur les Végétaux Fossiles renfermés dans les Grés de Hoer en Scanie;" par M. Ad. Brongniart. Annales des Sciences Naturelles, Février, 1825: —

"M. Mantell a découvert dans des couches dont l'époque de formation ne paraît pas beaucoup différer de celui des lieux que nous venons de citer, des fossiles végétaux qui confirment encore cette analogie du Grés de Hoer avec les divers membres de la formation jurassique. Ce géologue a en effet observé dans les couches désignées par les savans Anglais sons le nous de Iron sand, ou Sable ferrugineux, à Tilgate dans le Sussex, des débris de végétaux dont les uns sont évidemment des feuilles de fougères et dont les autres semblent être des tiges de Cycadées. Quant à la Craie dont les couches succèdent presqu' immédiatement à celles du calcaire jurassique, les plantes en petit nombre qu'on y a trouvées sont toutes d'origine marine: ce sont des fucus, des ulves, ou des conferves; on n'y a rien vu qui annonce une plante terrestre."

Wealden, as well as of the Shanklin sand, will be described and figured with so much accuracy and detail, in the masterly paper of Dr. Fitton, on the beds below the chalk, that I shall notice the subject in a very brief manner.

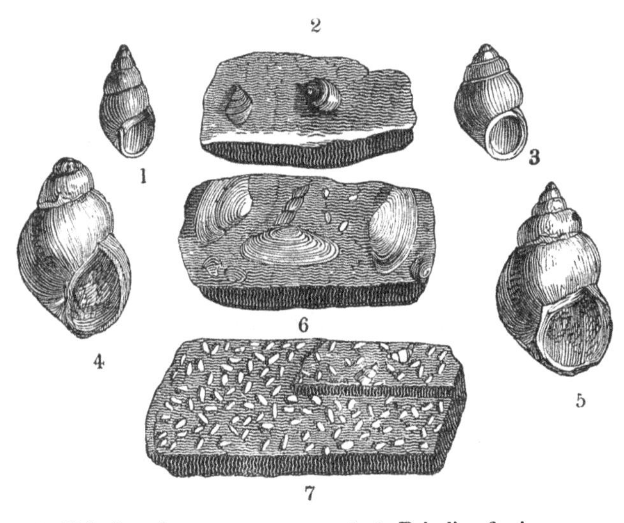

1. Paludina elongata.
2. Neretina Fittoni.
3. Paludina carinifera.
4, 5. Paludina fluviorum.
6. Psammobia.
7. Cypris faba.

Paludina fluviorum. — This freshwater univalve is the principal constituent of the Sussex marble: it occurs in the grit and sandstone, but more sparingly than in the upper limestone strata.

Paludina carinifera. — A small species found in septaria, in Resting-oak Hill, with *Cypris faba.*

Paludina elongata: in the grit of Tilgate Forest, and in the limestone, and sandstone at Hollington, near Hastings, and at St. Leonard's.

Neretina Fittoni. — The elegant little shell here

figured, occurs in the state of casts in the Tilgate grit, and is a well-marked, distinct species. I have named it in honour of Dr. Fitton.

Melanopsis. — Several shells apparently of this genus are found in the beds of shale at Pounceford, with the cyclades; they bear also considerable resemblance to *melania.*

Bulla. — Mr. Sowerby supposes that a small cast among some paludinæ belongs to this genus.

Succinea. — A cast of a small shell apparently of this genus, was observed with a group of paludinæ in the Tilgate grit; it was not sufficiently defined to admit of its characters being positively ascertained.

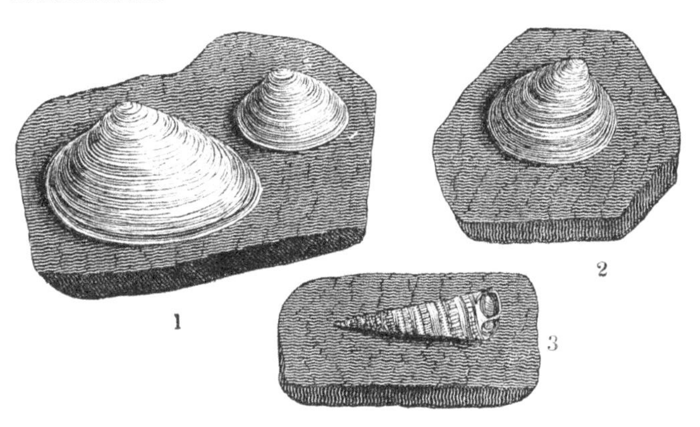

1. Cyclas membranacea.
2. ——— media.
3. Melanopsis, or Melania.

Cyclas. — On the shells of this genus from the shale of Pounceford, where entire beds are almost wholly composed of their broken remains, Dr. Fleming, the distinguished author of one of the most valuable works on zoology that has appeared in this or any other country, "British Animals," favoured me with the following observations.—"Of the propriety of referring these shells to the genus *Cyclas*, no doubt can exist. The two bifid hinge teeth, in some specimens, are distinctly displayed; even the *ligament* obviously exists, and sometimes the *cuticle*.

Cyclas media.—This is the strong shell with the concentric ridges coarser, and the posterior side-lip more produced, than the next: the *Cyclas cornea*, is, I suspect, the young of this species.

Cyclas membranacea. This is more inequilateral than the former, by the production of the hinder extremity; it appears to have been concentrically grooved."

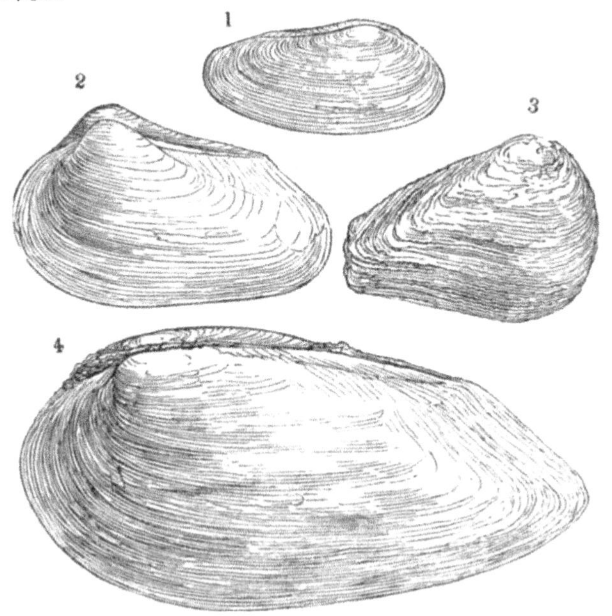

1. Unio antiquus. 3. Unio aduncus.
2. Unio compressus. 4. Unio porrectus.

Unio. Of this freshwater genus Mr. Sowerby has established five species from the specimens from Tilgate Forest, which are in my collection; namely,—

U. aduncus.
U. antiquus: this is the most common species.
U- porrectus.
U. compressus.
U. cordiformis.

Psammobia. An elegant little bivalve that is of rare occurrence in the cyclas limestone of Burwash, is referred by Dr. Fleming to this genus: it is represented fig. 6. of the woodcut, page 248.

Mytilus. Shells of this genus were found by myself and Mr. Lyell, in a bed of shale, that had been washed by a gentle stream which flowed over it, and by which many delicate shells were brought to view: they are too imperfect to be determined.

CRUSTACEA.

Cypris faba: fig. 7. of woodcut, page 248.

The Cypris is a crustaceous animal enclosed within two valves or shells of an oblong form; it has two antennæ terminated by a pencil of hairs; one eye and four legs; the head concealed, and the tail small. It is an inhabitant of fresh water only. The two little oblong shells, which appear like the two valves of a bivalve, are the only remains of this animal that occur in a fossil state, and which are profusely distributed throughout the strata of the Wealden. It is to the sagacity of Dr. Fitton that we are indebted for a knowledge of the various species, and of the important inferences to be drawn from their presence. That gentleman has discovered several species, which will be figured and described in the memoir already referred to. The preceding as well as the following observations are, in substance, those of Mr. Sowerby, in his Mineral Conchology:—

These minute shelly coverings are one tenth of an inch in length, and about half an inch wide: the surface is minutely punctated, the substance rather coriaceous, but brittle and very thin. They occur in the Sussex marble; in the Weald clay: and their casts in the ironstone: at Hollington in sandstone; in the Isle of Wight in clay, at Grange Chine; or in slaty clay in Sandover Bay. In France they have been found in the second freshwater formation above the chalk; at Puy-en-Velay along with gypsum, under lava and over clay, with Planorbes and Cyclostomæ.

FISHES.

The remains of fishes in the Wealden consist of detached bones, teeth, scales, and fins, with but two or three exceptions, where a portion of the scales and skeleton remain in juxtaposition; these belong to the fish which I shall first describe.

Lepisosteus Fittoni.—Scales of a dark-brown colour, of a rhomboidal form, possessing a high polish on the external surface, with bifurcating processes of attachment, are among the most abundant fossils of the Wealden: there is not a quarry in Kent or Sussex where they may not be found. Hemispherical teeth, the *bufonites* of the early geologists, either single, or disposed in rows on a bony plate, are almost equally common; both the scales and teeth belong to the same species of fish, as a beautiful specimen in my collection has established. A pectoral fin, and traces of the dorsal fin; some of the bony rays of the gills, and portions of the operculum; are the only other parts of this ichthyolite with which we are at present acquainted. The rhomboidal scales, &c. prove that the original was related to those species of pike which are comprised in the genus *Lepisosteus* of Lacépède, and I have given the specific name as a tribute of respect due to a gentleman by whose labours the establishment of the true geological relations of the Wealden formation has been principally effected.

The other species of fishes whose remains occur in these strata may be referred to the following:—

1. Fish, having tricuspid teeth longitudinally striated: these differ from the teeth found in the chalk, but resemble some from the Stonesfield slate.

2. Teeth slightly curved, with sharp entire edges.

3. Palate-teeth, approximating to a small palate found at Stonesfield.

4. Palatal bone of a fish allied to the *Ray*. Slightly curved and nearly flat pieces of enamel occupy the centre; and these are bordered by two rows of depressed sub-conical bones or teeth.

5. Radii, or dorsal defences, of a fish allied to the *Silurus*. These occur of three or four species, and very closely correspond to the bony fin of the recents Silurus.*

6. Minute scales and vertebræ: some of the argillaceous partings of the strata are full of these remains, but they are always in too mutilated a state to afford correct indications of the forms of the originals.

CHELONIAN REPTILES, OR TURTLES.

The disjointed bony skeletons are, with but few exceptions, the only remains of turtles that occur in a fossil state; and it is therefore by a careful comparison of the fossil with the recent bones that we can alone hope to obtain satisfactory conclusions respecting the forms and habits of the originals. Among the osseous remains of this order, which have been discovered in the strata of Tilgate Forest, none have been found in so perfect a state as to exhibit any considerable portion of the upper shell or *carapace:* but the detached ribs, vertebræ, portions of the bones of the sternum, pelvis, and extremities, which are abundantly distributed both in the Tilgate stone and in the friable sandstone, will, with the light which the illustrious Cuvier has thrown upon the anatomy of the recent animals, enable us to obtain highly interesting results.

* The *Silurus glanis* inhabits the rivers of Europe and the East; it is the largest of all freshwater fishes, being frequently twenty-four feet long and weighing 300 pounds. *Elements of Nat. Hist.*, vol. i. p. 368.

The bones of turtles, as well as those of the other oviparous quadrupeds of the Wealden, are generally of a dark brown colour, very heavy, brittle and strongly impregnated with iron. Their cellular structure is, in many instances, beautifully displayed, being injected, as it were, with limestone or white calcareous spar; and in numerous examples the medullary cavities of the long bones are filled with the same substance. There are, however, some remarkable exceptions, in which the bones are as light and porous as the osseous remains of diluvial deposits.

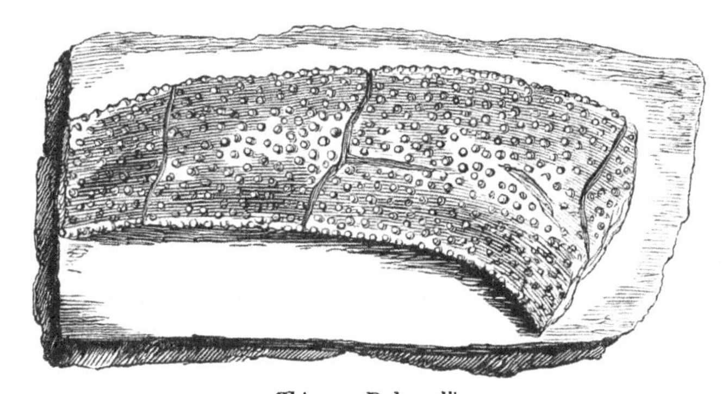

Trionyx Bakewelli.

Trionyx Bakewelli.—To the subgenus trionyx of M. Geoffroy, which contains freshwater turtles only, the specimen here figured bears the strongest affinity. The recent *trionices* are distinguished by the intervals between the ribs not being ossified throughout; by their extremities not being articulated to an osseous border; and by their surface being shagreened, and marked with little pits or hollows, that the soft skin, the only integument with which the trionyx is covered, may be firmly attached.* Being destitute of scales, their bones exhibit no traces of the furrows, or depressions, produced by their margins, as in the other sub-genera. The fossils we are about to describe, possess a shagreen surface, like the trionyx, but differ from the recent species in bearing decided marks of having had a scaly covering. In the rib above represented these impressions are clearly shown; and it is necessary to remark, that this bone, instead of being nearly of an equal width throughout, as in the fresh-water and marine species, gradually enlarges till the outer termination is nearly twice as wide as the inner. Such a character obtains in the ribs of land tortoises only†, and therefore presents another anomaly in the structure of the fossil animal. Portions of the sternum or *plastron*, and some of the scales belonging to this species, have been found. Large bones, covered with little pits, or fossæ, as in the soft turtles, have also been discovered; but they are too imperfec to be identified except by more experienced comparative anatomists than ourselves. From the slight degree of convexity of the rib, it is clear that the original was of a flattened form, like the common turtle, *Testudo mydas*; the shagreened surface proves its analogy to the trionyx; but the impressions of the scales show that it cannot be identified with any recent species. It may be mentioned, that among the numerous portions of the

* Oss. Foss. tome iii. p. 329.

† Oss. Foss., tome iii. p. 333. Edit. 1822.

osseous border of the carapace, found in Tilgate Forest, we have not observed any that have a shagreen surface; a negative proof, that the fossil, like the recent trionyx, was destitute of that appendage.

The specific name of this fossil trionyx is given as a tribute of respect to Robert Bakewell, Esq. a gentleman well known in the scientific world, and the author of decidedly the best "Introduction to Geology" in this or any other language.

Emydes. Freshwater Turtles.—In referring some of the bones of the Tilgate beds to this subgenus, we proceed on certain grounds; since the subject has already been elucidated by Baron Cuvier. Among the Sussex fossils which we transmitted to that celebrated naturalist some years since, were portions of the buckler, or carapace, of the same species as the fragments here delineated. These M. Cuvier determined to belong to a remarkably flat, unknown species of *Emys;* which occurs also in the Jura limestone, in the vicinity of Soleure*; they even correspond, in size, with a specimen figured in pl. xv. fig. 6. "Oss. Foss." tome v.; and which, if entire, would have exceeded twenty inches in breadth.

The species with smooth plates and ribs are by far the most abundant in the Tilgate strata; but, from our not having been so fortunate as to discover any united portions of the upper shell, we cannot, in every instance, determine whether the specimens presented to our notice belong to fresh-

* The specimens at Soleure are associated with the jaws and teeth of Spari, and other fishes; Terebratulæ, Echini, and other marine exuviæ.

water or marine sub-genera. The opinion of M. Cuvier is decisive as to the existence of an *emys* in these deposits; and, from the details we shall now offer, that of a marine turtle is more than problematical.

Cheloniæ, or *Marine Turtles.*—Ribs with a smooth surface, of nearly an equal width throughout, with pointed, striated extremities, and marked with impressions of scales; portions of a smooth osseous border; and sternal plates with radiated or dentated margins, are the remains which we propose to refer to this subgenus.

Some examples possess the striated projecting extremities, so characteristic of the ribs of the *cheloniæ;* but as they are not connected with an osseous margin, they might be supposed to belong to the trionyx. From this division of turtles they are, however, separated by the smooth surface and the impressions of scales.* Of this species we have specimens of the pointed extremities of two ribs, which, from the form of the portion of the osseous border adhering to them, are probably the first and second ribs of the left side; also the

* Mr. Clift, our first comparative anatomist, the Curator of the Museum of the Royal College of Surgeons, obliged me by comparing this specimen with the recent skeletons in the Museum, and stated that it resembled the third rib of *Testudo imbricata.*

Burtin (*Oryctographie de Bruxelles*) figures the extremity of a rib very like this fossil; and as M. Cuvier has referred the turtles of Melsbroeck to the Emydes, we at first entertained doubts whether our appropriation of this specimen to the *cheloniæ* were correct. Mr. Clift's remark, however, tends to confirm the opinion that it belongs to a marine turtle; and we are not aware, that the ribs of the recent Emydes possess such a character.

The largest ribs in our cabinet must have belonged to a turtle about 34 inches in length.

second or third sternal plates, which exhibit the curvature for the passage of the paws. The marine turtles have the plates of the breast-plate lobed and dentated, and we have very recently obtained a fine example of the third sternal plate of a turtle, from Tilgate, in which the dentated margin closely resembles that of the *Testudo imbricata*, or T. *carinata*, and of the fossil species found at Maestricht.* We have also part of a shoulder-bone, which is very like that figured in pl. xiv. fig. 6. *Ossemens Fossiles*, tome v. This specimen bears a greater affinity to the shoulder-bones of a marine turtle than to those of a land or freshwater species.

Our cabinet contains numerous other fragments of the carapace, sternum, ribs, humerus, clavicle, &c., but none sufficiently perfect to throw any additional light upon our investigations. From what has been advanced, we may, however, infer, that the strata of Tilgate Forest contain the remains of at least *three* very distinct kinds of turtles; viz.

1st. A freshwater species, allied to the *Trionyx*.

2dly. An unknown species of *Emys*, resembling a fossil freshwater turtle, found in the Jura limestone.

3dly. A marine species belonging to the subgenus *Chelonia*, and related to the fossil turtle of Maestricht.

* Oss. Foss., pl. xiv. fig. 3. tome v.

SAURIANS.

Of the Saurians, the bones and teeth of at least four genera have been discovered in the strata of Tilgate Forest; namely, the CROCODILE, PLESIOSAURUS, MEGALOSAURUS, and IGUANODON; but hitherto no connected portions of their skeletons have come under our observation.* The teeth, both in form and structure, present such striking differences, as to be readily distinguished from each other, and from those of existing species; but the bones possess so many characters in common, that when we consider the broken and detached state in which they occur, and their intermixture with the débris of turtles, of vegetables, of fishes, and of shells, the difficulty of the attempt to identify the bones of the respective animals seems almost insurmountable to observers so distant from any collection of comparative anatomy as ourselves. We therefore claim the indulgence of the reader, should the results of our investigations appear to be in some respects inconclusive and unsatisfactory; since, under such circumstances, rigorous conclusions must not be expected. We shall first describe the teeth, and such of the bones, as are referable, with but little doubt, to one or other of the above-mentioned genera; and afterwards notice those osseous remains, which we are unable to appropriate with any degree of certainty or probability.

* As it was considered advisable to print the memoir on the recently discovered fossil reptiles of the strata of Tilgate Forest without alteration, it became necessary to allow this portion of the text of the former work to remain, although some tautology is thereby incurred.

CROCODILE.—The remains of Crocodiles have been found in all the secondary formations of England, from the Oolite to the chalk inclusive; and also in the tertiary deposits. Their existence in the Wealden formation was first noticed in our former works; and imperfect as were the specimens then in our possession, we expressed our conviction that they approached very nearly to the crocodile with concave vertebræ found at Havre. M. Cuvier, in the new edition of the *Ossemens Fossiles*, confirms our conjectures, and also states, that the bones in question are almost identical with the fossil crocodile of Caen*, which belongs to the *Gavials*, a subdivision of the genus, characterised by the narrow, elongated, almost cylindrical jaws, which form an extremely lengthened muzzle. The teeth of the crocodile are distin-

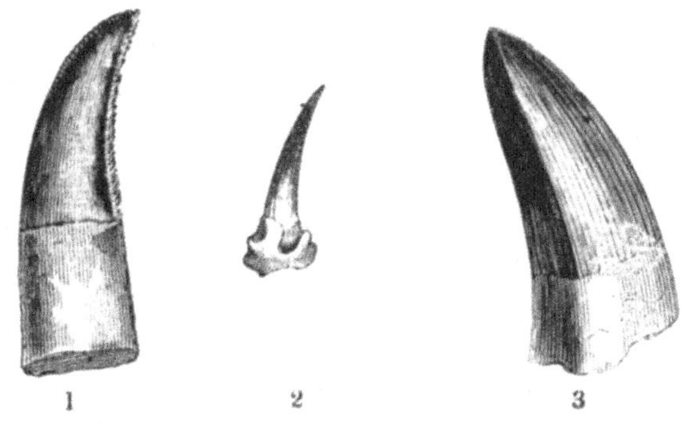

1. Tooth of the Megalosaurus.
2. Tooth of the fossil Gavial of Tilgate Forest.
3. Tooth of the fossil Crocodile of Tilgate Forest.

* The presence of Crocodilian remains in the Lias is not at present positively determined. Mr. Conybeare observes, that "the bones from Whitby are of the Ichthyosaurus, and Plesiosaurus." * Crocodiles'

* Parkinson's Introd. Org. Rem. p. 286.

guished by their conical form, striated surface, and lateral edges, but more particularly by their internal structure. Fig. 3. represents a fossil tooth from Tilgate Forest. They never become solid, but consist, as it were, of a series of cones enclosed within each other, the outer cone, or old tooth, being burst by the pressure of the included one, as the latter increases in size. Hence, at whatever age a tooth is removed, we find either in the socket or the cavity of the tooth, a small tooth ready to occupy the place of the old one, when the latter is destroyed by age or accident. This succession is repeated very frequently; and it is from this cause, that the fossil teeth of these animals are always so sharp and well defined; for although larger when old, they are not less perfect than in the young state. The lateral ridges are placed anteriorly and posteriorly.

Teeth of the *Crocodiles* of *Tilgate Forest.* (Figs. 2, 3. p. 261.)—These will be found to present all the essential characters of the teeth of the recent

---

teeth and bones occur in the Stonesfield slate, Purbeck limestone, &c. and in the Hastings formation: we have not observed them in the Shanklin sand, nŏr in the Galt; but they are found, though very rarely, in the lower chalk. M. Cuvier mentions, as a solitary instance, a crocodile's tooth in chalk from Meudon. It should be remarked, that there are occasionally found in the chalk, teeth of a conical form, and longitudinally striated, which, from their external appearance only, might be taken for those of crocodiles; but an examination of their internal structure readily distinguishes them. Mr. Parkinson describes the remains of a crocodile found in the London clay at Hackney, in which the vertebræ were concavo-convex, as in the recent species.* The head of an alligator was found in the London clay, of the Isle of Sheppey, during the last year (1832).

* Parkinson's Introd. Org. Rem. p. 387.

crocodiles; some (fig. 3.) are more obtuse than those of the crocodile of the Nile, and resemble the teeth of the second species of the fossil crocodiles of the Jura limestone, described by Baron Cuvier, vol. v. p. 142, *Oss. Foss.*: others (fig. 2.) are more slender, and possess a greater degree of curvature, bearing a close analogy to those from Caen.* The largest specimens in my possession must have belonged to an animal between twenty and thirty feet long.

Scales of a Gavial? — Strong, thick scales, of a dark-brown colour, and possessing a high polish, are very abundant in the Tilgate strata; those with bifurcating processes of attachment belong to the Lepisosteus Fittoni; but there are others of a different form, which we compared with the scales of a living alligator, and found them to resemble those which cover the legs of that animal. The scales of the fossil crocodile of Caen are described as being very thick, rectangular, thin towards the edge, and having the outer surface covered with little pits or hollows; we have not observed any with the characters here described, in the strata of Tilgate Forest.

Vertebræ. — In the recent crocodile, the vertebræ are convex posteriorly, and concave anteriorly; but those from Tilgate, like the vertebræ of the crocodile of Caen, and of one of the species of Havre, are, with but very few exceptions,

* We compared some of these fossils with the teeth of a crocodile from the banks of the Ganges, preserved in the museum of the East India Company, and could not detect any essential difference. Figs. 25, 26, 27. 30. pl. x. tome v. Cuvier's "Ossemens Fossiles," represent teeth of the Tilgate Gavial. Vide also *Foss. Tilg. Forest.*

slightly concave at both extremities. It may, perhaps, be observed that in the animals of this tribe, the epiphyses of the bones are cartilaginous in the young individuals, and that this circumstance may, in some instances, have given rise to this appearance; and we have ourselves remarked, among some detached crocodilian bones, sent by Sir Thomas Stamford Raffles to the College of Surgeons, vertebræ destitute of the convex articulating face, from this cause. But this character is too constant in the fossil vertebræ to be the result of such a circumstance, and can only have been produced by original conformation. The vertebræ are more contracted in the middle than those of the recent species, and are generally more or less compressed laterally. The caudal vertebræ, as in all the other lacertæ, are, of course, by far the most numerous.

Ribs.—Many of the ribs found in the Tilgate beds are decidedly those of crocodiles, presenting that double or bifurcating termination, so peculiar to the animals of the crocodilian family.

Bones of the extremities.—These, for the most part, occur in so mutilated a state, that but few examples can be identified with certainty: we have portions of the humerus, radius, and tibia.

Carpal and Metacarpal bones.—These resemble the bones of mammalia, and, in fact, cannot be readily distinguished from them: we have many examples, which, from their slightly depressed form, there is reason to conclude belonged to crocodiles.

Os frontis.—We have reserved for this place, a notice of a small bone (fig. 1. Plate II.) which, imperfect as it is, and partly obscured by the stone, bears a striking resemblance to the frontal bone of a small crocodile, about two feet in length, from the gypsum of Montmartre *: the inner surface only is exposed, and exhibits a channel which, if our conjecture be correct, formed the passage of the olfactory nerve to the nose; on the upper part the orbital arches are seen.

From the above remarks, it appears that the strata of Tilgate Forest contain the remains of at least two, if not four, species of crocodiles: that one of these (that with slender curved teeth) resembles the gavial of Caen, and probably was about twenty-five feet in length; and the other, with obtuse teeth, the fossil crocodile of Jura.

Megalosaurus Bucklandi.—Of this gigantic animal, whose remains have been discovered in the slate of Stonesfield, and were first described by Dr. Buckland, the teeth, ribs, vertebræ, and other bones, have been found in the strata of Tilgate Forest; the teeth occur also in the Purbeck limestone. Fig. 1. of the wood-cut, p. 261., represents a tooth from Tilgate Forest. No connected portion of the skeleton has been discovered, with the exception of part of the vertebral column, consisting of five anchylosed vertebræ; and a magnificent example of a fragment of the lower jaw, with several teeth remaining in their sockets. The bones which Dr. Buckland appropriates to the Megalosaurus, besides those above mentioned, are a perfect femur; a coracoid bone; ribs; a clavicle;

* Cuvier, Oss. Foss., tome iii. pl. lxxvi. fig. 8.

fibula; and a metatarsal bone; but, as M. Cuvier remarks, since these bones were found promiscuously intermingled with those of crocodiles and other oviparous reptiles, it is simply from their being discovered in the same stratum, and from their zoological characters, that we conclude they belong to the same kind of animal; a conclusion, which, as we shall hereafter have occasion to remark when treating of the iguanodon, must not always be regarded as unequivocal.

Teeth. (Fig. 1. p. 261.)—The teeth of the megalosaurus are compressed laterally, and slightly recurved backwards; their edges are finely serrated, the anterior edge being much thicker than the posterior, which is very sharp and thin. They bear a great analogy to the teeth of several species of the recent monitors; and the structure of the jaw of the megalosaurus indicates also a strong affinity to the animals of that genus.* In the jaw found at Stonesfield, the teeth are lodged in distinct alveoli, and do not adhere, as in the monitors, to the substance of the jaw, by any incorporation of the root or sides. In this respect they agree with the crocodile, but the outer edge of the jaw rises almost an inch above the inner margin, forming, as it were, a continuous lateral parapet to support the teeth externally, after the manner of the monitors. A series of triangular plates of bone arises from the inner edge, and constitutes a zig-zag buttress internally; and from the centre of each plate, a bony septum passes across to the outer parapet, by

* Vide Oss. Foss., tome v. pl. xvi. p. 276.

which the alveoli are formed. The new teeth rise in the angle between each triangular plate. None of the recent saurians have an analogous structure: in the monitors there is no inner alveolar ridge; the new teeth are formed in the substance of the gums between the bases of the old teeth: and they have no alveoli. The iguanas, which have the outer parapet very high, extending half-way up the teeth, are also destitute of any internal osseous border. Hence it may be inferred, that this monster of the ancient world belongs to an extinct genus of saurians, which partook of the characters of the crocodile and monitor; but was most nearly related to the latter.* From the straitness of the portion of the jaw found by Dr. Buckland, which is eleven inches long, it is evident that the jaws must have terminated in a flat, strait, and very narrow snout.

Vertebræ.—These differ from the vertebræ of any known recent monitors, but bear a great resemblance to those of the fossil crocodiles of Tilgate Forest, Havre, Caen, &c. They are generally a third longer than wide; slightly concave at both extremities; and more or less contracted in the middle; their annular part is united to the body by a suture, as in the crocodile and monitor; but there is a considerable lateral depression immediately beneath the annular part. It must, however, be confessed, that our knowledge of the osteological structure of the megalosaurus is at present too limited to enable us to determine the distinguishing characters of its vertebral column.

* The teeth are hollow in the young state, but become solid by ag

Coracoid bone.—This bone alone is sufficient to prove that the original animal was entirely distinct from the crocodiles, and approached very nearly to the monitors and iguanas. It differs, however, from the corresponding bone of existing species: the original is upwards of *two feet* in length.

Ribs, and supposed pelvis.—The ribs which Dr. Buckland has appropriated to this animal have a double articulation like those of the crocodile; but in the ribs of the monitors and other lizards, the spinal extremities are never divided into a head and tubercle; it is, therefore, probable, that the supposed ribs of the megalosaurus may have belonged to the crocodile, with whose remains they are associated; and that the ribs of the former have not yet been observed.

Extremities.—A remarkably perfect thigh-bone was found at Stonesfield, by Professor Buckland, and is two feet nine inches in length; the medullary cavity is filled with calcareous spar. We have numerous fragments of the femur from Tilgate Forest.

IGUANODON.—The discovery of the teeth and other remains of a nondescript herbivorous reptile, in the strata of Tilgate Forest, a reptile "encore plus extraordinaire que tous ceux dont nous avons connoissance *," is one of the most gratifying results of my labours. The first specimens of the teeth were found by Mrs. Mantell, in the coarse conglomerate of the Forest, in the spring of 1822 †;

* Cuvier, Oss. Foss., tome v. 2d part, p. 351.

† Vide Fossils of the South Downs, p. 54. No. 40, 41.

and a most interesting series has subsequently been collected, displaying every gradation of form, from the perfect tooth in the young animal, to the last stage, that of a mere bony stump, worn away by mastication. These teeth are comparatively rare; and the only locality in which they have hitherto been noticed, is in the immediate vicinity of Tilgate Forest: they have not been discovered in any other part of England. Their external form is so remarkable, and bears so striking a resemblance to the grinders of the herbivorous mammalia, that when a large partially worn tooth first came under my notice, its analogy to the incisors of the rhinoceros led me to suspect whether the deposit in which it was found might not be of diluvial origin. Subsequent discoveries proved that these teeth belonged to an unknown herbivorous reptile; but their structure was so extraordinary, that I determined to obtain the opinion of Baron Cuvier upon the subject: specimens were accordingly transmitted to Mr. Lyell, who was then residing in Paris, and by whom they were presented to M. Cuvier, who favoured me with the following remarks:—"Ces dents me sont certainement inconnues; elles ne sont point d'un animal carnassier, et cependant je crois qu'elles appartiennent, vu leur peu de complication, leur dentelure sur les bords, et la couche mince d'émail qui les rêvet, à l'ordre des reptiles; à l'apparence extérieure on pourrait aussi les prendre pour des dents de poissons analogues aux tetrodons, ou aux diodons; *mais leur structure intérieure est fort différente de celles là.* N'aurions-nous pas ici un animal nouveau, un reptile herbivore? et de même qu' actuellement chez les mammiferes terrestres, c'est parmi les herbivores que l'on trouve les espèces à plus grande taille, de meme aussi chez les reptiles d'autrefois, alors qu'ils étoient les seuls animaux terrestres, les plus grands d'entr'eux ne se seraient-ils point nourris de végétaux? Une partie des grands os que vous possidez appartiendrait à cet animal, unique jusqu'à présent dans son genre. Le tems confirmera ou infirmera cette idèe, puisqu'il est impossible qu'on ne trouve pas un jour une partie du squelette réunie à des portions de mâchoires portant des dents. Si vous pouviez obtenir de ces dents adhérentes encore à une portion un peu considérable de mâchoire, je crois que l'on pourrait

TEETH OF THE IGUANODON.

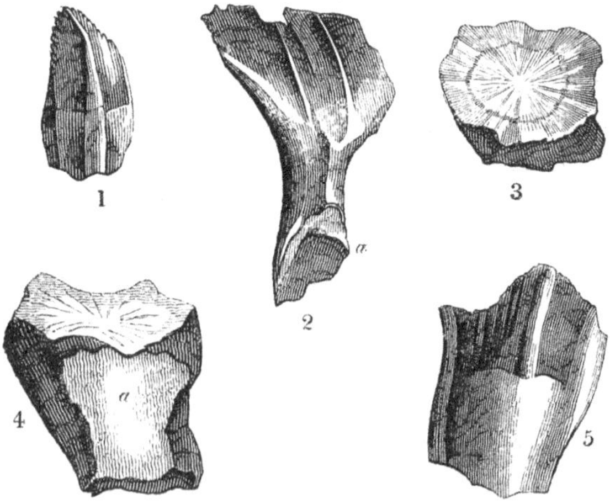

1. Tooth, with the apex perfect.
2. Tooth, with a cavity at the base *a*, caused by a new tooth.
3. Crown of a tooth worn down.
4. 5. Different views of a tooth partially worn down: *a*, the cavity formed by a supplementary tooth.

résoudre le problème." In the second part of the fifth volume of the *Ossemens Fossiles*, the learned author figures several of these teeth, and minutely describes their form and structure.* From the resemblance of the perfect specimens to the teeth of certain species of Iguana, I proposed to distinguish the fossil animal by the name of *Iguanodon;* and a memoir on the extraordinary dentature of the original was read before the Royal Society, and honoured with a place in its transactions.†

In the perfect teeth, and in those which have been but little worn, the crown is somewhat of a prismatic form; widest and most depressed in front; convex posteriorly, and rather flattened at the sides. As soon as the tooth emerges from the gum it gradually enlarges, and its edges approach each other and terminate in a point, making the upper part of the crown angular; the edges forming the sides of this angle are deeply serrated or dentated; see fig. 1. p. 270. The outer surface of the crown is covered with a thick enamel, but on the sides and back a thin coating of this substance only appears, as in the incisors of the gnawers. Dr. Buckland has pointed out to me, that the disposition of this enamel is not in straight lines, but in curves, so as to represent a series of gouges, as is beautifully shewn in fig. 1. p. 272.; by which the tooth was rendered more durable, and better capable of resisting mechanical injury. The ante-

* Oss. Foss., vol. v. p. 350.

† Notice on the Iguanodon, a newly-discovered fossil herbivorous reptile, from the sandstone of Tilgate Forest, in Sussex.—*Philosophical Transactions*, 1825, part i.

rior surface is divided longitudinally into three or more slightly concave parts, by obtuse ridges, of which the most prominent one is situated rather towards one side. In the young teeth, seldom more than one ridge occurs, dividing the surface into two unequal parts; but in some examples,

TEETH OF THE IGUANODON.

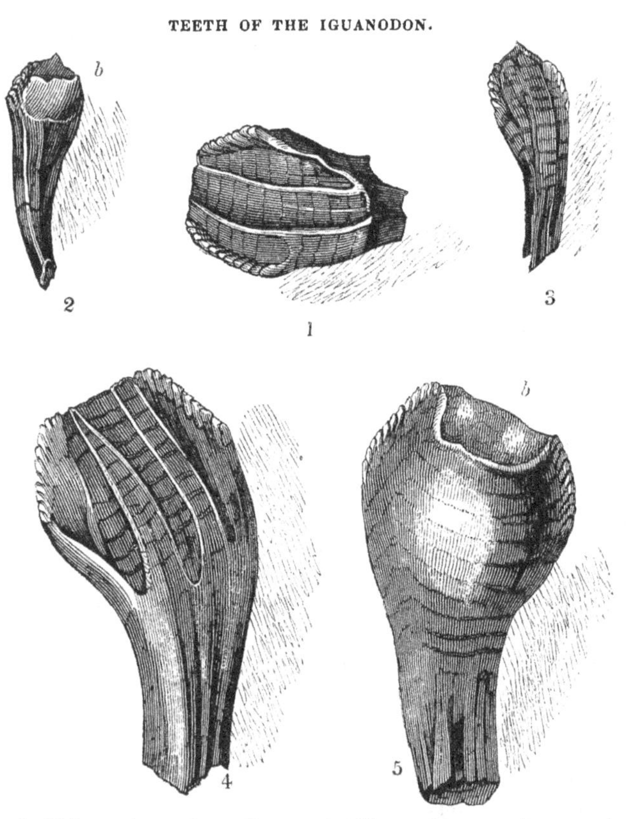

1. This specimen shows the crescent-like markings on the enamel, described above.

2, 3. The tooth of a young animal, in which the point is but little worn.

4, 5. A tooth of an adult, in which the crown of the tooth has suffered considerably from detrition.

several finer lines are observable: a specimen of the former is shewn in fig. 1. p. 270., and of the latter variety in the same cut, fig. 2., and in the figures in the above representation. The shank or fang of the tooth partakes of the form of the crown; it is slightly curved, rather flat anteriorly, and convex posteriorly, and much depressed at the sides; it gradually diminishes in size towards the base. The magnitude of the teeth is shown by the drawings, all of which are of the natural size: we have a young and almost perfect tooth in our possession, which is two inches and a half in length; being twelve times the length of the teeth of the recent Iguana: the largest specimens figured are twenty times larger than the teeth of that animal. All the teeth we have noticed possess traces of the characters above described; the vast difference of form they exhibit, depending on a cause which we shall now proceed to explain. In the elegant specimen, *fig.* 2. and 3. p. 272., we perceive the first variation from that angular form, which we have already stated to have been the original shape of the crown; and if we observe the internal surface at *b*, *fig.* 2., we see that the point of the tooth has been partially worn away, leaving an oblique triangular surface: if the reader compare this appearance with the surface of the perfect tooth, *fig.* 1. p. 270., he will form a correct idea of the change that has taken place. In a series of specimens, this wearing away of the crown of the teeth may be observed in every stage, from the slight appearance perceptible in *fig.* 2, 3. p. 272., to 4, 5. of the same, and *fig.* 2. of p. 270., in which the entire angular part of the tooth and its dentated edges disappear. If we trace the effects of this operation still farther, we find it going on as in *figs.* 4, 5. p. 270., till it reaches its maximum in *fig.* 3., which is merely a bony plate, without a fang, and with but a slight portion of enamel remaining on the anterior part. The process by which these changes have been effected is clearly that of mastication. So soon as the tooth performs its part in that operation, its point is, by degrees, worn away, till, by little and little, the serrated edges disappear, and the tooth assumes a truncated form; the masticating surface gradually becomes larger, and is always oblique (because the anterior coat of enamel is, from its superior thickness, less used than the rest of the tooth), till the tooth is worn down to the gum. During these changes in the crown of the tooth, the fang is suffering destruction from another but not less certain, process. A new tooth is formed at the base of the old one, which, increasing in size, occasions the absorption of the fang of the latter, till at length, when it is sufficiently developed, it displaces its predecessor, and occupies its place. The gradual progress of this change is shown in *fig.* 2. *a* p. 270.; and in *fig.* 4. of the same, in which the impression of the new tooth, or *dent de remplacement*, is marked *a*.

The teeth are hollow in the young animal, but become solid, almost throughout, in the adult state.

If we attempt to discover among the recent lizards a dentature at all analogous, we shall find among the Iguanas alone any kind of resemblance; yet even here, we cannot fail to remark, that in this, as in every other instance, if there be a general analogy, there are, also, striking and important differences, in the structure of the primitive animals of our planet, and of those which are its present inhabitants. Of the Iguana, there are several species; but the only skull we have had an opportunity of examining, is that of an individual from Barbadoes; we believe of *I. tuberculata.**

JAW AND TEETH OF THE RECENT IGUANA.†

* Ossemens Fossiles, vol. v. pl. xvi. figs. 24, 25.

The Iguanas are natives of many parts of America and the West India Islands, and are rarely met with any where north or south of the tropics. They are from three to five feet long, from the end of the snout to the tip of the tail. They inhabit rocky and woody places, and feed on insects and vegetables. Many of the Bahama islands abound with them; they nestle in hollow rocks and trees: their eggs have a thin skin like those of the turtle. Though they are not amphibious, they are said to keep under water an hour. When they swim, they do not use their feet, but place them close to their body, and guide themselves with their tails; they *swallow all they eat whole.* They are so impatient of cold, that they scarcely appear out of their holes but when the sun shines. — *Shaw's Zoology*, vol. iii. p. 199.

† I am indebted to my kind friend, Mr. Clift, of the College of Surgeons, for the original drawings of these parts of the recent Iguana.

Fig. 1. is a view of the inner surface of the right side of the upper jaw of this animal, of the natural size; and fig. 2. a portion magnified four diameters. The teeth are slightly convex externally, and have a ridge down the middle; they are slightly concave on the inner surface; the crown of a tooth, largely magnified, is shown, fig. 3. In the angular form of the crown, and its serrated edges, they strikingly resemble the fossil, fig. 1. page 270. The new teeth are formed at the bases of the old ones, and lie in a depression at the root of the fang, as is beautifully shown in the magnified drawing, fig. 2., at *a a*. The jaw throws up a lateral parapet on the outside of the teeth; but they have no alveoli, nor any internal protection but the gum. From the above observations, it appears that the fossil teeth bear a greater analogy to those of the recent Iguana, both in their form, and in the process by which dentition is effected, than to those of the crocodiles, monitors, and other living saurians. But, notwithstanding this general resemblance, the remarkable characters resulting from the act of mastication, separate the original animal from all known genera. None of the existing reptiles perform mastication; their food or prey is taken by the teeth or tongue, so that a movable covering of the jaws, similar to the lips and cheeks of the mammalia, is not necessary, either for confining substances subjected to the action of teeth as organs of mastication, or for the purposes of seizing or reaching food.* The herbivorous amphibia gnaw off the vegetable productions on which they

* Rees's Cyclopædia, art. Reptiles.

feed, but do not chew them; and the teeth, when worn, present an appearance of having been *chipped off*, and never like the fossil teeth of a *flat, ground, surface*; and it is certain that if the Iguanodon had *chopped off* its food like the Iguana, its teeth could never have presented the appearance which the illustrious Cuvier considered so extraordinary, and which led him to mistake the tooth of a reptile for that of a rhinoceros. Now, "as every organic individual forms an entire system of its own, all the parts of which mutually correspond, and concur to produce a certain definite purpose by reciprocal reaction, or by combining towards the same end," it follows from the peculiar structure of the fossil teeth alone, that the muscles which moved the jaws, and the bones to which they were attached, were widely different from those of any of the living lizards; and consequently, the form of the head of the Iguanodon must have been modified by these causes, and have differed from those of existing reptiles. Since the vegetable remains with which the teeth of the Iguanodon are associated, consist principally of those tribes of plants that are furnished with tough thick stems, and which, probably, were the principal food of the original animal, this peculiar structure of the teeth was evidently required, and beautifully adapted, to enable the animal to accommodate itself to the condition in which it was placed. Hereafter, perhaps, some more fortunate observers may discover a portion of the head or jaw, and be able to confirm or refute these conjectures.

Bones, supposed to be referable to the Iguanodon. We have already stated, that no united portion of the skeleton of this animal has been discovered: and since the bones of other gigantic reptiles are entombed with those of the Iguanodon, the attempt to identify the latter cannot, of course, be regarded as affording positive conclusions.

Vertebræ.—With the able assistance of the Rev. W. D. Conybeare, we examined our collection of vertebræ from Tilgate Forest, and after having separated those that appeared to belong to the Crocodile, Plesiosaurus, and Megalosaurus (or, rather, which resembled those from Stonesfield), several enormous vertebræ remained, which corresponded with each other, but differed from any we had previously noticed. The faces of the bodies of these vertebræ are almost flat on one side, and slightly depressed on the other; and are rather quadrangular. The spinous and lateral processes are very strong and thick. There is not that deep depression beneath the annular part of the vertebræ which is said to characterize those of the Megalosaurus. From their enormous magnitude, and from the circumstances above stated, we are induced to refer them to the Iguanodon, the other gigantic lizard of the Tilgate strata. We have one sacral vertebra, which is almost identical with that of the Monitor*, except that the body is slightly concave on both faces: this bone, if perfect, would measure ten inches from the extremity of one transverse process to the other. Some of the detached processes in our possession indicate vertebræ

* Oss. Foss. vol. v. pl. xvii. fig. 27.

of enormous proportions.* Our cabinet contains three very small vertebræ, which are anteriorly concave and posteriorly convex, as in the recent Iguana; and these, with the exception of two large examples, are the only instances of such a structure we have met with in the vertebræ of the Wealden. A *chevron-bone*, of a magnitude corresponding with the largest of the vertebræ, has also been found.

Extremities.— Fragments of thigh-bones, of a monstrous size, are occasionally found in the sandstone of the Forest: we have one specimen which is no less than 23 inches in circumference. Were it clothed with muscles and integuments of suitable proportions, where is the living animal with a thigh that could rival this extremity of a lizard of the primitive ages of the world?

Metatarsal and phalangeal bones.—Some of these are so large, that they appear more like the bones of mammoths or elephants, than of reptiles. M. Cuvier, with his usual candour, observes, "Des fragments d'os du métacarpe ou du métatarse sont si gros, qu'au premier coup-d'œil je les avois pris pour ceux d'un grand hippopotame."†

We have one metatarsal bone of the following gigantic proportions:—

Length of the bone, 4½ inches.

Circumference of the largest (*tarsal*) extremity, 13 inches.

We have shown that the teeth of the Iguanodon

* A spinous process, probably of a caudal vertebra, is 12 inches long.

† Oss. Foss. vol. v. p. 350.

are, at least, twenty times larger than those of the Iguana ; that the thigh-bone is of equally enormous proportions : and were we to calculate the probable magnitude of the original, from the data which this metatarsal bone affords, our readers might well exclaim, that the realities of Geology far exceed the fictions of romance. Even if we admit, what is, indeed, probable, that the linear dimensions of the extinct and living animals were not of the same relative proportions, still it must be allowed that the Iguanodon was one of the most gigantic reptiles of the ancient world ; and a colossus in comparison to the pigmy alligators and crocodiles that now inhabit the globe.

Horn of the Iguanodon.—We have now to request the attention of the reader to a very remarkable appendage, with which there is every reason to believe the Iguanodon was provided. This is no less than a *horn,* equal in size, and not very different in form, to the lesser horn of the Rhinoceros. This unique relic is externally of a dark brown colour ; and while some parts of its surface are smooth, others are rugose and furrowed, as if by the passage of blood-vessels. Its base is of an irregular oval form, and slightly concave. It possesses an osseous structure, and appears to have no internal cavity. It is evident that it was not united to the skull by a bony union, as are the horns of the mammalia. The nature of this extraordinary fossil was for some time undetermined ; and it is to the discrimination of Mr. Pentland, whose high attainments in comparative anatomy are well known, that we are indebted for the suggestion that it probably belonged to a saurian animal. It is well known that some reptiles of that order have bony or horny projections on their foreheads; and it is not a little curious, that, among the Iguanas, the horned species most prevail. The *Iguana cornuta*, which is a native of Saint Domingo, resembles the common Iguana in size, colour, and general proportions; on the front of the head, between the eyes and nostrils, are seated four rather large, scaly, tubercles; behind which rises *an osseous conical horn, or process, covered by a single scale.** That our fossil was such an appendage, there can be no doubt; and its surface bears marks of the impression of an integument by which it was covered, and probably attached to the skull. This fact establishes another remarkable analogy between the Iguanodon, and the animal from which its name is derived.

Plesiosaurus. The vertebræ, teeth, and other bones of this extraordinary genus, whose osteology has been so admirably and so fully described by Mr. Conybeare†, occur in the strata of Tilgate Forest.

* Shaw's Zoology, vol. iii. part i. p. 203.

† Geological Transactions, vol. v. and vol. i. new series. It may be interesting to the general reader to state, that the *Plesiosaurus* and *Ichthyosaurus* are two genera of fossil animals, supposed to have been oviparous, and to belong to the family of the saurians, but differing very essentially from all existing species, and in such particulars as evidently must have fitted them to live entirely in the sea. Their vertebræ are deeply cupped like those of fishes, and are as thin as those of the shark, so as to admit of a vibratory motion of the tail, to assist progression. The extremities terminate in four paddles, composed of a series of flat polygonal bones, greatly exceeding in number even the phalengic cartilages of the fins of fishes. The most wonderful animal of this division is the *Plesiosaurus dolichodeirus*, or long-necked plesiosaurus: the neck of this animal is equal to half the entire length of the body and tail united, and is composed of 35 ver-

The vertebræ agree with those from the *Lias*, and certainly belong to the *Plesiosaurus dolichodeirus*.

Among the bones which we are unable to assign to any known recent or fossil animal, is a humerus, which, from its shortness, thickness, and the extreme width of its cubital articulation, would appear to be adapted for swimming. It bears a greater analogy to the humeri of the *cetacea* than to those of the lizards: Dr. Buckland referred it to the whale; but there are essential differences between the humerus of the latter and this fossil.* It does not correspond with the humeri of the known plesiosauri; yet it may possibly belong to

---

tebræ; the back of 27, and the tail of 28; making a total of 90. The head is so small, that its length is not more than a fifth part of that of the neck.

* M. Cuvier, to whom this opinion was communicated, obliged us with the following observations on the subject:—

" Quant aux ossemens que vous croyez être de *cétacés*, ils méritent certainement d'être étudies avec attention, car ce serait la première fois que l'on trouverait des ossemens de ces mammifères dans les formations situées au-dessous de la craie. Mais il y a des vertèbres de grands reptiles, qui ressemblent si fort à celles de certains dauphins, qu'il est facile de s'y tromper; toujours faudrait-il examiner leur position avec beacoup de soin, et voir s'il n'y aurait point à cet égard de différence entre ces os de cétacés et ceux de reptiles. Rien n'empêcherait, ce me semble, que dans les couches de sables que se trouvaient former la surface du sol, la mer ne soit venue, dans ses nouvelles irruptions, apporter des dépouilles nouvelles, et les confondre avec les anciennes; mais je pense que soit dans leur position, soit dans leur état physique, on trouvera des différences charactéristiques de l'époque à laquelle ces dépouilles ont été enfouies."

From these remarks of this illustrious philosopher, we have hesitated to refer the fossils in question to the *cetacea*, until the discovery of more illustrative specimens shall establish or refute their supposed analogy, since of their geological habitat there cannot be the slightest doubt: it is probable they will hereafter be proved to belong to some of the saurian reptiles of the Wealden.

a new species of that genus. Among the gigantic vertebræ of Tilgate there are some which approach in their general form to those of the plesiosauri; but their bodies are concave on one side only, the other face being either flat, or slightly elevated; the largest are above six inches in their longest diameter.

Bones of Birds.—It is so rarely that the remains of birds are found in a fossil state, that, although a few mutilated bones are the only relics of these animals which the strata of Tilgate have hitherto afforded, the fact is too remarkable to be passed by unnoticed. We have, in our collection, remains of three of the larger bones of the extremities (either of the femur or tibia), of a wader larger than the common heron: they were supposed to be the bones of *Pterodactyles*, but M. Cuvier, to whom they were shown in 1830, did not hesitate to determine that they belonged to a bird, probably a species of *Ardea*.

In concluding this description of the organic remains of Tilgate Forest, we would repeat what we have elsewhere remarked, that the vast preponderance of the land and freshwater exuviæ over those of marine origin observable in these deposits, warrants the conclusion that the Wealden strata were formed by a very different agent to that which effected the deposition of the Portland beds below, and the sands and chalk above them. The seas, in the primitive ages of our planet, were inhabited by vast tribes of multilocular shells, which, however variable in their species, were not only of the same family, but also of the same genera, namely, *Belemnites*, *Ammonites*, and *Nautilites*. These shells, if we may draw any conclusions from our knowledge of the habits of the recent species of the only genus that still exists, were indisputably inhabitants of the ocean; and the presence of their remains, in any considerable quantity in a formation, affords a fair presumption that such formation was a marine deposit. The converse of this proposition, we conceive, must hold good in a case like the present, where not a vestige of these ancient marine genera can be traced, among innumerable remains of terrestrial vegetables and animals, and of freshwater testacea. The occasional occurrence of marine remains affords no grounds for a contrary opinion, since this fact is no more than might be expected under such circumstances, and is in strict accordance with what may be observed in the deltas and estuaries of all great rivers.

We cannot leave this subject, without offering a few general remarks on the probable condition of the country through which the waters flowed that deposited the strata of Tilgate Forest, and on the nature of its animal and vegetable productions. Whether it were an island, or a continent, may not be determined; but that it was diversified by hill, and valley, and enjoyed a climate of a higher temperature than any part of modern Europe, is more than probable. Several kinds of ferns appear to have constituted the immediate vegetable clothing of the soil: the elegant *Sphenopteris*, which probably never attained a greater height than three or four feet, and the beautiful

Lonchopteris, of still lesser growth, being abundant every where. It is easy to conceive what would be the appearance of the valleys and plains covered with these plants, from that presented by modern tracts, where the common ferns generally prevail. But the loftier vegetables were so entirely distinct from any that are now known to exist in European countries, that we seek in vain for any thing, at all analogous, without the tropics. The forests of *Clathrariæ*, and *Endogenitæ* (the plants of which, like some of the recent arborescent ferns, probably attained a height of thirty or forty feet), must have borne a much greater resemblance to those of tropical regions, than to any that now occur in temperate climates. That the *soil* was of a sandy nature on the hills and less elevated parts of the country, and argillaceous in the plains and marshes, may be inferred, from the vegetable remains, and from the nature of the substances in which they are enclosed. Sand and clay every where prevail in the Hastings strata; nor is it unworthy of remark, that the recent vegetables to which the fossil plants bear the greatest analogy, affect soils of this description. If we attempt to portray the animals of this ancient country, our description will possess more of the character of romance, than of a legitimate deduction from established facts. Turtles of various kinds, must have been seen on the banks of its rivers or lakes, and groups of enormous crocodiles basking in the fens and shallows.

The gigantic *Megalosaurus*, and yet more gigantic *Iguanodon*, to whom the groves of palms and arborescent ferns would be mere beds of reeds, must have been of such prodigious magnitude, that the existing animal creation presents us with no fit objects of comparison. Imagine an animal of the lizard tribe, three or four times as large as the largest crocodile; having jaws, with teeth equal in size to the incisors of the rhinoceros; and crested with horn: such a creature must have been the Iguanodon! Nor were the inhabitants of the waters much less wonderful; witness the Plesiosaurus, which only required wings to be a flying dragon; the fishes resembling *Siluri*, *Balistæ*, &c. &c.

From what has been advanced it seems obvious, that although, from the broken and rolled state of the greater proportion of the organic remains of the Wealden, it is manifest the large reptiles and vegetables have been subject to the action of river currents, yet dry land must have existed at no very great distance: some of the vegetables must have grown on the borders of a river or lake; and the habits of the recent species of reptiles most nearly related to the fossils warrant a similar conclusion, since they are well known to frequent the rivers and marshy tracts of tropical regions, in the sands and banks of which they deposit their eggs. Reflecting upon these extraordinary facts, may we not enquire with the illustrious Cuvier, "*At what period was it, and under what circumstances, that turtles and gigantic lizards lived in our climate, and were shaded by forests of palms, and arborescent ferns?*"

The discussion of this subject cannot, however,

be pursued in this place, without leading to the anticipation of facts hereafter to be examined; it will therefore be more convenient to reserve any further observations for the concluding chapter of this volume.

ON THE ANALOGY BETWEEN THE ORGANIC REMAINS OF THE TILGATE BEDS AND THOSE OF STONESFIELD, NEAR OXFORD.

In the course of this enquiry, allusion has been made to the fossils of the Stonesfield slate, and their general resemblance to those of the Tilgate strata: this correspondence in the organic remains of deposits, whose geological relations are so entirely dissimilar, is a fact sufficiently interesting to require some attention.

The Stonesfield limestone is supposed to belong to the inferior beds of the oolite, and has long been celebrated for the extraordinary character of its fossils, of which, however, no detailed account has yet appeared.

According to Dr. Kidd *, it contains crabs, birds, tortoises, and one or more large quadrupeds; and the Rev. W. Conybeare, in his highly interesting memoir on the *Plesiosaurus* †, mentions, that it also encloses the remains of "an immense saurian animal, approaching to the characters of the Monitor, and which, from the proportions of many of the specimens, cannot have been less than forty feet long:" this is the Megalosaurus.

* Geological Essays, by J. Kidd, M.D. 8vo. 1815. p. 38.

† Geological Transactions, vol. v. p. 592.

With the assistance of Mr. Lyell, and aided by an interesting collection of Stonesfield fossils, for which I am indebted to his liberality, I have been able to ascertain that the following organic remains occur in both deposits, viz.: —

The teeth, ribs, and vertebræ of the *Megalosaurus.*

Bones and plates of several species of *Turtle.*

Teeth of a species of *Shark,* and of a fish related to *Lepisosteus.*

Teeth of *Crocodiles.*

Scales of fishes.

Radii of *Siluri.*

A plant of the genus *Carex,* allied to recent species. But the resemblance extends no farther; the *Trigoniæ* and *Belemnites* of the Stonesfield slate do not occur in the Tilgate beds; and the *Paludinæ, Cyclades,* &c. of the latter, have not been discovered in the former deposit. It is scarcely necessary to add, that the strata of Tilgate Forest exhibit no traces of oolitic structure.

CHAP. X.

OBSERVATIONS ON THE FOSSIL REMAINS OF THE HYLÆOSAURUS, AND OTHER SAURIAN REPTILES DISCOVERED IN THE STRATA OF TILGATE FOREST, IN SUSSEX.

[Read by the Author before the Geological Society of London, December 5. 1832.*]

It is my object, in the observations which I have the honour to address to the Society, to give an account of the Saurian remains which have been discovered in the strata of Tilgate Forest, since my last publication on the subject; and more particularly, to describe a considerable portion of a skeleton imbedded in a block of indurated sandstone, which has recently been found, and is now, with other fossil bones from the same deposits, on the table of the Society.

Ten years have elapsed since the first notice of the fossil reptiles of Sussex appeared in the Illustrations of the Geology of the South Downs: — a paper on the Iguanodon, in the Philosophical Transactions for 1825—the description of a femur and some vertebræ of a reptile from the Wealden beds of the west of Sussex, by Mr. Murchison; and a memoir on the Megalosaurus of Stonesfield, by Dr. Buckland, both of which are in the Society's Transactions—the important remarks on these remains

* By the kind permission of the President, Roderick Impey Murchison, Esq. and the Council, this paper was allowed to be withdrawn.

in the *Ossemens Fossiles*—and the volume expressly dedicated to the fossils of Tilgate Forest, have subsequently been published, and comprehend all the contributions hitherto made towards the history of these extraordinary remains. I propose to offer a general view of what is already known on this subject, incorporating therewith the results of subsequent discoveries; to notice in a cursory manner the facts previously established, and dwell with requisite detail on the important additions to our anatomical knowledge of these extinct oviparous quadrupeds, which recent investigations have brought to light.

The geological characters of the country in which these remains occur are so well known, that it is only necessary to observe, that with the exception of alluvial deposits, and here and there insulated portions of the tertiary strata, the district is composed of two formations, namely, the Chalk, and the Wealden; the strata of the former containing marine exuviæ only, and having evidently been deposited in the tranquil depths of an ocean; while those of the latter teem with the remains of land and freshwater animals and vegetables, and are manifestly the sedimentary deposits which, in the earlier ages of our planet, constituted the delta of an immense river.

The mutilations which the carcases of the reptiles appear to have sustained, before they were imbedded and preserved in the silt and other sediments of the waters, and the promiscuous intermixture of the osseous fragments of various species, occasion the attempt to reconstruct their skeletons to be a work of considerable difficulty: and as I feel how much I shall require the indulgence of the Society, for the imperfect data on which, I fear, many of my conclusions will appear to rest, I would beg to cite the pertinent observations of M. Cuvier on the remains of the Megalosaurus of Stonesfield, and which are peculiarly applicable to those of Tilgate Forest:—"Par malheur ils ne sont pas trouvés ensembles, ni même (à l'exception des vertèbres) réunis deux à deux, ou trois à trois, de manière à rendre vraisemblable qu'ils soient provenus du même individu, et qui plus est, ce n'est que par leurs rapports zoologiques, et par leur existence dans les mêmes carrières, que l'on peut conclure qu'ils viennent d'une même espèce; encore ces rapports zoologiques sont ils d'une nature assez équivoque, et assez mélangée." (*Oss. Foss.* tom. v. p. 345.) In fact, with but very few exceptions, no two bones, or teeth, have been found together in their natural position; and not a fragment of the jaws, nor scarcely any portion of the skull, has been discovered; a circumstance truly remarkable, considering the vast numbers of detached teeth, and vertebræ, and bones of the extremities that have been collected.

From the almost entire absence of the bones of the head, we commence our enquiry with the teeth, some of which are very perfect, and highly characteristic, and belong to at least five genera of oviparous quadrupeds; namely, the *Plesiosaurus*, *Crocodile*, *Cylindricodon*, *Megalosaurus*, and the *Iguanodon*. I shall notice, in the order in which they are named, the teeth of these animals, and the bones which appear to be referable to them; as the examination of the large specimen on the table will, by this arrangement, come before us toward the conclusion of this paper, and we shall thus, by the previous consideration of the osteology of the other genera, be prepared for the investigation of its peculiar characters.

Plesiosaurus dolichodeirus.*—The teeth of this extraordinary animal, as shown in the admirable memoir of Mr. Conybeare, are "conical, very slender, curved inwards, finely striated on the enamelled surface, and hollow throughout the interior; they resemble those of the Crocodile, but are more slender." Many teeth of this kind have been found in the Tilgate strata, but the existence of the Plesiosaurus in these deposits does not rest on this indication alone: vertebræ occur which are identical with those of Lyme Regis; they belong to the *cervical*, *dorsal*, and *caudal*: no other bones have been identified.

Cylindricodon.—Under this name, Dr. Jæger, of Stutgard, has described the remains of a fossil reptile, of which almost the entire upper jaw, with the teeth, has been discovered by him in the *keuper* formation of Germany, near Wirtemburg. He considers the original to have been herbivorous, and describes it as characterised "par les dents latérales

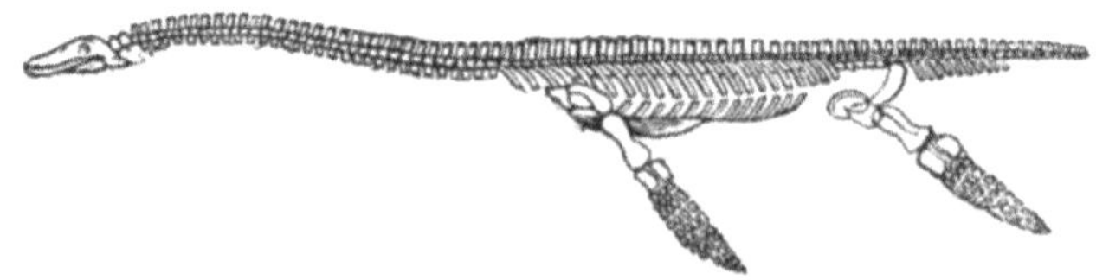

* This wonderful creature has been already described p. 281.: the annexed wood-engraving represents the skeleton, as restored from specimens discovered at Lyme Regis.

cylindriques, mais elle paroit avoir eue des dents aigues au bout de deux formes différentes." This eminent physician and naturalist had the kindness to send me drawings of these fossils, and I perceived considerable resemblance between the teeth and some collected in Tilgate Forest, and which are figured tab. xv. fig. 3, 4. in my work on the fossils of that district. (*Vide* Pl. II. fig. 2, 3, 4. of this volume; fig. 3. is intended to explain the mode in which the edges of the crown of the teeth are worn from contact with the corresponding edges of their antagonists.) These specimens I submitted to the examination of M. Cuvier, on his visit to this country in 1830, and that illustrious philosopher observed, that they were decidedly the teeth of an unknown reptile, but that their supposed identity with those of the *Cylindricodon* could not be determined from the figures of Dr. Jæger. The subject has since been elucidated by M. Boué, the distinguished editor of the "*Journal de Géologie*," who favoured me with a visit soon after his return from Stutgard, where he had carefully examined the originals. This gentleman expressed his conviction that the teeth, which I shall now describe, must have belonged to the same kind of saurian; their identification therefore rests upon his authority. These teeth are about an inch and a quarter in length, and commence with a subcylindrical shank, which gradually enlarges into a kind of shoulder, terminating in an obtuse angular apex, the margins of which are more or less worn, as if the teeth had been placed alternately so as to meet at their edges, as in Plate II. fig. 3. They are obscurely striated longitudinally, and have a thick coat of enamel: the crown of the tooth is solid, but the shank is more or less hollow. All the specimens appear as if they had been broken off close to the jaw; but they may have been separated by necrosis, occasioned by the pressure of the supplementary teeth. Meagre as are these details, they are not without value, if they shall be found to establish the identity of the animals to which these teeth belonged, since another fact will thus be added to our knowledge of the geological distribution of the ancient inhabitants of the globe.

Crocodile.—The teeth of the reptiles of this family are abundantly distributed throughout the strata of the Wealden; they present considerable variety of form, and are referable to two, if not more, distinct species. The first are very obtuse, much more so than the teeth of the crocodile of the Nile, and resemble those of the second species of the fossil crocodile of the Jura limestone; the second are more slender, possess a gentle curvature, and bear a close resemblance to the teeth of the fossil Gavial of Caen.*

Another species of this genus is indicated by the portion of a lower jaw imbedded in grit, having seven teeth in place, and two detached. This specimen is figured in the "Fossils of Tilgate Forest," and the teeth approximate so closely to those of the *Crocodilus priscus*, being alternately long and

* The first species is the *Crocodilus Mantelli* of Gray; the second is the *Gavialis Lamourouxii* of the same author.—*Synopsis Reptilium*, pp. 57. 61. With regard to their relation to the recent species, it may be observed, that the obtuse conical teeth resemble those of the Cayman or alligator, which inhabits Cayenne and North America: the more slender and recurved species, those of the Gavial, an inhabitant of the Ganges only.

short, that but little doubt can be entertained of the identity.* The existence of another species appears to be established by a small frontal bone, which, although, but 0·5 inches long, and partially obscured by the stone to which it is attached, is decidedly the os frontis either of a Crocodile, properly so called, or of an Alligator, and not of a Gavial. Plate 2. fig. 1. As it adheres to its matrix by its dermal aspect, the inner surface only is seen, and this displays a channel which served for the passage of the olfactory nerves to the nose: on the upper part, the orbital arches are displayed. This bone must have belonged to an individual not more than 18 inches in length. It is not a little remarkable, that, numerous as are the teeth of Crocodiles in the Tilgate strata, this minute frontal bone is the only vestige of the skull that has been noticed.

Vertebræ. — Until the discovery of the large specimen before us, I considered all those vertebræ, which had both extremities of the body nearly circular, and either flat or slightly concave, with the annular part united by suture, to belong to some one or more of the various species of fossil crocodiles whose teeth are above described. For although, in all the recent species the vertebræ are convex posteriorly and concave anteriorly, yet Cuvier, Soemmering, and other comparative anatomists, had shown that there were many fossil species with the vertebræ concave; and that in one species found at Havre, the vertebræ, although

* "Crocodilus priscus, rostro elongato cylindrico, dentibus inferis alternatim longioribus, femoribus dupla tibiarum longitudine." — *Oss. Foss.* tome v. p. 125.

concavo-convex, were reversed, the convexity being placed forwards, instead of backwards. M. Cuvier has figured and described several dorsal and caudal vertebræ from Tilgate Forest, which possess the first-mentioned characters (*Oss. Foss.* tom. v. pl. x. figs. 28. 31, 32, 33, 34.), and has approximated them to the fossil crocodile of Caen. That to follow in the steps of the immortal founder of fossil comparative anatomy is the only method by which we can obtain clear and satisfactory results in these investigations, no one can feel more persuaded than myself; yet, in the present instance, I am compelled, though with the utmost diffidence, and with feelings of respect amounting to veneration for the opinions of that illustrious philosopher, to hesitate in assigning *all* these vertebræ to the Crocodiles or Gavials, since, as will hereafter appear, the portion of the skeleton of a reptile on the table has similar vertebræ associated with

VERTEBRÆ OF REPTILES FROM TILGATE FOREST.

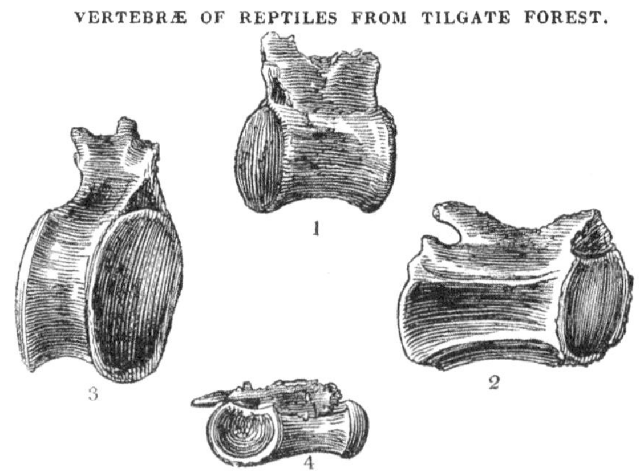

1. Crocodilian type.
2. Iguanodon.
3. Plano-concave.
4. Concavo-convex, of the lacertian type.

a sternal apparatus essentially different from that which characterises the animals of this family. It will materially elucidate our future remarks, if in this place we notice the several kinds of saurian vertebræ which have been found in the Wealden formation. They are referable to at least four systems, independently of the vertebræ of the Plesiosaurus above noticed. The first vertebræ are of the Crocodilian type (fig. 1.): these, as we have already remarked, have both extremities nearly flat, or slightly concave; and their facets are either circular or elliptical, according to the place they occupied in the column: the body is more or less arched, is somewhat contracted, smooth, and rounded in the dorsal, and ridged or keeled in the caudal (vide Tilg. Foss. pl. ix. figs. 1. and 11.). The vertebræ of the Megalosaurus of Stonesfield appear to be of this kind; they are described as much contracted in the middle, and *with a deep depression immediately beneath the annular part.* If the latter character should prove to be constant, it will serve to distinguish them; but among upwards of thirty vertebræ of this type before me, the depression is so variable, that it could not be assumed as a permanent character.*

* The specimen, pl. ix. 11. Tilgate Fossils, is described as a vertebra of a Megalosaurus; but it cannot, I now think, be separated from those figured in the same plate, as belonging to a crocodile. M. Cuvier describes the vertebræ of the Megalosaurus as resembling those of the crocodile of Honfleur, and other fossil species. I have lately seen a vertebra from the forest marble, in the collection of Mr. Channing Pierce, of Bradford, which has a deep sulcus immediately beneath the transverse process; and this character is so strongly developed, that it is probable it may prove to be constant.

In the second vertebral system (fig. 2.) the bodies are somewhat angular, and their articulating faces rather convex or flat on one side, and concave on the other; and in consequence of the angular form of the body they are obscurely hexagonal in the dorsal, and quadrangular in the caudal; the annular part is united by suture, and the transverse processes pass off horizontally beneath it.* The visceral aspect of the body does not, as in the caudal vertebræ last described, form a ridge, but is either flat or deeply sulcated: should this character be permanent, it will serve to discriminate this system from the preceding. The vertebræ from western Sussex, figured and described by Mr. Murchison in the Transactions of the Society (vol. ii. *second series*, tab. xv.) are of this kind; and I am induced here to quote the observations made upon them by the immortal Cuvier:—"They are the remains of a new species of large saurian, and I am not acquainted with any similar to them amongst either recent or fossil reptiles: nevertheless, as the different parts of the skeleton of the Iguanodon have not yet been found by Mr. Mantell, and these bones appear to be from the same formation, it is not impossible that some of them may belong to that animal. I can safely affirm, that the vertebræ now submitted to my inspection do not belong to the Plesiosaurus; but whether any of them are referable to the Megalosaurus, the specimens of the bones of that animal in my possession are too few to enable

* It is possible that this last-mentioned character, as well, indeed, as the others, may vary according to the situation the bones held in the vertebral column.

me to decide. The vertebræ are *lumbar*, *sacral*, and *caudal;* and there are two of the latter anchylosed, and two others disposed to become so; which is remarkable, for in all my skeletons of reptiles an analogous case is not observable, except where the tail has been broken, and then the bone becomes united by exostosis, which is not the case in these fossils: hence it would seem that these must have belonged to an animal making such feeble use of the tail, that the vertebræ were occasionally anchylosed together. One of the vertebræ is of the first caudal, viz. of those which support the largest chevron bones, for it has the articular tubercles of these bones very strongly marked." (*Geolog. Trans.* loc. cit.) Caudal vertebræ of this kind are figured, "Tilgate Fossils," pl. ix. 8. and 12., and pl. x. fig. 1.: the latter is a fine example, displaying the transverse apophyses, and the cavity for the medulla: the spinous process is broken off. These vertebræ are as numerous as those of the first system, and attain a large size; one in my cabinet has the following proportions: — height, 5 inches; length, 4 inches; transverse diameter of the extremities 4½ inches. The *first caudal* are deeply impres ed by the tubercles of the chevron bone: a *sacral* strikingly coincides with the corresponding bone of the Iguana and Monitor, and, if perfect, would measure 10 inches from the extremity of one transverse process to the other. The detached processes of these bones indicate an enormous magnitude; a spinous apophysis is 14 inches long; and the extreme strength of a similar process belonging to a sacral, is shown in fig. 1. pl. xii. Tilg.

Fossils. A chevron bone, which evidently belongs to these vertebræ, is 14 inches in length, and is identical with the similar bone in the Iguana. The vertebræ of the third system (fig. 3.) correspond in many respects with the above, but are more decidedly plano-concave, and the body is not sulcated, but is either smooth and rounded, or strongly keeled, as in the first system, from which they differ in their proportions: one of these, from my cabinet, is figured pl. xxi. 27. *Ossemens Fossiles*, tome v.; and is thus described by M. Cuvier, after alluding to the vertebræ of the Megalosaurus:—"Cette vertèbre, longue de 0·11 me paroît ressembler aussi fort exactement à celles de M. Buckland. Un caractère remarquable qu'on y observe c'est une arrête vive ou carène longitudinale à sa face inférieure: à la grosseur de ses apophyses on pourroit croire que c'est une vertèbre sacrée: elle a, comme celles de Stonesfield, beaucoup de rapports avec nos vertèbres de Honfleur." The dimensions of a large example are—length, 3 inches; height, 5 inches; transverse diameter, 6½ inches.

The vertebræ of the fourth system (fig. 4.) are very rare, only six or seven having come under my observation. They are of the true lacertian type, having the articular facets of the body convex posteriorly and concave anteriorly, and are *wider than high*, as in the Iguana and Monitors, and not in the reverse proportions, as in the recent Crocodiles. In two large but mutilated cervical, the admeasurements are as follow:—height of the concave extremity, 3½ inches; width of the same, 4½ inches; length of the body, 6 inches. It is not obvious whether the annular part be divided by suture or otherwise; the articular apophyses are horizontal, and very strong; the spinous process is destroyed. The other four vertebræ of this system are very small, and must have belonged to a young lizard. It must be admitted, that even if the above arrangement of the vertebræ should prove to be correct (which it is very probable it may not), still the appropriation of the bones to the various animals with whose teeth and other remains they are associated, cannot be satisfactory until we shall discover these relics in juxtaposition with other parts of the skeleton. We may, however, be permitted to hazard the conjecture, that the vertebræ of the first system, which approach in so many respects to those of the fossil Crocodiles, may have belonged to more than one genus or subgenus: some to the Crocodiles, which furnished the conical striated teeth; others to the Megalosaurus; and some to the animal whose osteological organisation will hereafter come under our notice.

Ribs.—Fragments of these bones are very abundant; and specimens sometimes occur which show the bifurcation of the head, for the articulation with the transverse process and body of the vertebra, which is so characteristic of the recent Crocodiles: they are but slightly arched. The same observation will apply to the ribs as to the vertebræ, and, in fact, to the other detached bones, namely, that they may have belonged to more than one genus of reptiles.

Bones of the extremities.—I shall only notice those fossil bones which approximate so closely to the recent, as to leave no reasonable doubt that they belonged to some species of Crocodile or Gavial. These are portions of the *omoplate*, a *humerus*, *radius*, *head of a tibia*, several *carpal*, and some of the *phalangeal* bones.

Megalosaurus;—or Great Lizard of Stonesfield. This enormous creature, which held an intermediate place between the Crocodiles and the Monitors, but is considered to have most nearly approximated to the latter, is computed to have attained a length of fifty feet. In the highly interesting memoir of Dr. Buckland, which first made us acquainted with the remains of this wonderful oviparous quadruped, the learned Professor has fully detailed its anatomical structure (so far as the remains of its skeleton hitherto found would admit): it will, therefore, be sufficient for my present purpose to notice very briefly its distinguishing characters. The principal bones described, consist, 1st, of a portion of the lower jaw, nearly a foot long, containing one tooth fully developed, and the germs of many others; 2dly, a femur; 3dly, a series of five vertebræ; 4thly, a coracoid bone; 5thly, a clavicle? 6thly, ribs; 7thly, an ischium; 8thly, metatarsal, or metacarpal bones. The teeth are very remarkable (vide fig. 1. p. 261.): they are compressed laterally, and recurved; their edges are finely serrated, the anterior edge being the thickest: the largest tooth is two inches long. They are lodged in distinct alveoli or sockets, but do not adhere to the jaw, as in the Monitor; although, like the latter, the outer edge of the jaw rises considerably above the inner margin, and thus forms a lateral parapet by which the teeth are supported. The vertebræ, as before remarked, belong to the first or crocodilian type; and the ribs have the double termination of head and tubercle. The thigh bone, which is entire, is 2 feet 9 inches long, and exhibits an intermixture of the characters of Crocodile and Monitor. The *coracoid bone* (fig. 4. p. 321.) is of a very extraordinary form; and I shall describe it minutely, that we may hereafter be able to decide in what respects it approximates to, or differs from, the coracoid of the fossil skeleton now before the Society. This bone is nearly flat, slightly concave on one side and convex on the other; thin towards its arched margin, but thick at its largest apophysis: one edge is slightly curved, the other is strongly contracted in the middle; it terminates in a point at one extremity, but on the other it is truncated, and divided into two processes. M. Cuvier remarks, that the only bone with which it can be compared is the coracoid of a saurian: it is two feet long, which, admitting the proportions of a Monitor, gives a total length to the animal to which it belonged of 65 feet. I would here observe, that in the coracoid bones of all the recent saurians, there is a foramen near the neck for the passage of blood-vessels, and which is not observable in this fossil.* In the strata of Tilgate Forest we have teeth of the same species of Megalosaurus; a femur identical with that figured by Dr. Buckland, and some other bones; so that no reasonable doubt can

* "Il y a toujours un petit trou pour les vaisseaux, percé au col de l'os entre ses apophyses et sa facette glenoide." — *Oss. Foss.* p. 290. tome v.

exist that the Megalosaurus was contemporary with the Iguanodon.

IGUANODON.—The undoubted remains of this gigantic herbivorous reptile of the ancient world, must be considered as having hitherto been found in the strata of Tilgate Forest only; for although some enormous saurian vertebræ and other bones have been discovered in the outlying strata of the Wealden, in the Isle of Wight, yet as neither the teeth, nor bones possessing decided characters, have been noticed, their relation to the Iguanodon cannot be positively determined.* The peculiar characters of the teeth of this animal, their analogy to those of the Iguana, and the conclusive evidence which they afford of the marvellous fact, that the original animal, unlike all other lizards either recent or fossil, masticated its food like the herbivorous mammalia, are now so well known, that it is unnecessary to trespass on the time of the Society by describing them. Assuming, then, that the Iguanodon bore a closer affinity to the Iguana than to any other of the recent genera of reptiles, an opinion in which I have the great satisfaction of knowing that the late Baron Cuvier fully concurred, it is inferred, that all those fossil

* Dr. Jæger, the eminent German naturalist, to whose interesting researches I had occasion to refer when speaking of the Cylindricodon, has found a tooth so like that of the Iguanodon, that he believes it to have belonged to the same animal. In a letter, dated August, 1829, he observes,—"J'ai une dent trouvée avec les restes du *Mastodonsaurus*, que je crois être d'un Iguanodon. Le 'sandstone of Tilgate Forest' nous étoit jusqu'alors inconnu; dernièrement on m'a envoyé une pièce avec des petrifications de reptiles qui lui ressemble parfaitement, mais je n'en connois pas encore les relations géologiques."

bones which are found associated with the teeth, and approximate more nearly to the corresponding bones of the Iguana than to those of the other saurians, belong to the Iguanodon ; and I now proceed to describe them: sooner or later, the discovery of some portions of the skeleton in connection with the jaws and teeth will establish or refute this conclusion.

Bones of the Head.—*A frontal bone,* $1\frac{1}{4}$ inch long, very closely corresponding to the os frontis of the Iguana, is figured Pl. II. fig. 6. ; it adheres to the sandstone by its dermal aspect, is very thick, and differs in some particulars from that of the recent animal; it must have belonged to a very young individual, whose length could not have exceeded five feet.

Ethmoïd bone. I have in my collection a mutilated bone, which M. Cuvier suggested might approximate to an ethmoïd bone of a lizard of enormous size, certainly larger than that of the *Geosaurus* of Manheim.*

Os tympani. This bone, which forms part of the auditory apparatus, and in reptiles unites the lower jaw with the skull, is placed vertically, and varies very much in form in the different saurian genera.† Two almost perfect bones of this kind have been discovered, which differ in many essential particulars from any previously noticed, yet approximate in some respects to the *os tympani* of

* *Vide* Mémoire sur quelques parties moins connues du Squelette des Sauriens Fossiles de Maestricht, par M. A. Camper, in which the *ethmoïd* of the Geosaurus is figured. Pl. I. fig. 1.

† This bone is analogous to the *os quadratum* in birds.

the Mososaurus.* In these bones the body bears some resemblance to a vertebra, but the large cells or hollows which pervade it throughout, readily distinguish it; it forms a thick pillar or column, which is contracted in the middle, and terminates at both extremities in an elliptical and nearly flat surface: two lateral processes, or *alæ*, pass off obliquely, and are small in proportion to the size of the column; on placing these bones beside the os tympani of an Iguana, we at once perceive that the relative proportions of these parts are reversed; for in the recent animal the pillar is small, and the lateral processes large. From the great size of the body in the fossil, and the extreme thinness of its walls, the *tympanic cellulæ* must have been of considerable magnitude, and have constituted a large portion of the auditory cavities. Pl. II. fig. 1. accurately represents the most perfect specimen in my cabinet; it is 6 inches high, and 5½ inches wide at the longest diameter of the extremity of the body. It exceeds in magnitude the corresponding bone of the Mososaurus, and is 14 times as large as the same bone in an Iguana four feet long.†

Vertebræ. The somewhat angular vertebræ de-

* The os tympani is beautifully preserved in the head of the Mososaurus, in the Jardin des Plantes; and its form is perfect in a splendid model of that specimen presented to me by the late M. Cuvier: yet this bone does not appear in any of the numerous published representations of that celebrated fossil; and has not, I believe, even been described.

† It is due to my excellent and distinguished friend, Dr. Hodgkin, to mention, that he suggested to me long since, that these fossil bones approximated, in some respects, to the os quadratum in birds; but

scribed in a former part of this memoir, as appearing to constitute a second system, I should be disposed, from their number, and from their so commonly occurring in the localities where the teeth of the Iguanodon most abound, to refer to that animal; it must, however, be mentioned, that the concavo-convex vertebræ which correspond so entirely with those of the Iguana, and Monitor, would seem to offer a more probable approximation; yet the extreme rarity of the latter renders it questionable, since there appears no reason why the vertebræ should not have been preserved in as considerable numbers as the teeth. The chevron bones already described as belonging to the second vertebræ, serve to add to the probability that they are referable to the Iguanodon.

Ribs. Portions of enormous ribs of a prismatic form occur, which must have belonged to the skeleton of a colossal reptile, but they present no other remarkable character.

Sternum. A small sternal bone has been discovered, which corresponds to the sternum of an Iguana; it is represented of its natural size, Pl. III. fig. 4.

Clavicles. The clavicles in the Iguana and Monitor are very simple, forming a gentle curvature, which is attached by one extremity to the sternum, and by the other to the omoplate; a small bone of this kind has been found, which is

their true relations were not ascertained until the opportunity was afforded me of comparing them with the os tympani of the Mososaurus.

about four times the size of the clavicle of the recent species.

Under this head it is my intention to notice a bone of so singular a form, that even M. Cuvier declared it was the most extraordinary that he had ever seen, and that he was utterly at a loss as to what animal, and even as to what place in the skeleton, it should be assigned. "It might be a clavicle; but if it were, it did not resemble that of a reptile, nor indeed of any other living creature." The bone figured and described by Dr. Buckland as a clavicle (in the memoir on the Megalosaurus, to which reference has already been made,) appears to approximate to it; but it is broken at its widest extremity, so as to admit the possibility of a difference in the form of the apophyses. One of my specimens is entire, and two views of it are given, Pl. IV. fig. 1, 2.: it is 28 inches long; several fragments of other similar bones have been collected, and I have a portion of one that is one third larger, and if perfect would be 35 inches in length. Dr. Buckland's specimen is 21½ inches long, and if admitted to be the clavicle of a lizard, indicates a total length of 55 feet for the original animal; the large fragment in my possession must, therefore, according to such calculation, have belonged to an individual 80 feet long. To point out in a still more striking manner the extraordinary anomalies which this fossil bone exhibits, I may mention that our first comparative anatomist, Mr. Clift, among all the osteological treasures which the museum of the College of Surgeons places at his command, could discover no bone at all approaching it, except the first rib of an ostrich, which has processes bearing a distant resemblance to the apophyses observable at the wider extremity.

The drawing so accurately represents the original, that a brief description will suffice. The bone is slightly arched, of a prismatic form in the middle, and enlarged and flat at both extremities: from the smaller end the bone contracts, then widens, and at not quite one third from the other extremity sends off a small flat apophysis; it then becomes enlarged, and terminates in two unequal flat processes, which are equal in width to one third of the whole length. If we consider the form of this bone, it appears that the only place it can hold in the skeleton must be either the thorax or lower extremities; it may be a fibula, a rib, or a clavicle; and that it is a clavicle of some extraordinary extinct reptile is certainly most probable. Notwithstanding the analogy which the specimen from Stonesfield bears to this bone, I am disposed to refer the Sussex fossil to the reptile which exhibits the most striking differences from all known types of saurian organisation — the Iguanodon.

Bones of the extremities.—Humerus, radius, and *ulna.* Of the first I have two large examples, which bear considerable resemblance to the corresponding bones of the Iguana, and Monitor. Of the *ulna* and *radius* I have seen no decided examples, but a fragment which is 5 inches long, and was considered by M. Cuvier as resembling the head of a *radius.*

Femur, tibia, and *fibula.* The fortunate discovery of a considerable portion of a femur in juxtaposition with the corresponding bones of the leg, in a bed of soft clay, which admitted of the bones being cleared without injury, has afforded certain data for the restoration of a limb of the colossal Iguanodon. A considerable portion of the shaft of the thigh bone, much broken and compressed, was found at a little distance from the inferior extremity; but the latter, with its condyles, is in good preservation; and the leg bones, though somewhat distorted, are very entire. These bones present the true lacertian characters, and approximate more nearly to the Iguana, than to the Monitor and other related genera. The inferior portion of the *femur* is represented Pl. IV. figs. 3, 4.; the circumference at the condyles is 24 inches. The *tibia* is figured Pl. II. fig. 8.; it is 31 inches long. Fig. 7. Pl. II. represents the *fibula*, which is 27½ inches in length. If these be compared with the corresponding bones of the recent Iguana, they give to the original animal a total length of 60 feet. A specimen of about one third of the shaft of a femur, in my cabinet, is still more gigantic, being 23 inches in circumference.

Bones of the feet and toes. Some very large tarsal and carpal bones, corresponding to those of the crocodile, rather than to those of the lizards, have been discovered, yet their enormous size seems to render it probable that they are derived from the Iguanodon, or Megalosaurus: the largest in my possession is 4½ inches long; circumference of the tarsal extremity, 13 inches. Of the phalangeal bones two very distinct kinds are met with; the one resembling the phalanges of the crocodile, the other those of the Iguana, and Monitor. A group of four bones has lately been discovered of the last mentioned species; and although they are displaced and mutilated, yet they form a highly interesting specimen: they give the following admeasurements:—

1st, or longest bone, 8 inches long.

2d; this is imperfect, but was probably as large as the first.

3d, 6½ inches long.

4th, 3½ inches long.

There can be no doubt that these bones formed part of the foot of an Iguanodon, or similar reptile; they are about ten times the size of the similar bones of a recent Iguana 5 feet long.

Unguical or *claw-bone*. The distal phalangeal bone of these reptiles constitutes a claw, and is protected by a horny covering, or nail: a most interesting bone of this kind has been found in the Tilgate grit, and is figured Pl. III. fig. 1.; and the corresponding bone of the Iguana, and the nail which covered it, are represented of the natural size in the same plate, figs. 2, 3.: the analogy between the recent and fossil bones is most perfect, the latter differing but in magnitude, being about sixteen times larger, thus indicating a total length of 80 feet. A smaller but more perfect claw-bone has recently been found by the Marquess of Northampton, and through his Lordship's liberality is added to my collection; it is 3·7 inches long, and shows the distal termination of the claw in a very beautiful manner; two other specimens only have been noticed.

Horn of the Iguanodon. Among the recent genera of lizards, the Iguanas are distinguished by their exuberant dormal appendages; many of the species have enormous serrated processes on the back; others on the tail and guttural pouch; while some have warts or horny protuberances on the head; these, however, are so small, the horn in the most favoured species, the *I. cornuta* (vide wood-cut, p. 325.), being scarcely a quarter of an inch high in an animal 5 feet long, that no one could have imagined that the corresponding process of an extinct reptile would have been preserved through countless ages, and submitted to our examination. Such, however, is the case; like the claw-bone, it was discovered imbedded in the conglomerate of Tilgate Forest. This relic is of so extraordinary a nature, that although it has been noticed in my former work, I am anxious to be permitted to dwell on it in this place, that I may introduce the remarks of M. Cuvier, by whom it was examined during his last visit to London, and at whose suggestion a more accurate representation has been made than that published in the fossils of Tilgate Forest. (Vide Pl. III. fig. 5.) This horn is of a conical form, moderately compressed laterally, and slightly recurved; its surface is corrugated by the integument by which it was covered: it is 4 inches high; but as the apex is broken off, it must have been at least half an inch higher when entire, thus exceeding by eighteen times the magnitude of the horn of the recent animal. The base is of an irregular elliptical form, and slightly concave; its longest diameter 3·2 inches; its shortest 2·1 inches. In this fossil we have another most interesting analogy between the Iguanodon and the Iguana.

Having thus shown, or rather rendered highly probable, that the reptile whose teeth have furnished the characters on which the genus Iguanodon is founded, possessed also a skeleton, which, except in its gigantic proportions, bore a general resemblance to that of the recent Iguana, it will be interesting to take a summary view of what has been advanced, and consider what parts of its osseous fabric we have collected together, and what data they afford from which to determine the size, and form, and habits of the original. The *teeth* occur in so perfect a state, and in all the variety of form produced by the different periods of growth, and the respective situations which they occupied in the jaws, as to furnish conclusive evidence that the animal not only was herbivorous like the Iguana, but that, unlike all other known reptiles, it was capable of performing mastication like the herbivorous mammalia. The *os tympani* affords us some information respecting the auditory organs; and the *horn* exhibits another striking resemblance to the recent genus. The *vertebral column* approaches in some respects to that of the fossil crocodiles, unless we select the large concavo-convex vertebræ of the true lacertian type: the question, however, as to which of these systems should be referred to the Iguanodon, cannot, I think, in the present state of our knowledge, be positively determined. The ribs are too imperfect to furnish any certain data;

but they offer very gigantic proportions. Of the sternal apparatus nothing certain is known, except that a small *sternum* so closely resembles that of a lizard, that it may be inferred the coracoids and omoplates possessed a similar analogy; but no characteristic portions of the latter have been observed. The supposed clavicle is very peculiar, and of a most extraordinary character. The bones of the extremities approach very nearly to those of the Iguana; and the phalanges terminate in claw-bones, as in the recent animals.

To assist the attempt to estimate the size of the original animal, I have constructed the following table, which exhibits a comparative view of the dimensions of the corresponding bones in an Iguana 5 feet long, and in the Iguanodon: —

BONES.	RECENT IGUANA FIVE FEET LONG.	IGUANODON.		Length of the original, as indicated by the comparison.
			Exceed the recent by	
Teeth . .	. . .	. . .	20 times.	100 feet.
Horn . .	¼ inch high.	4½ inch.	18 —	90 —
Os tympani	0·6 inch high.	6 inch.	10 —	50 —
Clavicle .	1½ inch long.	30 inch.	20 —	100 —
Femur .	Circumference of the shaft 1½ inch.	23 inch.	15 —	75 —
	Length of the bone 3½ in.	4 feet.		
Tibia . .	2·8 inch.	31 inch.	11 —	55 —
Claw-bone	. . .	. . .	16 —	80 —

From this table it appears, that a comparison of seven different parts of the skeleton, give to the Iguanodon an average length of 70 feet, which is equal to fourteen times that of the Iguana.* It is not, of course, pretended that such a calculation can afford more than a very distant approximation to the truth; yet it may be confidently affirmed, that a reptile, which required a thigh bone larger than that of the largest elephant to support it, could not be of less colossal dimensions. In truth, I believe that its magnitude is here under-rated; for, like Frankenstein, I was struck with astonishment at the enormous monster which my investigations had, as it were, called into existence, and was more anxious to reduce its proportions, than to exaggerate them. If the conclusions above advanced, be admitted to be legitimate deductions from the facts which the fossil remains before us appear to have established, we can scarcely err in assuming, that the living Iguanodon bore considerable resemblance in form to the Iguana of the present day; and if we attempt its restoration, by investing it with muscles and integuments, still keeping the recent animal as the standard of comparison, the following admeasurements will be obtained: —

* Should subsequent discoveries prove that the original more nearly corresponded in the length of the tail with the crocodilian family than with the lizards, its total length would, of course, be much less than is here inferred; and from the form of some of the fossil *carpal* and *phalangeal* bones, it appears highly probable that the original animal was more bulky in proportion to its length than the existing lacertæ.

Length of the animal, from the snout to the tip of the tail,	70 feet.
head - - - - - -	$4\frac{1}{2}$ feet.
body - - - - -	13 feet.
tail - - - - - -	$52\frac{1}{2}$ feet.
Height from the ground to the top of the head - -	9 feet.
Circumference of the body - - - - -	$14\frac{1}{2}$ feet.
Length of the thigh and leg - - - -	8 ft. 2 in.
Circumference of the thigh - - - - -	$7\frac{1}{2}$ feet.
Length of the hind foot, from the heel to the point of the long toe - - - - - -	$6\frac{1}{2}$ feet.

The separate bones which appear to have belonged to the reptiles, the character of whose teeth were previously determined, having, however imperfectly, been referred to their several genera, I now proceed to describe the large specimen before us, which exhibits a considerable portion of the skeleton of a saurian animal; and I shall endeavour, by a careful investigation of its osteological structure, to determine what relation it bears to the known recent, and fossil species.

In the summer of the present year (1832), some workmen employed in quarrying the Tilgate grit in that part of Sussex where it is used as a road material, and who had been instructed to preserve whatever objects of interest their labours might discover, blasted a large block of the hardest variety of the stone to pieces, and perceiving traces of bones in some of the fragments, placed them aside for my inspection, and informed me of the circumstance. Upon repairing to the quarry, the considerable number of pieces into which the block was broken, the extreme hardness of the stone, and the unpromising appearance of the fragments of bone that were visible, seemed to render the attempt to dissect it alike hopeless and unprofitable. I resolved, however, to collect the scattered fragments together; and after much labour succeeded in reducing the specimen to the state in which it now appears. It is 4½ feet long, and 2 feet 3 inches broad at its widest extremity; but not more than two thirds of the original mass remain; of the other third a considerable part was lost, and the pieces that were preserved could not be made to fit together. The portion of the skeleton displayed in the specimen, consists of a series of ten vertebræ, five *cervical* (*a*, Pl. V.) and five *dorsal* (*b*, Pl. V.), adhering to the stone by their spinous processes; and of three other dorsal vertebræ, which are dislocated, but lie near to each other: there are also two more dorsal in other parts of the block. Several ribs, more or less displaced, are situated on each side of the vertebral column; and at the end of the fifth dorsal vertebra, two *coracoid* bones (*g*, Pl. V.) and two *omoplates* (*f*, Pl. V.) are seen: these are somewhat displaced, the left coracoid overlying the right, and concealing one third of its sternal portion; in fact, the left omoplate and coracoid, appear as if they had been driven with great violence against the vertebral column, and over the opposite bone, and had occasioned the removal of the four dorsal vertebræ from their place. On the left side of the column is a series of bony processes (*h*, Pl. V.), of the form of an isosceles triangle; they are irregularly disposed, yet seven of these are placed somewhat in a parallel line with the vertebral column; but three of the largest and most remarkable in form (*h** *h** *h** Pl. V.) lie near to each other, and in a direction at right angles with the former, and above the level of the coracoid bones. Fragments of ribs and other bones, with two of the osseous bases of the scales, and here and there traces of lignite and vegetables, and casts of freshwater shells, comprise all the other organic remains observable on the face of the stone which is presented to view. On the opposite side (which it was necessary to place in cement), the ends of the spinous processes of the vertebræ were visible; there were also some interesting vegetable remains, and a fine portion of the stem of Clathraria Lyellii was extracted, as well as several of the seed-vessels which M. Ad. Brongniart conceives to have belonged to the same plant. Among this mass of vegetables were many shells of the genera Unio and Paludina. Such is a general description of this remarkable fossil; and I must now solicit the indulgence of the Society, while I enter upon those anatomical details, which, however dry and tedious they may appear, can alone afford a clue to guide us through the labyrinth of fossil zoology, and lead to those interesting and imposing results, which the genius of Cuvier first taught us how to obtain.

*Vertebræ.** These approximate very closely to those of the fossil crocodiles, comprised in the first system previously described. Of the cervical (*a*, Pl. V.), the remains of five are distinguishable. The confused bony mass at the extremity of the

* By a reference to Plate V. the reader will easily follow this anatomical description.

column may be supposed to be the remains of two; then follows one that is compressed, but its outline is defined; the succeeding vertebra is entire; it has two tubercles for the attachment of supplementary transverse processes; the form of the latter unfortunately cannot be discovered: the transverse processes are short and very strong: this bone is 1·7 inch long and 2 inches wide. The seventh cervical is much crushed; it has traces of the left lateral apophysis. The first dorsal (*b**, Pl. V.) is entire; it is 2 inches long, and 2·2 inches wide at the extremities; it is depressed laterally, and the anterior part of the body rounded; the tubercle for the attachment of the rib is well marked; and the transverse process short and strong: its rib is near it, and shows the deep bifurcation of the head, as in the crocodile. The 2d, 3d, 4th, and 5th dorsal succeed, and differ but little from each other; except that the 3d is more keeled; the 5th (*c*, Pl. V.), which is 2½ inches long, has its left rib near it; and the latter, unlike the four other ribs, has no strongly marked bifurcation, but its process sinks into a mere tubercle, as in the corresponding rib of the crocodile; and it is this character which, in the absence of other evidence, has been taken to determine the respective situations in the column to which the vertebræ belonged. The 2d, 3d, 4th, 5th, and 6th ribs are seen on the left side of the vertebræ; on the right, the 1st, 2d, and 4th only remain; the face of the 6th dorsal vertebra is shown beneath the sternal margin of the left coracoid, and one of its transverse processes appears near the 4th rib (*d*, Pl. V.). The 7th,

8th, and 9th dorsal are displaced, and lie in a hollow formed by the extremities of the left ribs and the corresponding coracoid. Another vertebra of the back, 2·8 inches long (perhaps the 10th, *e*, Pl. V.), is thrown to the left side of the stone, resting on portions of two ribs; the body is smooth and rounded, slightly arched, and its extremities, which are flat, are nearly circular, and 2 inches in diameter.

It is scarcely necessary to observe, that the ribs and vertebræ above described are decidedly of the fossil crocodilian structure; the union of the annular part by suture occurring, as M. Cuvier observes, in the living reptiles, in the crocodiles and turtles only; and if these bones were the only data from which to form an opinion of the nature of the original animal, we could not hesitate to assign it to some one of the fossil crocodiles mentioned in a former part of this memoir. But the bones we have next to describe incontrovertibly prove that the animal could not have belonged even to the same family; and they afford another striking example of that union and blending, as it were, of the different generic characters, which geology is constantly presenting to the comparative anatomist.

In the Crocodile (fig. 5. in the opposite page), the sternum consists of a long, slender, flat bone, pointed both before and behind; and this is supported on each side of the middle of its lateral edges by a coracoid of an elongated form, which has a thick neck near the humeral extremity, that enlarges into a plane and wide portion, to attach itself to the sternum. The omoplate or scapula is not unlike the coracoid; its plane forms a narrow isosceles triangle; its neck is subcylindrical and curved internally, and widens to present a face to the coracoid: on the external edge of this is an articular apophyses, which, with

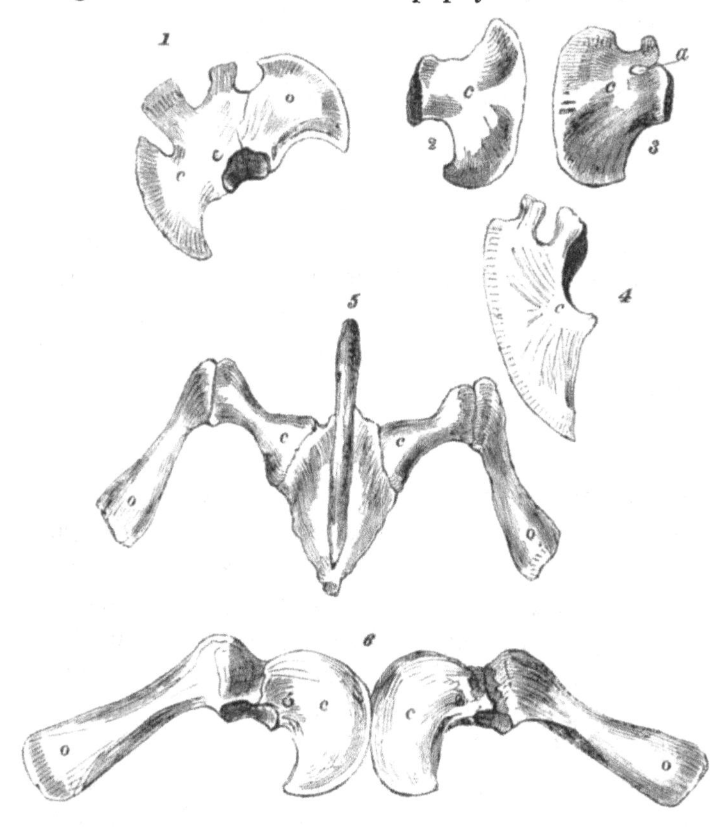

1. Coracoid bone and omoplate of the Monitor.
2. Coracoid bone of the Ichthyosaurus.
3. Coracoid bone of an Ichthyosaurus, in which there is a foramen near the neck, formed by a division in the bone.
4. Coracoid bone of the Megalosaurus.
5. Sternal apparatus of the Crocodile.
6. Coracoids and omoplates of the Hylæosaurus, the newly discovered fossil reptile of Tilgate Forest.

the corresponding one of the coracoid, forms the cavity for the reception of the head of the humerus. There are no clavicles ; the coracoid alone forming a buttress against the sternum.

In the fossil before us, the *omoplates* (of which restored outlines are given in p. 321. fig. 6, O.) correspond in many particulars with those of the crocodile above described, and unite with the coracoid in a similar manner. The head of the bone is 6 inches wide, and very thick; it is contracted at the neck, and passes off into a flat and wide extremity: the articular facet is 2½ inches high, and 3½ wide. These bones are about twelve times larger than the omoplates of a crocodile 3 feet long (they are marked *f f* in Plate V.). The *coracoid bones* (*c c*, fig. 6. p. 321. and Plate V. *g g*) are totally distinct from those of the crocodile; they are like the coracoids of the true lizard, hatched-shaped; but they are not emarginated, and have no apophyses corresponding to those of the recent Monitors (fig. 1.) and Iguanas, and the fossil Megalosaurus (fig. 4.). The longest diameter of the coracoid is 7 inches; the transverse diameter 5·1 inch: it presents a large articular fossa for the glenoid cavity, which is formed mutually by this bone and the omoplate. Near the neck of the bone is a large foramen for the passage of vessels, which invariably is the case in the lizards, but does not occur in the Plesiosaurus, Ichthyosaurus (fig. 2.)*, Megalosau-

* The only known exception is a coracoid of an Ichthyosaurus, of which a reduced sketch is given fig. 3. p. 321 : here the humeral extremity is seen to throw off a kind of apophysis, which unites at the upper end of the glenoid cavity, and thus produces a foramen at *a*. This interesting specimen was in the matchless collection of Ichthyosauri and

rus, or Crocodile. There are no traces of sternum, or clavicles.

We have next to direct our attention to the bony processes which are distributed on the left of the vertebral column, and three of which hold so conspicuous a place near the base of the left omoplate; these are marked *h h h* in Plate V. Of these bones there are no less than ten, more or less perfect, and varying in size from 5 to 17 inches in length, and from 3 to 7½ inches wide at the base; they differ much in form, but they approach to an isosceles triangle. They are so accurately represented in the drawing, that I shall only particularly notice the three large examples which lie near the anterior part of the fossil (*h** *h** *h**, Plate V.). The longest of these is 17 inches in length, and but 3½ wide, at six inches from the base, which is broken, and is 4 inches thick; this bone differs from the corresponding ones in this respect, and more nearly resembles the displaced bone (at *i*, Plate V.) that has a semicircular hollow at its largest extremity, as if for articulation with another bone. The middle process is 13½ inches long, and 4·7 inches wide at the base, and is flat, and slightly depressed in the centre: the third is also very flat; it is 11 inches long, and the base imperfect. What the nature of these processes may be, it must be ac-

---

Plesiosauri of Thomas Hawkins, of Glastonbury, Esq., F.G.S., who, with his accustomed generosity, presented it to me, on learning that it might assist in the elucidation of my investigations. This gentleman has in the press a Memoir on the Ichthyosauri and Plesiosauri, with superb plates, in folio, which will shortly appear before the public.

knowledged, is a question exceedingly embarrassing. They are evidently apophyses, and their number is so considerable as to suggest at once, that if they were attached to other bones, it must have been to some part of the vertebral column; yet they can be neither spinous nor transverse processes, for we have examples of both of these still attached to the vertebræ, and besides, some of these bones are so large that none of the vertebræ could support them. At first I supposed them to be *chevron bones* or inferior spinous processes; and, as above remarked, the detached bone at *i*, Plate V. has a hollow, now filled with stone, as if for attachment to a vertebra; but they differ most essentially from all the other chevron bones of the Tilgate strata, as will appear at once if reference be made to what has already been stated respecting those of the Iguanodon and Crocodile. They cannot have been supplementary transverse processes to the cervical vertebræ, (as in the Crocodile,) for they are both too numerous and too large. Upon taking the dimensions of their bases, it appears that the united length of the whole is four feet. It is therefore evident that if they belonged to the vertebral column, it must have been to the vertebræ of the tail; but then this difficulty arises, that not even the large vertebræ which we have ascribed to the Iguanodon have such enormous processes; for the base of some of these is 7½ inches wide. Now the whole length of the five dorsal and two cervical vertebræ in the fossil is 17½ inches; and of the whole series of fourteen vertebræ but 3 feet. The same number of corresponding vertebræ constitute in the Crocodile about one-sixth of the length of the entire animal; in the Monitor nearly the same; and in the Iguana one-ninth. If, therefore, the original animal bore the linear proportions of the first and second, it would not exceed 20 feet in length, and if of the Iguana not 30; and although, from the mixed characters of the skeleton, and the size and strength of the omoplates and coracoids, it seems probable that it was larger than the ordinary Crocodiles, yet its total length cannot reasonably be estimated at more than 25 feet; and from the annular sutures in the vertebræ being almost obliterated, there is reason to conclude that the individual was an adult. This creature, then, did not approach in magnitude to the Iguanodon, or Megalosaurus, by more than one-half; and yet the processes, if assumed to be vertebral, indicate vertebræ much larger than any that have been discovered of those animals. Another conjecture has occurred to me, and, extravagant as it may appear, it seems the most probable. It is known that many of the lizards, particularly the Iguanas, have large cartilaginous processes with horny coverings, which form a sort of dermal fringe along the back: in an animal 5 feet long, these spines are about an inch in height. (A sketch of the recent *Iguana cornuta* is here introduced, for illustration.) Now the

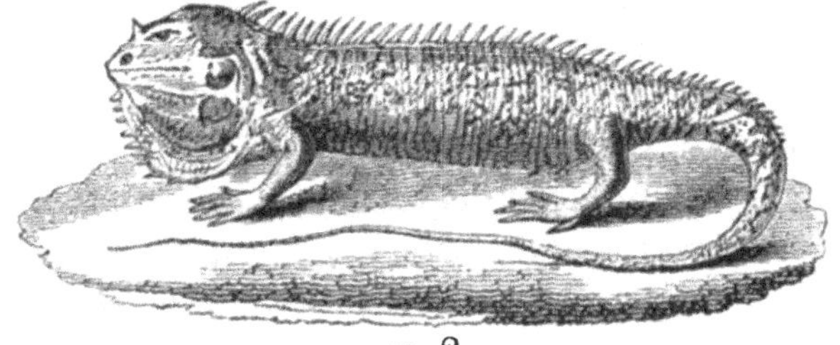

situation which the processes in the fossil occupy, is precisely that which the remains of such a dermal appendage may be supposed to have occupied, had the animal been pressed down on its back, and the scaly integument turned to the left side. It may also be remarked, that if, on the other hand, it be assumed that the bones, the nature of which is so perplexing, belonged to the tail, it is very extraordinary that they should have been forced into their present situation, associated with the bones of the anterior part of the skeleton, and maintaining a certain degree of parallelism with the line of cervical and dorsal vertebræ; and as the bones which I have next to describe render it certain that the original animal was covered in some parts of its body with enormous scales, or horny appendages, the supposition that these enigmatical processes are the remains of dermal spines, seems to be the more probable. It may, too, be added, that the base of one of these apophyses that was detached, has been carefully cleared from the stone, and that the extremity has a *small sulcus* only, thus bearing no resemblance whatever to that of a chevron bone; *in that sulcus a small dermal bone was found.*

Dermal bones.—Among the mass of vegetable matter removed in clearing the skeleton, many small bones were cut through and destroyed, until their peculiar structure attracted attention, and induced me to preserve the next which came under my observation. Two only remain attached to the specimen; one which is broken is in juxtaposition with the first right rib, *k*, Pl. V.: the other is near the vegetable impression in the centre of the stone: this last is of the size and form of a small patella, and is composed of minute linear spiculæ of bone which decussate each other almost at right angles: this structure prevails in all the specimens: one bone of this kind, of an irregular form, is 2 inches in diameter.

DERMAL BONES OF THE HYLÆOSAURUS.

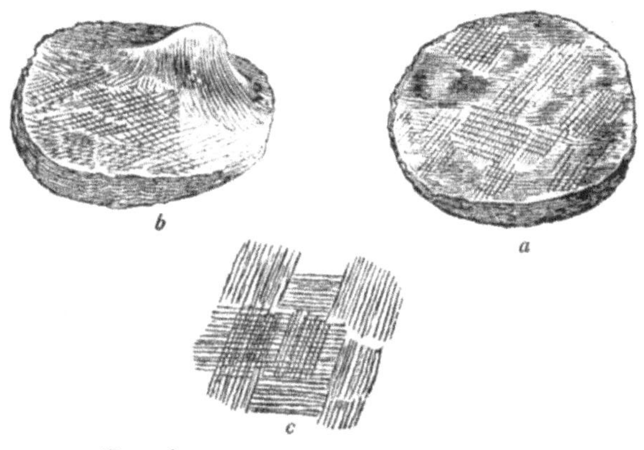

a. Dermal aspect.
b. Surface attached to the scale.
c. Magnified view of a part of the surface.

There cannot exist a doubt that these are dermal bones to which the scales of the animal were attached.

Such are the anatomical developements which the fossil skeleton before us offers for our consideration, and it remains to determine what place it should hold in the saurian family. It is unnecessary here to repeat the reasons which prove, that although the vertebral column and ribs approximate it to the crocodile, the sternal apparatus shows it to be more nearly related to the monitors, from which it is distinguished (as well as from the Megalosaurus), by the entire margin of the coracoid, and the figure of the omoplate; and as the teeth and other bones of the Iguanodon possess so many of the peculiar characters of the lacertæ, (not to mention the disparity in size,) it cannot, it is presumed, in the present state of our knowledge, be referred to that animal. To avoid the confusion, therefore, which would necessarily arise from a mistaken appropriation of this skeleton to the teeth and other bones found in the Tilgate strata, and from the conviction that its osteological organisation is peculiar and important, I venture to suggest the propriety of referring it to a new genus of saurians. The generic characters would, of course, be the peculiarity of the sternal apparatus, and the remarkable processes which are distributed around it; and I propose to distinguish it by the name of *Hylæosaurus**, to indicate the locality in which these remains were discovered. The original animal in all probability differed as much in its external form, as in its skeleton, from known species; we are certain that it was covered with scales, and there appears every reason to conclude that either its back was armed with a formidable row of spines, constituting a dermal fringe, or that its tail possessed the same appendage, and was enormously disproportionate to the size of the body, as in some of the recent lacertæ; the specific name of *armatus*, in either case, would not be inappropriate.

It is not my intention to trespass longer on the indulgence of the Society, for although many large

* From ὕλη, *sylva*, wood, *weald*, or forest; and *saurus*, lizard; the WEALDEN LIZARD, or *Fossil Lizard of Tilgate Forest.*

and characteristic bones have been discovered in the strata of Tilgate Forest, which cannot have belonged to any of the animals above described, it would extend this paper to too great a length were I to notice them on this occasion. I would, however, solicit permission to offer a few remarks on what has been advanced. It has been shown that the Wealden deposits have afforded the remains of several species of crocodiles, and gavials; of animals holding an intermediate place between the lizards, and crocodiles, the *Hylæosaurus*, and *Cylindricodon;* and of the two colossal monsters of the reptile world, the Megalosaurus, and the Iguanodon: with these are associated the remains of tortoises, and turtles; of river fishes, and river shells; of plants which grew in marshes, and of others which grew on the land, and which, like the Dracæna, and the Cycas, and the palms, and arborescent ferns, could not have existed but in a climate of a high temperature. I will not presume, to repeat here, the observations which I formerly made on the probable condition of the country which produced such a fauna and flora, and through which the river flowed that brought down and deposited the remains which offer such an inexhaustible field for geological enquiry. Yet I may be permitted to state, that all which I have since observed has served to confirm me more and more in the opinion, that the strata of the Wealden were a vast delta formed by some mighty river, and that the few marine stragglers which occur serve to support rather than to weaken such a hypothesis. But I do not think there is reason to conclude that the reptiles, and the terrestrial plants, like the zoophytes, and mollusca, and fishes of the chalk, lived on the very spots where we now find them entombed; for, with the exception of the beds of shells, and among the vegetables, perhaps, the Equiseta Lyellii, (which are so local, and occur in such quantities, and with but so little an intermixture of other plants,) all bear marks of having been transported from a distance by water. But although three fourths of the bones which we discover are more or less broken and rolled; the teeth detached from the jaws; the vertebræ and bones of the extremities, with but very few exceptions, disjointed and scattered here and there; the stems of the plants torn to pieces; and all these intermixed with pebbles of quartz, and flinty slate, and jasper, all concurring to prove that these heterogeneous materials have been subject to the action of water, yet it is manifest that the action was fluviatile, not littoral. The pebbles, though smooth from their angles having been worn away, are not rounded into beach or shingle; they have been subject to the operation of currents and torrents, but not to attrition from the waves of the ocean. And if we reflect upon the immense strength which the tendons and ligaments of the joints possess, even in the pigmy lizards that are our contemporaries, we must be convinced that the gigantic limbs of the Iguanodon, and Megalosaurus, could not have been dissevered from their sockets without great violence, or by the decomposition of their tendons by long maceration; and if the latter were alone the cause, we should not find the bones broken and separated, but lying in juxtaposition, as in the specimen of a fossil fox in the collection of the President; this condition of things seems to indicate that they were transported from a considerable distance. The state of the skeleton of the *Hylæosaurus* is in this point of view highly interesting: many of the vertebræ and ribs are crushed and splintered, yet the fractured portions remain near each other; the bones are more or less dislocated, yet they maintain a situation bearing some relation to the place they occupied in the recent animal; the humeri, or in other words the fore-legs, have been torn from their sockets, and this, too, must have taken place before the specimen was imbedded in the mud and sand, for the glenoid cavities were filled with stone: these circumstances seem to show, that the carcase of the original must have suffered injury, and mutilation, before its bones were reduced to a skeleton; and that the dislocated and broken parts were kept somewhat together, as we now see them, by the muscles and integuments; in this state the headless trunk was borne down the stream, and at length sank into the mud of the delta, and formed, as it were, a nucleus around which the stems and leaves of Clathrariæ, palms, and ferns were deposited, and river shells became intermingled with the general mass. The phenomena here contemplated cannot, it appears to me, be explained upon any other supposition than that which implies a considerable period of transport; the carcasses of the large reptiles must have long been exposed to such an agency; and the river which flowed through the country of the Iguanodon, must have had its source far distant from the delta which it deposited; the course of that river, the extent of that delta, and the situation of that country, will probably for ever remain unknown.

In concluding these observations, I cannot but express how deeply I have felt in every stage of my labours, and what all must feel who engage in enquiries of this kind, the irreparable loss which the science of fossil comparative anatomy has sustained by the death of that illustrious philosopher, Baron Cuvier, whose powerful mind, and enlightened genius, could, like the fabled wand of the sorcerer, cause to pass before us the beings of former ages, and from the relics which have descended to us from the eternity of the past, construct anew the forms of organisation, which had ceased to exist before the creation of the human race. Could I upon this, as upon former occasions, have obtained his invaluable assistance, the interpretation of the interesting records before me would have been clear and satisfactory, and an apology would not then have been necessary, as I now feel it to be, for the obscure and imperfect manner in which my investigations have been conducted.

Note.—Since the above was written we have obtained, through the liberality of R. Trotter, Esq., F.G.S., a series of the first six caudal vertebræ of a reptile, in a slab of Tilgate-stone, from the same quarry as the remains of the *Hylæosaurus*. The bodies of these vertebræ, like those of the newly-discovered reptile, are slightly concave at both extremities; the visceral aspect is smooth, but gently ridged. The spinous processes are almost perfect, and are 15 inches high; but, although so large, they bear no resemblance whatever to the extraordinary bony apophyses of the Hylæosaurus, but rather afford a negative proof that those appendages belonged to the back of that animal. The transverse processes are very short. The *chevron bones* are like those of the Iguana; three of them remain almost entire. The width (or rather height) of the tail, at this part, must have been, at least, 27 inches, and its entire length 22 feet.

CHAP. XI.

RESULTS OF THE GEOLOGICAL INVESTIGATION OF THE SOUTH-EAST OF ENGLAND.

In this chapter we propose to lay before the reader a summary of the geological phenomena described in the preceding pages, and the inferences resulting therefrom; bearing in mind the admirable remark of a distinguished philosopher, that "the language of theory can never fall from our lips with any grace or fitness, unless it appear as the simple enunciation of those general facts, with which, by observation alone, we have become acquainted."* Happily, the evidence of the great physical mutations, and important changes in organic life, which have taken place in this part of the earth during the geological periods to which our researches refer, is so clear and satisfactory, that even the general reader will perceive that our deductions, extraordinary as they may appear, naturally result from the facts themselves.

The several formations or groups of strata, previously described, may be regarded as geological chronometers, marking certain distinct epochs or

* Professor Sedgwick's Annual Address to the Geological Society of London. 1830.

periods; the lowermost or most ancient of which (as we have already noticed) is of fluviatile origin, and reposes on the *Oolite*, a marine formation of great extent, that forms an important feature in the physical structure, not only of England, but also of the Continent. The *Portland Limestone* constitutes the uppermost division of the Oolite, and contains marine remains only; it is succeeded by the *fresh-water strata of the Isle of Purbeck*, which may be considered as the lowermost deposits of the Wealden; for although the Purbeck marble, composed of a small species of *paludina*, has not been observed within the wealds of Kent and Sussex, yet the shales and limestones of Ashburnham and Brightling bear a close resemblance to some of the slaty strata of Purbeck, and the organic remains in all are of the same fluviatile character.

But there is a fact connected with the history of the Portland and Purbeck beds, so highly interesting, and which illustrates in so striking a manner the nature of one of those grand geological mutations which have taken place in the south of England, that it will be necessary to notice it here, although it occurs without the limits of the district, which it is the professed object of this work to describe.

In the island of Portland, the oolitic limestone is extensively quarried for architectural purposes, and supplies most of the cities and towns in the south-east of England. On these oolitic strata are placed deposits of a totally different character. Immediately on the uppermost marine stratum (which abounds in *ammonites, terebræ, trigoniæ*, &c.) is a bed of limestone, much resembling, in appearance, some of the tertiary lacustrine limestones. Upon this stratum is a layer of what appears to have been an *ancient vegetable soil; it is of a dark brown colour, contains a large proportion of earthy lignite, and, like the modern soil on the surface of the island, many water-worn stones.* This layer is called the *dirt-bed* by the quarrymen; and in, and upon it, are a great number of silicified trunks of coniferous trees, and plants allied to the recent *cycas* and *zamia.* Many of the stems of the trees, as well as the plants, are still erect, as if petrified while growing undisturbed in their native forest; the former, having their roots in the soil, and their trunks extending into the superincumbent strata of limestone. On a late visit to the quarries, a large area of the surface of the *dirt-bed* having been cleared, preparatory to its removal for the purpose of extracting the building-stone from beneath, several stems, from two to three feet in height, were exposed, each standing erect in the centre of a mound or dome of earth, which had evidently accumulated around the base and roots of the trees; presenting an appearance as if the trees had been broken, or torn off, at a short distance from the ground. Portions of trunks and branches were seen, some lying on the surface, and others imbedded in the dirt-bed; many of these were nearly two feet in diameter, and the united fragments of one tree measured upwards of thirty feet in length. The silicified plants allied to the *cycas* are found in the intervals between the trees; and I dug up from the dirt-bed several that were standing erect, evidently upon the very spot on which they grew, and where they had remained undisturbed amidst all the revolutions which had subsequently swept over the surface of the earth.* "The dirt-bed extends through the north of the Isle of Portland, and traces of it have been observed in the coves at the west-end of Purbeck; and a stratum, with bituminous matter and silicified wood, occurs in the cliffs of the Boulonnois, on the opposite coast of France, occupying the same relative situation with respect to the Purbeck and Portland formations. A similar bed has also been discovered in Buckinghamshire, and in the Vale of Wardour, proving that the presence of this remarkable stratum is coextensive with the junction of the Portland and Purbeck strata, so far as they have hitherto been examined."†

Above the *dirt-bed* are thin layers of limestone, the total thickness being about eight feet, into which the erect trunks extend, but no other traces of organic remains have been noticed in them. These limestone beds are covered by the modern vegetable soil, which scarcely exceeds in depth the ancient one above described; and instead of giving support, like the latter, to a tropical forest, can

* These highly interesting facts were first pointed out by Mr. Webster (Geological Transactions, 2d series, vol. ii.). The fossil plants have been beautifully illustrated by Dr. Buckland (Geol Trans. 2d series, vol. ii.), who, in conjunction with Mr. De la Beche, has subsequently laid before the Geological Society a masterly paper on the geological phenomena of the Island of Portland, and of the vicinity of Weymouth.—*Geol. Proceedings*, 1829, 1830.

† Vide Geology of Hastings, p. 76. *et seq.*

barely maintain a scanty vegetation, there being scarcely a tree or shrub on the whole island.*

Here, then, we have recorded in characters which cannot be mistaken the nature of the changes which took place in this part of the globe, after the sea of the oolite had deposited the marine strata of Portland. A portion of the bed of that sea was elevated above the surface of the waters, and became clothed with a vegetation, which, reasoning from the close resemblance of the fossil plants to the recent *Cycadeæ*, must have enjoyed a climate of a much higher temperature than is known in these latitudes at the present day. How long this island, or continent, (for of its extent no correct estimate can be formed,) remained above the level of the ocean, cannot be conjectured; but that it was dry land for a considerable period, is manifest from the number and magnitude of the petrified trees which remain. It is equally evident, that it was submerged before the Purbeck and Wealden strata began to be deposited; for the dirt-bed, and its contents, are covered by the fresh-water limestone of the former. The tropical forest of Portland must, therefore, have gradually and

* The appearance of the large quarry on the northern brow of the Island of Portland was, at the time of my visit (in July, 1832), peculiarly interesting; and although prepared by a perusal of the excellent Memoirs of Mr. Webster, and Dr. Buckland, for the phenomena presented to my view, I was struck with astonishment at the extraordinary scene; the floor of the quarry was literally strewed with fossil wood, and before me were the remains of a petrified tropical forest, the trees and the plants, like the inhabitants of the city in Arabian story, being converted into stone, yet still maintaining the places which they occupied when alive.

tranquilly subsided (like many subterranean forests of the modern epoch*) beneath a body of fresh water, sufficiently profound to admit of the accumulation of the limestone and fluviatile strata that compose the Wealden. What contemporaneous changes took place in other parts of Europe, it would be foreign to our purpose, and perhaps, in the present state of our knowledge, in vain to enquire; but we may remark, that the submergence of so extensive a tract of country, probably produced in other regions important mutations in the relative level of the land and water.† At this epoch, then, the land and its tropical forest sank to the depth of many hundred feet, and became the bed of a vast lake or estuary, into which we have the clearest evidence that a river flowed, and formed a delta, made up of the debris of the rocks which composed its bed, intermixed with the remains of the animals and vegetables of the country from whence its waters were derived; for, as Mr. Bakewell has sagaciously remarked‡, a river that

* Vide p. 18.

† M. Elie de Beaumont has shown, that the higher ridges in eastern France, of the Côte d'Or, and Mount Pilas, and a portion of the Jura chain, were all elevated *after the formation of the Oolite was completed*, and *before* the deposition of the chalk; these elevatory movements therefore took place at the Iguanodon period, and may have been connected with the physical mutations of the south-east of England mentioned in the text. The immense area over which the effects of an earthquake may extend, was remarkably exemplified in that of Lisbon, in 1755. It was most violent in Portugal, Spain, and the north of Africa; but the whole of Europe, and even the West Indies, felt the shock on the same day. *Consult Mr. Lyell's "Principles of Geology,"* vol. i. p. 439.

‡ In his correspondence with the Author.

could form a delta of such extent as the Wealden, it must have required the drainage of a vast continent to supply.* An interesting question here arises; what was the nature of the rocks that formed the boundaries of this ancient river, lake, or estuary? It would seem most probable, that these were composed, in part at least, of the oolite, lias, and other secondary strata, and which were worn down into the sands and clays of the Wealden: but, if this were the case, we should expect to find in the conglomerate of Tilgate Forest some traces of the peculiar fossils of those formations, and some of the harder portions of the rocks in the state of pebbles; whereas, these strata contain rolled masses, and gravel, of quartz, jasper, and flinty slate, which must have resulted from the disintegration of primary, rather than of secondary formations: it seems, therefore, most probable, that the Wealden is made up of the debris of rocks of various ages.†

The proofs of the Wealden having been the delta of some ancient river, are so fully stated in the preceding chapter, that it is unnecessary to dwell upon the subject. Of its original extent, our conjectures must necessarily be extremely vague: Dr. Fitton has, however, ingeniously instituted a comparison

* Had the fossil vegetables of the Wealden been identical with those of the Isle of Portland, it might have been supposed that the latter was dry land at the Iguanodon period: but although the vegetable remains in both deposits indicate the floras of tropical climates, they are totally distinct from each other, and belong to different species and genera.

† Dr. Fitton has a fragment of an ammonite in limestone, from the Wealden of the Isle of Wight: oyster shells occur in some of the Purbeck strata.

between the known superficial surface of the Wealden, and the deltas of some modern rivers. Assuming that the occurrence of the Wealden strata at Beauvais is established, this eminent geologist computes that the remains of the delta of the Iguanodon period, are from west to east, or from Lulworth Cove, to the boundaries of the Lower Boulonnois, about 200 miles; and from north-west to south-east, or from Whitchurch to Beauvais, 220 miles; the total depth or thickness being about 2000 feet.* This but little exceeds the modern deltas of the Ganges, and the Mississippi; and is not equal to that of the Quorra, or Niger, which forms a surface of 25,000 square miles, being equal in extent to one half of England.†

We have no data from which to calculate the probable duration of the Iguanodon epoch; it is, however, manifest that no brief period could have sufficed for that profuse evolution of animal life, of which we have such positive evidence in the organic remains. It may here, too, be remarked, that the vegetables and animals of this era, like the forest of Portland, denote a tropical climate, and belong to species and genera, wholly unknown; and, as we have elsewhere observed, the fossil bones of the oviparous quadrupeds are so enormous, that it is even difficult to believe the evidence of our senses, when

* Geology of Hastings, p. 58.

† Ibid. Strata containing marine remains, with iron ore, have been observed between the upper beds of the oolite and the chalk, in the department of the Hante Saone; near Cander, in the Brisgau; and in Poland: they are supposed, by Mr. De la Beche, to have been formed in the sea during the Iguanodon era.—*Man. Geolog.* p. 303.

we attempt, from these remains, to restore the forms of the extinct monsters of the ancient world.

The next great change is the subsidence of the Wealden into the abyss of that extensive and profound ocean which deposited the chalk formation. Whether this mutation were effected suddenly, or by slow degrees; whether the Wealden subsided entire, or were broken up previously to its submergence; or whether, like the Isle of Portland, it constituted dry land at some remote period, antecedently to its being buried beneath the sea, we have no data to enable us to decide. The principal lines of elevation of the Wealden are clearly referable to those movements which up-heaved the chalk and incumbent strata: but we may observe, that the deeper beds exhibit traces of extensive faults and dislocations, which seem to belong to previous disruptions, for the fissures and chasms are filled up with broken shale, and clay, and sand, the debris of the Wealden, and contain no intermixture whatever of the marine deposits which may be supposed to have once covered them.

The ocean of the chalk appears to have been of vast extent; it buried beneath its waters a considerable part of Europe; and, probably, like the Atlantic, its waves reached the western world, and covered a portion of the continent of North America.* The nature of the strata, and the organic

* The occurrence of the remains of the *Mososaurus*, that extraordinary reptile of the Maestricht beds, in the strata of the United States, previously mentioned, is a remarkable fact in corroboration of such an inference. See Dr. Morton on the *Ferruginous Sand Formation of North America*, 8vo. 1 vol. with plates. Philadelphia. 1833.

remains which they enclose, prove that the chalk was deposited in the tranquil depths of a profound ocean; the abundance of *Ammonites*, *Nautili*, and other multilocular shells that inhabit the bottom of the deep*; the almost entire absence of pebbles and gravel†; the perfect state in which the fishes and other perishable organic bodies occur—not as in the Wealden, crushed, and disjointed, but as perfect as if they had been enveloped by a soft paste when living, or even while in a state of progression—all bear evidence in favour of such a conclusion.‡

There are but few, if any remains of terrestrial animals and plants, to throw light on the nature of the climate during the cretaceous epoch: we may, however, infer from the nautili and other tropical shells, as well as from the presence of the stony *polipidoms*, or corals, that the temperature was not much inferior to that of the Iguanodon period, for this

* "The recent *Nautilus pompilius* is a ground-dwelling animal, and its principal sphere of action is at the bottom of the sea, as is proved by the nature of its food. The peculiar structure of its shell enables it to rise to the surface; but to float there at ease, an additional volume of air is probably taken into the dwelling chamber; in which case the act of sinking would be accomplished by simply reversing the shell." In several ammonites and nautili from the Sussex chalk, the siphunculus, or air-tube, which communicates with the concealed chambers of the shell, is beautifully preserved.— Consult the highly interesting and scientific *Memoir on the Pearly Nautilus*, by Richard Owen, Esq., of the Royal College of Surgeons of London. 4to. 1832.

† Small quartz pebbles occur in the chalk of Kent; and, very rarely, in that of Sussex. I have a flat pebble from Steyning, 3 inches in diameter, to which are attached a small caryophillia, and several minute parasitical shells.

‡ These remarks, of course, refer to the chalk of this part of England only; the chalk ocean, like other seas, must have had its shallows, and its limits; and, probably, sooner or later, traces of its shores, and beds of shingle, and sand, with littoral shells, will be discovered.

division of zoophytes is not known to exist in low latitudes, in our modern seas.* The *cretaceous* strata of the chalk, with their nodules and veins of flint, have more the character of a chemical production, than of a mere mechanical deposit; and may perhaps owe their origin to precipitation from thermal waters. The shells and crustaceous coverings of the echini are invariably changed into calcareous spar; and in many instances the *terebratulæ* are twisted and contorted in every direction, without the shells exhibiting a single fracture; changes which probably resulted from the influence of a high temperature under considerable pressure.

With the exception of the *pentacrinus*, the teeth of fishes resembling those of the shark, the teeth of crocodiles, and perhaps a few shells†, the organic remains of the chalk differ entirely from all known existing species, as well as from the fossils of other formations. The thickness of the chalk, which is estimated at upwards of 1200 feet, and the immense variety and numbers of its organic remains, evince that the agents which produced it were in full activity through a long period of time.‡ Although

* M. Lamouroux observes, that in the colder latitudes the *Cellarias*, and Sertulariæ alone are to be found; with a few closely woven sponges, and a small number of alcyonia. The minute *Pentacrinus Europæus*, recently discovered by Mr. Thompson in the Cove of Cork, is an exception; but the recent *Pentacrinus Caput Medusæ*, to which the pentacrinal stems that occur in the chalk bear considerable analogy, is found in the sea off the West India Islands.

† Even these few exceptions are very equivocal, and probably the species will hereafter prove to be distinct from their supposed analogues.

‡ The fossils of the chalk of Sussex are enumerated in the Catalogue in the Appendix to this work. In the list of the organic remains of

we have no satisfactory evidence to determine whether the chalk were deposited over the entire surface of the Wealden (as seems most probable), or whether the latter were undergoing elevation during the deposition of the chalk, and were but partially covered by the cretaceous strata, yet there can be no doubt that the chalk originally very much exceeded its present limits. It is true that gravel, and partially rolled flints, occur but rarely on the Wealden, the diluvial covering of the latter chiefly consisting of its own debris; yet this fact may have resulted from the action of the sea during the elevation of the strata, or many other causes, and cannot be admitted as affording conclusive evidence that the Wealden was never wholly covered by the chalk. Our limits will not allow us to examine this interesting question in all its bearings, which will be fully elucidated in the 3d volume of Mr. Lyell's "Principles of Geology," now in the press; and we proceed to the consideration of the next geological era—that in which the older *tertiary* strata were deposited.

The epochs we have already noticed are marked by immense mutations in the relative situation of the land and sea; yet these changes appear to have been effected in such manner as to have occasioned comparatively but little derangement in the strata, and to have been succeeded by periods

---

the cretaceous strata of Europe, given by M. De la Beche, there are, of Reptiles, 6 or 8; Fishes, 10 or 12; Crustacea, 15; Mollusca, 225; Conchifera, 285; Annulata, 110; Radiaria, 90; Vegetable remains, 20 species, 16 of which are marine.

of repose of long duration. In the tertiary era, on the contrary, it is manifest that the disturbing forces were in frequent and violent action, and produced elevations and subsidences, and enormous dislocations and fissures, throughout the whole mass of the strata of the south-east of England. In the anticlinal axis of the Forest ridge, from whence the strata diverge to the south-east in Sussex and the north-west in Kent, we have evidence of a force having acted from beneath, in a direction from east to west, by which the Wealden beds have been elevated above the chalk formation, and the cretaceous strata broken up, and swept away from the whole central area of Kent and Sussex. On these phenomena Dr. Fitton observes that, "whether the fractures and upheavings took place entirely beneath the sea, or after the strata were in part or wholly raised above its surface, at once or at distant epochs, we have no facts to enable us to decide; it is, indeed, not impossible that the very act of rending the strata may itself have effected their protrusion from beneath the waves."* If, however, we consider that the chalk was upwards of 1200 feet in thickness, and extended over the whole southern denudation, it seems probable that elevation and destruction were going on simultaneously. So soon as the first ridge of chalk on the anticlinal line protruded above the surface of the ocean, it would become exposed to the action of the waves; and as elevation proceeded, degradation would

* Geology of Hastings, p. 83.

proceed also, until the whole of the chalk strata were carried away, and the Wealden beds in their turn became exposed to the same destructive agency. The debris of both formations would thus become intermixed and deposited in the hollows of the chalk, giving rise to those accumulations of transported materials of which the tertiary strata are principally composed. During these important and extensive changes, the tertiary ocean which then covered the south-east of England must have been studded with islands, formed by the most elevated portions of the chalk and Wealden*; the marshes of the then existing continent were peopled with tribes of extinct animals allied to the *Tapir* (the *palæotherians*), and the lacustrine formations of Hampshire and the Isle of Wight were deposited.

The organic remains of the tertiary epoch differ entirely from those of the chalk upon which in the south-east of England they repose. In the Isle of Wight, in the Paris Basin, and many contemporaneous deposits on the continent, they consist of alternations of marine and freshwater shells, indicating the existence of lakes communicating with the sea. The ammonites, and other ancient pelagian shells entirely disappear, and a small proportion of recent species occurs in the most ancient, and a much more considerable number in the newer deposits. With these are associated the

* Vide the "Principles of Geology," vol. ii. In the map illustrating the extent of the tertiary sea, or seas, it will be seen that Mr. Lyell has delineated a range of chalk islands in the south-eastern part of England, agreeably to this theory of the gradual elevation of the land.

remains of the Palæotheria, of crocodiles, turtles, birds, and fishes; and the stems and leaves of palms, and other vegetables characteristic of an equatorial climate. In the tertiary strata of the south-east of England, no traces of mammalia have been discovered; the organic remains consisting of shells, the bones and teeth of fishes, and the leaves and stems of vegetables.

The next era is marked by the existence of the fossil elephant, or mammoth, in these latitudes, having for contemporaries a species of deer, ox, and horse; and in other parts of England, the rhinoceros, hippopotamus, &c. The teeth which have been found in Sussex belong to a species nearly allied to the Asiatic elephant, and the deposits in which they occur are decidedly of a more recent date than those above described, for they contain boulders of tertiary sandstone, and breccia; while, in the older tertiary, the remains of the elephant have not been discovered. The perfect state of the teeth in the deposits at Brighton, forbids the supposition that they were transported from a distance: and we have, too, the remarkable fact, that while the shingle on which the elephant bed reposes, is composed not only of chalk pebbles, but of boulders of granite, porphyry, and other primary rocks which must have been brought from a distant part of the country, and of tertiary sandstone and breccia, and the sand beneath contains the bones of whales, no remains of elephants have been found therein. It would seem, therefore, that the sand, and the shingle, were formed in an estuary, and that when the upper beds were deposited, all communication with the ocean was cut off; for neither the bones, nor the materials of which the bed is composed, appear to have suffered from attrition, nor is there any intermixture of marine exuviæ. These deposits were evidently of considerable extent: there are outlying patches on the chalk along the coasts of Sussex and Kent, and also at Etables, and other points on the opposite shores of France. Similar beds occur on the banks of the Loire, and probably the same series is represented by the *Crag*, overlying the London Clay, on the eastern shores of England; facts which tend to prove that the estuary once extended over a considerable portion of the area now occupied by the British Channel.* The geological relations of this group of deposits are as yet but imperfectly known. The zoological characters which distinguish them from the older tertiary strata, are the absence of the *palæotheria*, and the occurrence of the remains of the mammoth, rhinoceros, and other mammalia, whose bones are so constantly found in the superficial gravel of Europe, intermixed with those of recent species.

To this epoch we may probably refer the existence of hyenas, tigers, and other carnivorous animals, whose skeletons are entombed in such immense numbers in caverns, and fissures, and in beds of superficial gravel, in various parts of England,

* Mr. Samuel Woodward, of Norwich, the author of the "Synoptical Table of British Organic Remains" (a work indispensable to the practical geologist), states, that in the crag on the coast of Norfolk, the remains of Mammoths are so abundant, that on the oyster-ground off Harborough, the grinders of these animals which have been found must have belonged to upwards of 500 individuals.

and the continent. One solitary instance only is known of the occurrence of remains of this kind in the south-east of England. The lower jaw and a few fragments of other bones of a hyena were discovered, a few years since, in a chasm in a stone quarry at Boughton, near Maidstone.

The next era is that during which the *Crag*, and the tertiary strata, and the chalk on which they repose, were lifted up to their present situations; the channel which separates England from France was broken through, and the transverse valleys of the north and south downs were produced or enlarged; for, although these valleys are now river courses, yet it is obvious that they originated in disruption, for the strata, in every instance which I have observed, diverge from the line of fracture.* We should doubtless err in assigning all these mutations to one and the same period; the phenomena are extremely complicated, and an appearance which may seem to have been produced at the same time, and by a single operation, may have been the result of many and varied changes. There is, however, one fact respecting which there can be no hesitation, namely, that the disturbing forces which have broken up the tertiary deposits, came into action *after the elephant epoch*. These elevatory movements and convulsions were manifestly of great intensity, and ma-

* Mr. Woodward arrives at the same conclusion from an examination of the chalk valleys of Norfolk. "These," he observes, "are *valleys of disruption;* that is, they were formed by the elevation of the chalk and its consequent fracture, as is evident from the strata of chalk and flints on each side the valley being now found to decline from the line of elevation." — *Correspondence with the Author*.

terially changed the physical geography of the south-east of England, and the contiguous parts of the continent, and occasioned the vertical position of the strata in the Isle of Wight and Hampshire.* These alterations in the surface of the country, must, too, have been attended with great changes in the hydrography of Hampshire, Surrey, Kent, and Sussex; the waters resulting from the drainage of the land, and which, before the existence of the transverse fractures, probably flowed through the longitudinal valleys towards the east, would be thrown into different channels, and find their way to the ocean by the existing river courses.† Traces of these revolutions remain in the boulders and superficial loam and gravel, which occupy the valleys and low elevations of the south-east of England.

Subsequently to these last-mentioned changes the surface of the country appears to have under-

* In other parts of Europe we have also proofs of great vicissitudes of land and sea, and oscillations of level since the commencement of the tertiary period, and that earthquakes have taken place by which the land has not only been lifted from above the waters, but to an elevation of several thousand feet above the level of the sea. Thus the Alps have acquired an additional altitude of from 2000 to 4000 feet; and the Apennines owe a great part of their height, from 1000 to 2000 feet, to subterranean convulsions which have happened within the same epoch. Vide *Principles of Geology*, vol. ii. p. 308. In Mr. Bakewell's *Travels in the Tarentaise, and various Parts of the Grecian and Pennine Alps* (2 vols. 8vo. Longman and Co. 1823), the reader will find many highly interesting remarks on the physical structure of the Alps, and the elevatory movements which they have sustained; as well as on the extinct volcanoes of central France. The Pyrenees have acquired the whole of their present altitude, which in Mount Perdu exceeds 11,000 feet, since the origin of some of the newest members of our *secondary* formations.

† The Cliff Hills, near Lewes, afford an interesting example of the manner in which a large proportion of the chalk valleys have been

gone no material alteration; the ordinary effects of the atmosphere, the degradation of the shores by the action of the sea, the erosion by river currents of the strata over which they flow, and the formation of deltas, and the silting up of valleys, being the only physical changes that have taken place in the south-east of England during the modern epoch, and which are still in active operation.

The existing rivers in this district are producing on a small scale the same effects as the mighty river of the Iguanodon period; bringing down from the interior the debris of the strata over which they flow, mixed with the bones of animals, and the trunks, branches, and leaves of vegetables, and imbedding a portion in the chalk valleys in a deposit of mud or silt, and transporting the remainder to form deltas at their entrance into the ocean.

---

formed. The annexed figure represents the actual section presented by the face of this range towards the town.

CLIFF HILLS, NEAR LEWES.

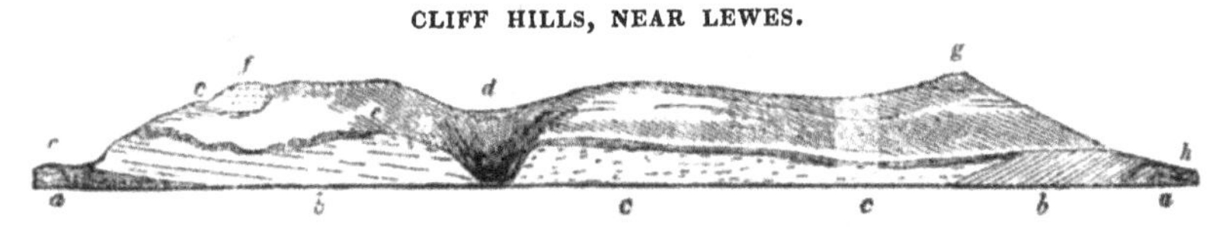

a. a. Chalk marl. *b. b.* Lower chalk. *c. c. c.* Flinty chalk.
d. The Coombe. *e.* Malling road.
f. Bridgwick chalk pit, consisting of the upper chalk.
g. Mount Caburn. *h.* Southerham Corner.

At Malling road, on the north, the chalk marl is seen, inclined towards the south-east, and coming out from beneath the lower chalk. Approaching the deep ravine called *the Coombe*, the *lower chalk* forms the lower two thirds of the cliff, the upper portion being composed of the *flinty chalk*. The southern side of the valley, on the

The levels near Lewes, described in a former part of this volume, afford so interesting an illustration of the silting up of the disrupted valleys of the chalk, during a comparatively very recent period, that we subjoin the following summary of the sequence of events which they record.* First, there was a salt-water estuary peopled for many years by marine testacea identical with existing species, and into which some of the large cetacea, as the sea-unicorn, and porpoise, occasionally entered. Secondly, the inlet grew more shallow, and the water became brackish, or alternately salt and fresh, so that freshwater and marine shells were mingled in the blue argillaceous sediment at the bottom. Thirdly, the shoaling continued until the river water prevailed, and was no longer habitable by marine testacea, but fitted only for the abode of fluviatile species and aquatic insects. Fourthly, a peaty swamp or morass was formed, into which trees and terrestrial animals, as deer, were occasionally drifted by land-floods. Lastly, the soil, being only subject to periodical inundations from the river, became a verdant plain, through which the narrow Ouse now winds its way to the British

contrary, is made up entirely of the *flinty chalk*, proving that the valley coincides with a line of fault. The flinty chalk continues, with a slight southern declination, to near Southerham corner, where the *lower chalk* is again brought to view, and is seen tilted up in the chalk-pit near the turnpike-gate, and dipping at a considerable angle towards the *north;* the *chalk-marl* appearing beneath it. This section is a beautiful illustration, on a small scale, of the faults and dislocations produced by elevations and subsidences; and the *Coombe* is a fine instance of a valley formed by the derangement of the strata.

* Principles of Geology, vol. ii. p. 276.

Channel. It is in alluvial deposits of this kind *that the remains of man first appear:* human skeletons, and the rude instruments of a half-civilised race, are found associated with the bones of animals which still inhabit this country, and in some instances intermixed with the osseous remains of a few species that appear to have been extirpated by man.

Such are the results which a review of the geological phenomena of the south-east of England offers to our consideration. We have evidence of great physical mutations of the surface of the earth,—of vast changes in the temperature of the climate; and we perceive that these revolutions were accompanied by a corresponding alteration in the forms of organic life: these are general conclusions, which cannot be disputed, although the laws that governed these co-existing phenomena may be concealed from our view. It is, however, obvious, that the great changes which have taken place in the relative proportion of the land and water, must have materially influenced the temperature of the climate, and consequently the geographical distribution of animals and vegetables. Mr. Lyell has treated this question in a very luminous and admirable manner, and has shown that there is every reason to conclude that since the commencement of the tertiary period, the dry land in the northern hemisphere has been increasing; not only because it is now greatly in excess beyond the average proportion which land generally bears to water on our planet, but because a comparison of the secondary and tertiary strata affords indications throughout the space occupied by Europe, of a transition from the condition of an ocean, interspersed with islands, to that of a large continent: and to this increase of the land in the northern hemisphere we may probably attribute, in a great measure, that gradual diminution of temperature which the organic remains of the different periods denote. "The climate was hottest when the northern hemisphere was for the most part occupied by the ocean; and the refrigeration did not become considerable until a very large proportion of that ocean was converted into land, and replaced in some parts by high mountain chains: nor did the cold reach its maximum until these chains attained their greatest elevation, and the land its utmost extension."*

* Lyell's Principles of Geology, vol. i. chap. vii. The remarks of one of our most eminent astronomers on this beautiful theory of Mr. Lyell are so pertinent, that I cannot omit them. "The fact of a great change in the general climate of large tracts of the globe, if not of the whole earth, and of a diminution of general temperature, having been recognised by geologists, from their examination of the remains of animals and vegetables of former ages enclosed in the strata, various causes for such diminution of temperature have been assigned." After observing on the theories which attempt to explain the present decrease of temperature, by supposing that the whole globe has gradually cooled from a state of absolute fusion; and that, in former ages, there was an immensely superior activity in volcanic action, Sir John Herschel proceeds to remark:—"Neither of these can be regarded as real causes in the sense here intended; for we do *not* know that the globe has so cooled from fusion, *nor* are we *sure* that such supposed superior activity of former than of present volcanoes really did exist. A cause, possessing the essential requisites of a *vera causa*, has, however, been brought forward by Mr. Lyell, in the varying influence of the distribution of land and sea over the surface of the globe: a change of such distribution, in the lapse of ages, by the degradation of the old continents, and the elevation of new, being a demonstrated fact; and the influence of such a change on the climates of particular regions, if not

The changes that have taken place in the forms of the animal and vegetable kingdoms, are not less striking than those which we have above described in the inorganic world. The animals and plants of the more ancient strata, are not only such as could not now exist in the latitudes which they formerly inhabited, but almost all the species, and very many of the genera, are no longer to be found in any part of the known globe. In the newer deposits, on the contrary, we perceive an intermixture of existing with extinct species; the proportion of the former increasing according to the more recent formation of the strata, till, in the deposits of the modern era, the remains of existing species alone are discovered, and, as we have already remarked, in these accumulations of débris, the skeletons of man, and traces of the works of art of the early tribes of our race, are sometimes found imbedded.

The extinction of whole genera of animals and plants has, no doubt, depended on various causes. In the earlier revolutions, the vicissitudes of climate, and the mutations of land and water, were, pro-

---

of the whole globe, being a perfectly fair conclusion, from what we know of continental, insular, and oceanic climates by actual observation. *Here, then, we have, at least, a cause on which a philosopher may consent to reason;* though, whether the changes actually going on are such as to warrant the whole extent of the conclusion, or are even taking place in the right direction, may be considered as undecided till the matter has been more thoroughly examined." Another possible cause of the refrigeration of the earth is to be found in the "diminution of the excentricity of the earth's orbit round the sun; and which, as a general one, affecting the mean temperature of the whole globe, and as one of which the effect is both inevitable, and susceptible, to a certain degree, of exact estimation, deserves consideration."—*Discourse on the Study of Natural Philosophy*, p. 146, 147.

bably, the principal agents of destruction: but since man became the lord of the creation, his necessities and caprice have occasioned the extirpation of many tribes, whose relics are found in the same superficial strata with those of species concerning which all human history and tradition are silent.*

The obliteration of certain forms of animal life (and perhaps the creation of new ones) appears, therefore, to be dependent on a law in the economy of nature, which is still in active operation. Of this we have a remarkable instance in the case of the *Dodo*, which has been annihilated, and become a denizen of the fossil kingdom, almost before our eyes. The Dodo was a bird of the gallinaceous tribe, larger than the turkey, which existed in great numbers in the Mauritius and adjacent islands, when those countries were first colonised by the Dutch, about two centuries ago. This bird was the principal food of the colonists; but it was incapable of domestication, and its numbers soon became sensibly diminished. Stuffed specimens were sent to the museums of Europe, and paintings of the living animal were executed, and copied into the works on natural history. *The Dodo is now extinct:* it is no longer to be found in the isles where it once flourished, and even all

* In Great Britain, we may instance, as belonging to species which formerly existed in this country, and are still living in other parts of the globe, the *beaver*, *bear*, *wolf*, *hyena*, &c.; and, as wholly extinct, the *Irish Elk*, and *Mammoth*, with whose bones existing species of shells are sometimes found associated. Consult *Dr. Fleming's British Animals*, 1 vol. 8vo. 1828: also an excellent Memoir, by the same author, in the Edinburgh Philosophical Journal, No. xxii.

the stuffed specimens are destroyed; the only relics that remain being the head and foot of an individual in the Ashmolean museum at Oxford, and the leg of another in the British museum. To render this history complete, the fossilised remains were alone wanting, and these have actually been found beneath a bed of lava in the Isle of France, and are now in the museum of the Jardin des Plantes at Paris; affording the most unexpected and conclusive evidence of the truth of what was formerly considered one of the most startling propositions in modern geology.*

THE DODO INEPTUS,

from a painting of the original.

* See an excellent paper on the dodo, by Mr. Duncan, *Zoological Journal for January*, 1828: also, " Contributions towards the History of the Dodo (*Didus ineptus*), by J. V. Thompson, Esq., *Mag. Nat. Hist.*, vol. ii. p. 442.; and Mr. Lyell's Principles of Geology, vol. ii. p. 151.

THE FOOT OF THE DODO

which is still preserved in the British Museum.

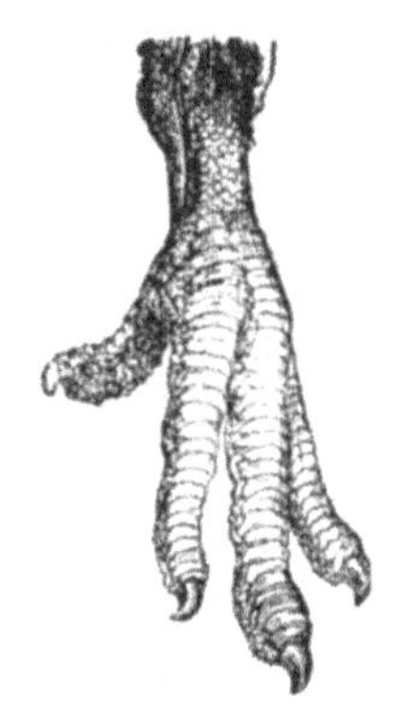

Another highly interesting and important fact is proved by the phenomena that have been presented to our examination, namely, the comparatively recent period at which man became an inhabitant of the earth, and exercised dominion over the animal creation; a fact in strict accordance with those sacred records which reveal the moral obligations and destiny of the human race.

With these observations I conclude this volume; entreating the indulgence of the geologist for much prolixity on subjects with which he was already familiar, but which without such detail would have presented but little interest to many; and assuring the general reader who may feel desirous of further information, that the more he becomes acquainted with the nature and objects of geological enquiries, the more he will find them to possess in an eminent degree the charms and advantages which are so eloquently described by Sir John Herschel, as being inseparably connected with the study of every branch of natural philosophy. "To the natural philosopher there is no natural object unimportant or trifling. From the least of nature's works he may learn the greatest lessons. The fall of an apple to the ground may raise his thoughts to the laws which govern the revolutions of the planets in their orbits; or *the situation of a pebble may afford him evidence of the state of the globe he inhabits, myriads of ages ago, before his species became its denizens*....Accustomed to trace the operation of general causes, and the exemplification of general laws, where the uninformed and unenquiring eye perceives neither novelty nor beauty, he walks in the midst of wonders: every object which falls in his way elucidates some principle, affords some instruction, and impresses him with a sense of harmony and order; while the observation of the calm, energetic regularity of nature, the immense scale of her operations, and the certainty with which her ends are attained, tends, irresistibly, to tranquillise and re-assure the mind, and render it less accessible to repining, selfish, and turbulent emotions. And this it does, not by debasing our nature into weak compliances and abject submission to circumstances, but by filling us, as from an inward spring, with a sense of nobleness and power which enables us to rise superior to them; by showing us our strength and innate dignity, and by calling upon us for the exercise of those powers and faculties by which we are susceptible of the comprehension of so much greatness, and which form, as it were, a link between ourselves and the best and noblest benefactors of our species, with whom we hold communion in thoughts, and participate in discoveries, which have raised them above their fellow-mortals, and brought them nearer to their Creator."*

* Discourse on the Study of Natural Philosophy, p. 14—17.

APPENDIX.

A Tabular Arrangement of the Organic Remains of the County of Sussex.

Alluvial Deposits.

NAMES.	REFERENCES.*	LOCALITIES.
Subterranean Forests.	Geol. Suss. 288, 289.	Felpham, near Bognor. Pevensey Levels.
PEAT.		
(*a*) Consisting of the remains of fresh-water and marsh plants, trunks and branches of trees, hazle-nuts, &c.	Ibid. 287—290.	Lewes and Arundel Levels. The Wish, near Eastbourn. Little Horsted, Isfield.
(*b*) Consisting of the remains of marine plants, confervæ, fuci, &c.	. . . .	Pevensey Levels.
Blue Clay or Silt.	Ibid. 286.	Lewes Levels.

* The fossils which are not in the possession of the author, and those not examined by him, are marked by an asterisk.

CLASS.	GENUS.	SPECIES.	REFERENCES.	LOCALITIES.
FRESH-WATER.				
Insecta.	Phryganea.[a]		. . . .	Lewes Levels.
Conchifera.	Cyclas	cornea.	Geol. Suss. p. 287.[b]	Ibid.
Mollusca.	Succinea	amphibia.		
——	Planorbis	carinatus.		
——	——	corneus.		
——	Limnea	stagnalis.		
——	——	palustris.		
——	——	limosa.		
——	Valvata	piscinalis.		
——	Paludina	impura.		
MARINE.				
Conchifera.	Lutraria	compressa.	Ibid. 287.[c]	Ibid.
——	Tellina	solidula.		
——	Cardium	edule.		
Mollusca.	Turbo	ulvæ.		
Mammalia.	Cervus.		Cuv. Oss. Foss. iv. Pl. 3. f. 16. . . .	In sand several feet beneath the bed of the Ouse, Lewes Levels; in alluvium.[d]

[a] The indusia or cases of the larvæ of this genus of insects, with minute shells of the genera Planorbis, Limnea, &c. adhering to them, are very abundant in the silt or blue clay.

[b] Still inhabit the rivers and ditches.

[c] Exist in the neighbouring ocean.

[d] The entire skeleton. A species allied to the Canadian, figured by Cuvier.

CLASS.	GENUS.	SPECIES.	REFERENCES.	LOCALITIES.
Mammalia.	Monodon	Monoceros.	. . . .	Lewes Levels.[e]
———	Delphinus	Phocœna.	. . . .	Mouth of the Cuckmere.[f] Beeding Levels.[g]

Diluvium.

No vegetable remains have been noticed in these deposits.
Testacea very rare.

CLASS.	GENUS.	SPECIES.	REFERENCES.	LOCALITIES.
Conchifera.	Mytilus.		Geol. Suss. 277.	Cliffs between Brighton and Rottingdean.[h]
———	Modiola.			
———	Nerita.			
Mollusca.	Trochus.			
Mammalia.	Balæna.		. . . .	Base of the cliff near Kemptown, Brighton.[i]
———	Bos.		. . . .	Brighton Cliffs.[k]

[c] Portion of the skull.

[f] The skull eighteen inches long; dug up at a depth of ten feet in blue clay.

[g] Human skeletons, in coffins of very rude workmanship, have been found in the silt at the depth of several feet; the bones and teeth were of a deep chocolate colour like those of the deer, &c. above mentioned, and coated in parts with blue phosphate of iron.

[h] In a confused mass, resting on the shingle bed.

[i] Portion of the lower jaw (12 feet long, 32 inches in circumference) in the sand beneath the shingle bed, and resting on the chalk.

[k] Teeth.

CLASS.	GENUS.	SPECIES.	REFERENCES.	LOCALITIES.
Mammalia.	Cervus.		. . . .	Brighton Cliffs. Copperas Gap.
———	Equus.		Geol. Suss. 277.	Ibid.[m]
———	Elephas	primigenius.	Ibid. 203.	Peppering near Arundel.[n]
			Ibid. 278.	Brighton Cliffs. Brick-yard, near Hove.[o]

Tertiary Formations. *Térrains de sédiment superieurs.*

London Clay. *Calcaire Grossier. Prémier Térrain Marine.*

1. *Blue Clay of Bracklesham.* Geology of Sussex, p. 268.

CLASS.	GENUS.	SPECIES.	REFERENCES AND SYNONYMS.	LOCALITIES.
Annelides.	Dentalium.[p]		. . . .	Bracklesham Bay.
Conchifera.	Crassatella	sulcata.	Min. Con. Tab. 345. f. 1.	Ibid.
———	———	compressa.	Lam. Coq. Foss. Env. de Paris, Pl. 18. f. 1.	
———	Corbula.			
———	Sanguinolaria	Hollowaysii.	Min. Con. Tab. 159.	
———	Cytherea	nitidula.	Lam. Coq. Foss. Env. de Paris, Pl. 21. f. 1. 2.	
———	Venericardia	planicosta.	Min. Con. Tab. 50.	
———	———	acuticosta.	Lam. Coq. Foss, Env. de Paris, Pl. 20. f. 2.	
———	Cardium	semigranulatum.	Min. Con. Tab. 144.	
———	Pectunculus	pulvinatus.	Lam. Coq. Foss. Env. de Paris, Pl. 16. f. 9.	
———	Chama	squamosa.	Min. Con. Tab. 348.	

[l] Jaw, teeth, and bones.
[m] Bones and teeth.
[n] The skeleton.
[o] Bones and teeth.
[p] Resembles D. entale.

CLASS.	GENUS.	SPECIES.	REFERENCES AND SYNONYMS.	LOCALITIES.
Conchifera.	Ostrea	Flabellula.	Min. Con. Tab. 253. Chama plicata, Brander Foss. Hant. f. 84, 85.	Bracklesham Bay.
———	———	undetermined.		
Mollusca.	Melania	sulcata.	Min. Con. Tab. 39.	Ibid.
———	———	costellata?	Lam. Coq. Foss. Env. de Paris, Pl. 12. f. 2.	
———	Ampullaria	patula.	Min. Con. Tab. 284. (Two middle figures.)	
———	———	sigaretina.	Ibid. Tab. 284. (Two lower figures.)	
———	Natica	similis.	Ibid. Tab. 5. (Two middle figures.)	
———	Scalaria	acuta.	Ibid. Tab. 16. f. 4, 5. (Two lower figures.)	
———	Solarium	canaliculatum.	Brander Foss. Hant. f. 7, 8.	
———	Trochus	agglutinans.	Min. Con. Tab. 98. f. 1. (Two smaller figures.)	
———	Turritella	multisulcata.	Lam. Hist. Nat. Anim. sans Vert. vii. 562.	
———	———	conoidea.	Min. Con. Tab. 51. f. 1. 4.	
———	———	brevis.	Ibid. Tab. 51. f. 3.	
———	———	elongata.	Ibid. Tab. 51. f. 2.	
———	Cerithium	Cornucopiæ.	Ibid. Tab. 188. f. 1. 3, 4.	
———	Pleurotoma[q]			
———	Fusus.	longævus.	Ibid. Tab. 63.	
———	Pyrula	bulbiformis?	Fusus bulbif. Min. Con. Tab. 291. Murex Pyrus, Brander Foss. Hant. f. 52, 53.	
———	———	lævigata.	Lam. Coq. Foss. Env. de Paris. Pl. 4. f. 7.	
———	Murex	argutus.	Brander Foss. Hant. f. 13.	
———	Voluta	Luctator.	Min. Con. Tab. 115. f. 1.	

[q] Too imperfect to determine the species.

CLASS.	GENUS.	SPECIES.	REFERENCES AND SYNONYMS.	LOCALITIES.
Mollusca.	Voluta	Bicorona.	Lam. Hist Nat. Anim. sans Vert. vii. 351.	Bracklesham Bay.
———	Ancilla	aveniformis.	Min. Con. Tab. 99. f. 1, 2. (Middle figures.)	Ibid.
———	———	Turritella.	Ibid. Tab. 99. f. 3, 4. (Larger figures.)	
———	———	canalifera.	Lam. Coq. Foss. Env. de Paris, Pl. 2. f. 6.	
———	Conus	Dormitor.	Min. Con. Tab. 301.	
———	Nummularia	lævigata.[r]	Ibid. Tab. 538. f. 1.	
Pisces.	Raia.[s]		Brander Foss. Hant. f. 117.	
			Ibid. f. 109.[t]	

2. *Arenaceous Limestone or Sandstone of Bognor.*

CLASS.	GENUS.	SPECIES.	REFERENCES AND SYNONYMS.	LOCALITIES.
Polypi.[x]			. . . .	The rocks on the Coast near Bognor.
Annelides.	Dentalium	planum.	Min. Con. Tab. 79. f. 1.	
———	Serpula.		Geol. Trans. 1st series, ii. 205.	
———	Vermetus	Bognoriensis.	Geol. Suss. 272. Hist. Suss. vol. iii. Pl. 1. f. 3. Min. Con. Tab. 596. f. 1, 2, 3.	
Conchifera.	Fistulana	personata.	Lam. Coq. Foss. Env. de Paris, Pl. 24. f. 6, 7, 7*a*, 7*b*. Teredo antenautæ, Min. Con. Tab. 102.	
———	Pholadomya	margaritacea.	Min. Con. Tab. 297. f. 1. 2, 3.	
———	Panopæa	intermedia.	Ibid. Tab. 76. f. 1. and Tab. 419. f. 2.	
———	Lutraria ?	oblata.	Ibid. Tab. 534. f. 3.	

[r] The young shell, of the size of a millet-seed, occurs in immense quantities. [s] Palates.
[t] Vertebræ. [u] Dicotyledonous wood perforated by Fistulana personata occurs occasionally in large masses.
[x] A ramose zoophyte, genus undetermined.

CLASS.	GENUS.	SPECIES.	REFERENCES AND SYNONYMS.	LOCALITIES.
Conchifera.	Venericardia	Brongniarti.[y]	Not figured.	
———	Cardium?			Bognor Rocks.
———	Pectunculus	breverostris.	Min. Con. Tab. 472. f. 1.	Ibid.
———	———	decussatus.[z]	Ibid. Tab. 27. f. 1.	
———	Modiola	elegans.	Ibid. Tab. 9. f. 5.	
———	Pinna	affinis.	Ibid. Tab. 313. f. 2.	
———	*Ostrea	edulis?	Webster, Geol. Trans. 1st series, ii. 205.	
———	Anomia	lineata.	A. striata, Min. Con. Tab. 425.	
———	Lingula	tenuis.	Min. Con. Tab. 19. f. 3.	
Mollusca.	Calyptræa	trochiformis.	Trochus apertus, Brander Foss. Hant. f. 1, 2.	
———	Ampullaria	patula.	Min. Con. Tab. 284. (Two middle figures.)	
———	———	sigaretina.	Ibid. Tab. 284. (Two lower figures.)	
———	Natica	similis.	Ibid. Tab. 5. (Two middle figures.)	
———	Pyrula.		Brander, Foss. Hant. f. 52, 53.	
———	Murex	Smithii.	Min. Con. Tab. 578. f. 1, 2, 3.	
———	Rostellaria	Sowerbii.[a]	R. Parkinsoni, Min. Con. Tab. 349. f. 1. 3, 4.	
———	Voluta	Luctator.	Min. Con. Tab. 115. f. 1.	

[y] The specific name is in honour of M. Alex. Brongniart.

[z] As it differs from the recent P. decussatus, (see Turton's Brit. Bivalves, 173,) a different specific name should be imposed.

[a] This Rostellaria was figured and described by the late Mr. Sowerby as R. Parkinsoni of the Geol. of Sussex; it is, however, perfectly distinct, and it becomes necessary to adopt a different specific name; that of Sowerbii is here given as a tribute of respect to the present scientific editor of the Mineral Conchology of Great Britain. The R. Sowerbii occurs in the tertiary formations only; R. Parkinsoni in the chalk marl and Shanklin sand.

CLASS.	GENUS.	SPECIES.	REFERENCES AND SYNONYMS.	LOCALITIES.
Mollusca.	Conus.			Bognor rocks.
———	Nautilus	imperialis.	Min. Con. Tab. 1.	
Pisces.	Squalus ?[b]			

3. *Sand on Emsworth Common.*

CLASS.	GENUS.	SPECIES.	REFERENCES AND SYNONYMS.	LOCALITIES.
Annelides.	Dentalium	cylindricum.	Min. Con. Tab. 79.	
Mollusca.	*Nummularia	elegans.	Ibid. Tab. 538. f. 2.	

PLASTIC CLAY. *Argile Plastique. Prémier Terrain d'eau douce.* Castle Hill, Newhaven.

Leaves; remains and impressions. Geol. Suss. Tab. 8. f. 1, 2, 3, 4. Brit. Min. Tab. 500. These are supposed by Mr. Sowerby to resemble the larger foliage of Platanus orientalis, Geol. Suss. p. 262.
"Fruit of a species of Palm?" Webster, Geol. Trans. 1st series, vol. ii. p. 191.
Wood; dicotyledonous. Occurs in small fragments in the reddish brown sandy-marl. Geol. Suss. p. 257. No. 7.

CLASS.	GENUS.	SPECIES.	REFERENCES AND SYNONYMS.	LOCALITIES.
Conchifera.	Cyclas.		Brit. Min. Tab. 500. iv. 185. . . .	Castle Hill, near Newhaven.
———	Cyrena.		Geol. Suss. 264.	
———	Cytherea	convexa.	Ibid. Pl. 25. f. 2. Desc. Geol. Env. de Paris, Pl. 8. f. 7. p. 282. (Edit. 1822.)	
———	Unio.		Brit. Min. Tab. 500.	
———	Avicula	media.	Min. Con. Tab. 2. f. 2.	
———	Ostrea.		Geol. Suss. 264.	

[b] A small tricuspid tooth.

[c] Very rarely perfect.

CLASS.	GENUS.	SPECIES.	REFERENCES AND SYNONYMS.	LOCALITIES.
Mollusca.	Hælix	lævis.[d]	Geol. Suss. Tab. 18. f. 19, 20. H. lævis, Fleming Brit. Anim. 265.	Castle Hill, near Newhaven.
———	Cerithium	funatum.	Geol. Suss. Tab. 14. f. 4. Min. Con. Tab. 128.	
———	———	politum.[e]	C. melanoides, Geol. Suss. Tab. 18. f. 3. Min. Con. Tab. 147.	
Pisces.	Mustelus ?[f]		Geol. Suss. 264.	

Chalk Formation.

1. *Chalk with Flints.* 2. *Chalk without Flints.* (Craie blanche.)

CLASS.	GENUS.	SPECIES.	REFERENCES AND SYNONYMS.	IN SUSSEX.	ELSEWHERE.
Agamia.	Confervites	fasciculata[g]	Ad. Brong. Hist. Verg. Foss. Pl. 1. f. 1. .	Lewes. Steyn-[ing.	Isle de Bon-[holm.
———	———	undetermined.	Geol. Suss. Pl. 9. f. 12.	Ibid.	
———	Fucoides	Brongniarti.[h]	Ibid. Pl. 9. f. 1.	Ibid.	
———	———	undetermined.	. . . .	Ibid.	
Phanerogamia. (Dicotyledonous)	Wood.[i]		Geol. Suss. 157.	Ibid.	

[d] Bath is named by Dr. Fleming as its locality, evidently by mistake.

[e] In Min. Con. vol. iv. p. 48., Mr. Sowerby expresses a doubt whether these shells should not be referred to Potamides: Dr. Fleming, however, retains them in Cerethium, see Brit. Anim. p. 358.

[f] Teeth resembling those of this species.

[g] In flint and chalk: very rare.

[h] Specific name in honour of M. Adolphe Brongniart, author of the Hist. Veget. Foss. *Vide* p. 96.

[i] Sometimes in flint nodules, and perforated by Fistulanæ or Teredines: in chalk, in the state of a brown friable mass; rare.

CLASS.	GENUS.	SPECIES.	REFERENCES AND SYNONYMS.	IN SUSSEX.	ELSEWHERE.
Polypi.	Flustra	utricularis.[k]	Lam. Hist. Nat. Anim. sans Vert. ii. 224. König. Icon. Foss. Sect. f. 61.	Lewes.	
———	———	undetermined.			
———	Orbitolites	lenticulata.	Geol. Suss. Tab. 16. f. 22—24. Desc. Geol. Env. de Paris. Pl. 9. f. 4. (Edit. 1822.)	Ibid.	
———	Caryophyllia	centralis.	Madrepora centralis, Geol. Suss. Tab. 16. f. 2. 4.	Ibid.	Heytesbury.
———	Madrepora.		Geol. Suss. 160.	Brighton.	
	Spongia	ramosa.[l]	Geol. Suss. Tab. 16. f. 11.	Lewes.	Warminster.
———	———	lobata.	Parkin. Org. Rem. ii. Pl. 7. f. 6. Fleming, Brit. Anim. 526.	Ibid.	
———	Spongus	Townsendi.[m]	Geol. Suss. Tab. 15. f. 9.	Lewes.	Ibid.
———	———	labyrinthicus.	Ibid. Tab. 15. f. 7. S. hemisphærica, Flem., Brit. Anim. 526.	Ibid.	Heytesbury.
———	Polypothecia.	several species[n] undetermined.	Miss Benett's Cat. Foss. Wilts.	South Downs.	Wilts.
———	Siphonia.[o]		Parkin. Introd. Org. Rem. 52.	Ibid.	
———	Alcyonium.		Geol. Suss. Tab. 15. f. 4, 5. Tab. 16. f. 17, 18.	Ibid.	
———	Choanites	subrotundus.	Ibid. Tab. 15. f. 2.	Ibid.	
———	———	Königi.[p]	Ibid. Tab. 16. f. 19—21.	Ibid.	Warminster.

[k] Attached to Echini. Ventriculites quadrangularis, Geol. Suss. Tab. 15. fig. 6. probably belongs to this genus. There are also several undetermined species.

[l] Common in flints.

[m] This species is probably distinct from the Pewsey cup-corals.

[n] Vast numbers of the flints derive their forms from the sponges they enclose. Common in flints.

[p] Radiated spicula are observable in some specimens.

CLASS.	GENUS.	SPECIES.	REFERENCES AND SYNONYMS.	IN SUSSEX.	ELSEWHERE.
Polypi.	Choanites	flexuosus.[q]	Geol. Suss. Tab. 15. f. 1.	Lewes.	
———	Ventriculites	radiatus.[r]	Ibid. Tab. 10, 11, 12, 13. Linn. Trans. vol. xi. Mantellia radiata, Parkin. Introd. Org. Rem. 53.	Ibid.	
———	———	alcyonoides.	Geol. Suss. 176. Smith's Strata, Tab. 3. f. 1. Ocellaria König, Icon. Foss. Sect. f. 98, 99.	Ibid.	Ibid.
———	———	Benettiæ.	Geol. Suss. Tab. 15. f. 3.	Ibid.	
Radiaria.	Apiocrinites	ellipticus.	Miller, Hist. Crin. p. 34. Bottle Encrinite, Parkin. Org. Rem. ii. Pl. 13. f. 31. 34. 75.	Ibid.	Northfleet.
———	Pentacrinites.[s]				
———	Marsupites	Milleri.[t]	Geol. Suss. Tab. 16. f. 6. Mil. Hist. Crinoid. 133.	Brighton.	Ibid.
———	Pentagonaster	semilunatus.[u]	Parkin. Org. Rem. iii. Pl. 1. f. 1. . . .	Lewes.	
———	———		. . . .	Ibid.[x]	
———	Cidaris	cretosa.[y]	Ibid. iii. Pl. 4. f. 3. Pl. 1. f. 11.	Ibid.	Ibid.
———	———	variolaris.	Desc. Geol. Env. de Paris, Pl. 5. f. 9. .	Ibid.	
———	———	corollaris.	Geol. Suss. Tab. 17. f. 2. Parkin. Org. Rem. iii. Pl. 1. f. 7.	Ibid.	
———	Echinus	saxatilis.	Parkin. Org. Rem. iii. Pl. 3. f. 1. . .	Ibid.	
———	———	Königi.	Geol. Suss. 189. Parkin. Org. Rem. iii. Pl. 1. f. 10.	Ibid.	

[q] In flints.

[r] Immense quantities of flints owe their forms to this genus of Polypi.

[s] Portion of a stem resembling that of P. Caput-Medusæ. Mill. Hist. Crinoid. p. 46. probably distinct.

[t] Tortoise Encrinite of Parkinson.

[u] Very rare.

[x] Detached ossicula too imperfect to admit of determination.

[y] It differs essentially from C. papillata of the Oolite.

CLASS.	GENUS.	SPECIES.	REFERENCES AND SYNONYMS.	IN SUSSEX.	ELSEWHERE.
Radiaria.	Echinus	Spines belonging to four or more species.	Geol. Suss. Tab. 17. f. 12—14. Parkin. Org. Rem. ii. Pl. 4. f. 19, 20. . .	Lewes.	
———	Spatangus	Cor anguinum.	Desc. Geol. Env. de Paris, Pl. 4. f. 11. Parkin. Org. Rem. iii. Pl. 3. f. 11. . .	Ibid.	Meudon.
———	———	rostratus.	Geol. Suss. Tab. 17. f. 10. 17. . . .	Brighton.	
———	———	planus.	Ibid. Tab. 17. f. 9, 21.	Lewes.	
———	———	Prunella ?	Ibid. Tab. 17. f. 22, 23.	Brighton.	
———	Conulus	Albogalerus.	Ibid. Tab. 17. f. 8. 20. Galerites albogalerus, Desc. Geol. Env. de Paris, Pl. 4. f. 12.	Lewes.	Dieppe.
———	———	vulgaris.[z]	Parkin. Org. Rem. iii. Pl. 2. f. 3. . .	South Downs.	
———	———	subrotundus.	Geol. Suss. Tab. 17. f. 15. 18. . . .	Lewes.	
———	Echinocorys	scutatus.	Parkin. Org. Rem. iii. Pl. 2. f. 4. Ananchytes, Lamarck	Ibid.	Meudon.
———	———	ovatus.	Desc. Geol. Env. de Paris, Pl. 5. f. 7. Ananchytes ovata, Lamarck	Ibid.	Ibid.
———	———	hemisphæricus.	Desc. Geol. Env. de Paris Pl. 5. f. 8. . .	South Downs.	Toigny.
Crustacea.	Astacus	Leachii.	Geol. Suss. Tab. 29, 30, 31. . . .	Lewes.	
———	———	Sussexiensis.	Ibid. Tab. 30. f. 3.	Ibid.	
Crustacea.	Pagurus	Faujasii ?	Ibid. Tab. 29. f. 3. Brongniart, Hist. Crust. Foss. Pl. 11. f. 2.	Lewes.	Mæstricht.
———	Scyllarus	Mantelli.	Brongniart, Hist. Crust. Foss. 130. . .	Ibid.	
———	Eryon.[a]		Ibid. 128. Geol. Suss. Tab. 29. f. 2. .	Steyning.	
Annelides.	Serpula	ampullacea.	Geol. Suss. 196. Min. Con. Tab. 596. f. 1.	Lewes.	

[z] Quere if specifically distinct ?

[a] Too imperfect for the species to be ascertained.

CLASS.	GENUS.	SPECIES.	REFERENCES AND SYNONYMS.	IN SUSSEX.	ELSEWHERE.
Annelides.	Serpula.	Plexus.	Geol. Suss. 196. Min. Con. Tab. 598. f. 1.	Lewes.	
———	Spirorbis.		Desc. Geol. Env. de Paris, 251. . .	Ibid.	Meudon.
Cirripeda.	Pollicipes	sulcatus.	Min. Con. Tab. 606. f. 7. Geol. Suss. Tab. 33. f. 11.	Ibid.	Norwich.
Conchifera.	Fistulana	personata.	Geol. Suss. Tab. 18. f. 23. Lamarck. Coq. Foss. Env. de Paris, Pl. 24. f. 6, 7, 7*a*, 7*b*, Teredo antenautæ, Min. Con. Tab. 102. .	Ibid.	
———	Inoceramus	Cuvieri.	Geol. Suss. Tab. 27. f. 4. Catillus C., Desc. Geol. Env. de Paris, Pl. 4. f. 10. (Edit. 1822.)	Ibid.	Meudon.
———	———	Brongniarti.	Geol. Suss. Tab. 27. f. 8.	Ibid.	Warminster.
———	———	Lamarckii.	Ibid. Tab. 27. f. 1. Geol. Trans. 1st series, v. Pl. 1. f. 3.	Ibid.	Dover.
———	———	mytiloides.	Geol. Suss. Tab. 28. f. 2. Mytaloides labiatus, Desc. Geol. Env. de Paris, Pl. 3. f. 4. (Edit. 1822.)	Ibid.	Bougival.
———	———	cordiformis.	Min. Con. Tab. 440.	Ibid.	Gravesend.
———	———	latus.	Geol. Suss. Tab. 27. f. 10. Min. Con. Tab. 582. f. 1.	Ibid.	Norfolk.
———	———	Websteri.	Geol. Suss. Tab. 27. f. 21.	Ibid.	
———	———	striatus.	Ibid. Tab. 27. f. 5. Min. Con. Tab. 582. f. 2.	Ibid.	Heytesbury.
———	———	undulatus.	Ibid. Tab. 27. f. 6.	Ibid.	Ibid.
———	———	involutus.	Min. Con. Tab. 583.	Ibid.	
———	Plagiostoma	spinosum.[b]	Geol. Suss. Tab. 26. f. 10. Min. Con. Tab. 83.	Ibid.	Meudon, Rouen.

[b] One of the most characteristic shells of the chalk; p. 125.

CLASS.	GENUS.	SPECIES.	REFERENCES AND SYNONYMS.	IN SUSSEX.	ELSEWHERE.
Conchifera.	Plagiostoma	Hoperi.	Geol. Suss. Tab. 26. f. 2, 3. 15. P. Mantelli, Desc. Geol. Env. de Paris, Pl. 4. f. 3. (Edit. 1822.)	Lewes.	Rouen.
——	——	Brightoniensis.[c]	Geol. Suss. Tab. 26. f. 15.	Brighton.	
——	Pecten	quinquecostatus.	Ibid. Tab. 26. f. 14. 20. Min. Con. Tab. 56. f. 4. 8.	Lewes.	Meudon.
——	——	nitidus.[d]	Geol. Suss. Tab. 26. f. 4. 9. . . .	Ibid.	Dieppe.
——	——	undetermined.	Ibid. Tab. 25. f. 4.	Ibid.	
——	Dianchora	lata.[e]	Ibid. Tab. 26. f. 21. Podopsis, Desc. Geol. Env. de Paris, (Edit. 1822.) . .	Ibid.	Le Havre.
——	——	obliqua	Geol. Suss. Tab. 25. f. 1. Tab. 26. f. 12. .	Brighton.	Ibid.
——	Ostrea	vesicularis.[f]	Desc. Geol. Env. de Paris, Pl. 3. f. 5. (Edit. 1822.) Gryphæa globosa, Min. Con. Tab. 392.	Lewes.	Meudon.
——	——	semiplana.	Geol. Suss. Tab. 25. f. 4. Min. Con. Tab. 489. f. 3.	Ibid.	
——	——	canaliculata.	Min. Con. Tab. 135. f. 1.	Ibid.	Cromer.
——	Crania	Parisiensis.	Ibid. Tab. 409. f. 1. Desc. Geol. Env. de Paris, Pl. 3. f. 2. (Edit. 1822.)	Brighton.	Meudon.
——	Terebratula	subrotunda.[g]	Min. Con. Tab. 15. f. 1, 2.	Lewes.	
——	——	carnea,	Desc. Geol. Env. de Paris, Pl. 4. f. 7. .	Ibid.	Ibid.
——	——	ovata.	Min. Con. Tab. 15. f. 3. . . .	Ibid.	Rouen.

[c] Very rare.

[d] P. cretosus, Min. Con. Tab. 394.; and P. Arachnoides Desc. Geol. Env. de Paris, Pl. 3 f. 8. (Edit. 1822) are but varietes of this species.

[e] Podopsis striata, Desc. Geol. Env. de Paris, Pl. 4. A. B. p. 83. (Edit. 1822), is probably a distinct species.

[f] A characteristic fossil of the chalk; very common.

[g] Very common.

CLASS.	GENUS.	SPECIES.	REFERENCES AND SYNONYMS.	IN SUSSEX.	ELSEWHERE.
Conchifera.	Terebratula	undata.[h]	Min. Con. Tab. 15. f. 7, 8, 9. . . .	Lewes.	Meudon.
———	———	elongata.	Ibid. Tab. 435. f. 1, 2.	Ibid.	Norwich.
———	———	plicatilis.[i]	Ibid. Tab. 118. f. 1, 2. Tab. 83. f. 6. . .	Ibid.	Dieppe.
———	Magas	pumila.	Min. Conch. Tab. 119.	Brighton.	Norfolk.
———	Terebratula	subplicata	Geol. Suss. Tab. 26. f. 5, 6. 11. . .	Offham.	
Mollusca.	Trochus	Basteroti.[k]	Desc. Geol. Env. de Paris, Pl. 3. f. 3. .	Lewes.	Meudon.
———	Cirrus	depressus.	Geol.Suss. Tab.18. f.18.22. Min.Con.Tab.428.	Ibid.	Warminster.
———	———	perspectivus.	Ibid. Tab. 18. f. 21. 12. Min. Con. Tab. 428.	Ibid.	Ibid.
———	———	granulatus.[l]	Ibid. 195.	Southerham, [near Lewes.	
———	Dolium	nodosum.	Ibid. 196. Min. Con. Tab. 426. . .	Clayton, near [Hurst.	
———	Belemnites	mucronatus.[m]	Ibid. Tab. 16. f. 1. Desc. Geol. Env. de Paris, Pl. 3. f. 1. (Edit. 1822.) B. Allani, Fleming, Brit. Anim. 240. Min. Con. Tab. 600. f. 1.	Brighton.	Meudon. Scandinavia. Near Giant's Causeway, Ireland.
———	———	granulatus.	Min. Con. Tab. 600. f. 3. 5.	Lewes.	Mæstricht.
———	———	lanceolatus.	Ibid. Tab. 600. f. 8, 9.	Steyning.	Ibid.
———	Hippurites	Mortoni.	The first discovered in Great Britain. . .	Lewes.	

[h] T. undata, T. subundata, T. intermedia, T. semiglobosa of Min. Con. are included, being considered as varieties only.

[i] T. plicatilis, T. octoplicata, T. concinna of Min. Con.

[k] Occurs in the form of casts only.

[l] Lower chalk.

[m] A beak or mandible has lately been discovered, which probably belonged to some species of Belemnites, or Nautilus.

CLASS.	GENUS.	SPECIES.	REFERENCES AND SYNONYMS.	IN SUSSEX.	ELSEWHERE.
Mollusca.	Baculites	Faujasii.	Min. Con. Tab. 592. f. 1.	Lewes.	Mæstricht.
———	Nautilus	elegans.[n]	Geol. Suss. Tab. 20.	Ibid.	
———	Ammonites	varians.[o]	Ibid. Tab. 21. f. 2. 5. 7.	Ibid.	
———	———	Woollgari.[p]	Ibid. Tab. 21. f. 16. Tab. 22. f. 7. Min. Con. Tab. 587.	Ibid.	
———	———	navicularis.	Geol. Suss. Tab. 22. f. 5. Min. Con. Tab. 555. f. 2.	Offham.	Guildford.
———	———	catinus.	Ibid. Tab. 22. f. 10.	Southerham.	
———	———	Lewesiensis.	Ibid. Tab. 22. f. 2. Min. Con. Tab. 358. Hist. Mont St. Pierre, Pl. 31.?	Lewes.	Mæstricht.
———	———	peramplus.	Geol. Suss. 200. Min. Con. Tab. 357.	Ibid.	
———	———	rusticus.	Min. Con. Tab. 177.	Ibid.	
———	———	undatus.[q]	Ibid. Tab. 569. f. 2.	Ibid.	
———	Scaphites	striatus.[r]	Geol. Suss. Tab. 22. f. 3, 4.	Brighton.	
———	Hamites	armatus.[s]	Ibid. Tab. 23.	Lewes.	
Pisces.[t]	Muræna	Lewesiensis.	Ibid. Tab. 39. f. 11. Tab. 40. f. 2.	Ibid.	
———	Zeus	Lewesiensis.	Ibid. Tab. 35, 36.	Ibid.	Gravesend.
———	Salmo ?	Lewesiensis.	Ibid. Tab. 33. 40.	Ibid.	
———	Esox	Lewesiensis.	Ibid. Tab. 41. f. 1, 2. Tab. 25. f. 13.	Ibid.	
———	Amia ?	Lewesiensis.	Ibid. Tab. 37, 38.	Ibid.	
———	Squalus.[u]		Ibid. Tab. 32. f. 1. Sq. Cornubicus	Ibid.	

[n] Rare in the white chalk. [o] Very rare. [p] Lower chalk. [q] Upper chalk: unique.

[r] Rare; p. 160. [s] Exceedingly rare. [t] This arrangement and nomenclature of the fishes of the chalk must be considered only as temporary; the greater part will require the establishment of new genera for their reception.

[u] Teeth resembling those of several recent species.

CLASS.	GENUS.	SPECIES.	REFERENCES AND SYNONYMS.	IN SUSSEX.	ELSEWHERE.
Pisces.	Squalus.	———	Geol. Suss. 32. f. 2, 3. 6. Sq. Mustelus . .	Lewes.	Prevail throughout the chalk.
			Ibid. Tab. 32. f. 4. 7, 8. Sq. Zygœna . .	Ibid.	
			Ibid. Tab. 32. f. 12. 14, 15. Sq. Galeus .	Ibid.	
			Ibid. Tab. 33. f. 10.	Ibid.[x]	
———	Balistes ? [y]		Ibid. Tab. 33. f. 5, 6.	Ibid.	
———	Diodon ? [z]		Ibid. Tab. 32. f. 18. 20. . . .	Ibid.	
			Ibid. Tab. 39, 40, 41.	Ibid.[a]	
			Ibid. Tab. 42. Hist. Mont St. Pierre, Tab. 29.	Ibid.[b]	Mæstricht.
Reptiles.[c]	Mososaurus	Hoffmannii.[d]	Ibid. Tab. 33. 41. Hist. Mont. St. Pierre .	Ibid.	Ibid.
Amio-copros.[e]			Ibid. Pl. 9. f. 3. 6. 9, 10. . . .	Ibid.	Ibid.

3. *Chalk Marl.*

CLASS.	GENUS.	SPECIES.	REFERENCES AND SYNONYMS.	IN SUSSEX.	ELSEWHERE.
Agamia.	Confervites	fasciculata.	Ad. Brongn. Hist. Veget. Foss. Pl. 1. f. 1. .	Hamsey.	
Phanerogamia.† (Dicotyledonous)	Wood.		. . . .	Ibid.	

[x] Vertebræ. [y] Defence of.

[z] The palates resemble those of Diodon Histrix; but, from the numbers often found grouped together, the mouth of the original appears to have been paved with them: they probably belong to an extinct species of shark.

[a] Radii or fin bones of unknown fishes allied to Balistes. These radii are entirely distinct from those of the Lias, and belong to three or more species. [b] Jaw with teeth and bones of an unknown fish.

[c] Bones, teeth, portions of the mandibles, &c. of several reptiles and fishes too imperfect to be determined.

[d] Dorsal and caudal vertebræ. [e] The name given to these bodies by Dr. Buckland, who has proved them to be fœcal remains. See the Professor's paper on Coprolites.

† Wood. In the state of a brown friable mass.

CLASS.	GENUS.	SPECIES.	REFERENCES AND SYNONYMS.	IN SUSSEX.	ELSEWHERE.
Polypi.	Flustra.[f]			Hamsey.	
——	Millepora	Fittoni.[g]	Geol. Suss. Tab. 15. f. 10.	Ibid.	
——	Spongia	several undetermined.[h]	Ibid. 107.	Ibid.	
——	Alcyonium ?	pyriforme.	Ibid. 105.	Ibid.	
Radiaria.	Cidaris	claviger.	König	Ibid.	
——	Echinus	saxatilis ?	Geol. Suss. Tab. 17. f. 1.	Ibid.	
——	Spatangus	cordiformis.	Ibid. 108.	Middleham.	
——	Conulus	Hawkinsii.[i]		Hamsey.	Guildford.
	Spines of several species.				
Crustacea.	Astacus	Sussexiensis.[k]	Ibid. Tab. 30. f. 3.	Southerham.	
Annelides.	Serpula	Plexus.[l]	Min. Con. Tab. 598. f. 1.	Hamsey.	
——	Vermicularia	umbonata.	Geol. Suss. Tab. 18. f. 24. Min. Con. Tab. 57.	Ibid.	
——	——	Sowerbii.	Ibid. Tab. 18. f. 14, 15.	Ibid.	
Conchifera.	Venus	Ringmeriensis.	Ibid. Tab. 25. f. 5.	Middleham.	
——	Astarte ?		Ibid. 126.	Ibid.	
——	Venericardia.		Ibid. 126. Min. Con. Tab. 259.	Ibid.	
——	Cardium	decussatum.	Ibid. Tab. 25. f. 3. Min. Con. Tab. 552. f. 1.	Hamsey.	
——	Cucullæa.[m]			Middleham.	

[f] There are probably several species.

[g] The specific name is in honour of Dr. Fitton, P.G.S.

[h] A flexuose species in masses of an oval form is very common.

[i] A remarkable species found in the chalk marl only, hitherto neither figured nor described. Diameter of the base two inches and a half, height two inches; base nearly circular, flat; vent placed in the base two thirds from the mouth, and one third from the margin. Specific name in honour of John Hawkins, Esq. F.R.S., &c. of Bignor Park.

[k] Rare.

[l] Two small masses, very rare.

[m] M. Brongniart sent me a similar cast from Rouen.

CLASS.	GENUS.	SPECIES.	REFERENCES AND SYNONYMS.	IN SUSSEX.	ELSEWHERE.
Conchifera.	Arca	2 or 3 species undetermined.	. . . .	Ringmer.	
———	Chama.[n]		. . . .	Hamsey.	
———	Avicula	2 species undetermined.[o]	. . . .	Ibid.	
———	Inoceramus	tenuis.	Geol. Suss. 132. No. 65. . . .	Ibid.	
———	———	Cripsii.	Ibid. Tab. 27. f. 11.	Ibid.	
———	Plagiostoma	elongatum.	Ibid. Tab. 19. f. 1. Min. Con. Tab. 559. f. 2.	Ibid.	Folkstone.
———	———	asper.	Ibid. Tab. 26. f. 18.	Ibid.	
———	Pecten	Beaveri.	Ibid. Tab. 25. f. 11. Min. Con. Tab. 158. .	Ibid.	
———	———	triplicatus.	Ibid. Tab. 25. f. 9.	Ibid.	
———	———	quinquecostatus.[p]	Ibid. Tab. 25. f. 10.	Ibid.	
———	———	orbicularis.	P. laminosus, Geol. Suss. Tab. 26. f. 8. Min. Con. Tab. 186.	Ibid.	
———	Plicatula	inflata.	P. spinosa Geol. Suss. Tab. 26. f. 13. 16, 17. Min. Con. Tab. 409. f. 2. . . .	Ibid.	
———	———	pectinoides.[q]	Min. Con. Tab. 401. f. 9. . . .	Ibid.	
———	Terebratula	subrotunda.[r]	Ibid. Tab. 15. f. 1, 2.	Ibid.	
———	———	undata.	Ibid. Tab. 15. f. 7.	Eastbourn.	
———	———	striatula.	Ibid. Tab. 536. Geol. Suss. Tab. 25. f. 7, 8. 12. T. Defrancii, Desc. Geol. Env. de Paris, Pl. 3. f. 6. (Edit. 1822.) . .	Hamsey.	

[n] A subglobose shell, not uncommon.
[o] The shells very thin and fragile.
[p] Probably a distinct species.
[q] Cambridgeshire, in galt.
[r] Rare.

CLASS.	GENUS.	SPECIES.	REFERENCES AND SYNONYMS.	IN SUSSEX.	ELSEWHERE.
Conchifera.	Terebratula	Mantelliana.	Min. Con. Tab. 537. f. 5. T. sulcata, Geol. Suss. 130.	Hamsey.	
——	——	Martini.	T. Pisum, Min. Con. Tab. 536. f. 6, 7. Geol. Suss. 131.	Ibid.	Folkstone.
——	——	rostrata.	Min. Con. Tab. 537. f. 1, 2.	Ibid.	
——	——	squamosa.[s]	Geol. Suss. 132. No. 64.	Hamsey.	
Mollusca.	Auricula	incrassata.	Ibid. Tab. 19. f. 2, 3. 34. Min. Con. Tab. 163	Stoneham.	Blackdown, [Devon.
——	Ampullaria?		Ibid Tab. 18. f. 11.	Hamsey.	
——	Trochus	linearis.	Ibid. Tab. 18. f. 17.	Ibid.	
——	——	agglutinans?[t]	Ibid. Tab. 18. f. 7.	Ibid.	
——	Rostellaria	Parkinsoni.	Ibid. Tab. 18. f. 1, 2. 4. Min. Con. Tab. 558. f. 3	Ibid.	Ibid. in Shank- [lin sand.
——	Cassis	avellana.[u]	Desc. Geol. Env. de Paris, Pl. 6. f. 10. (Edit. 1822.)	Ringmer.	Rouen.
——	Eburna?		Geol. Suss. Tab. 18. f. 13.	Hamsey.	
——	Voluta	ambigua?[x]	Ibid. Tab. 18. f. 8.	Ibid.	
——	Baculites	Faujasii.	Min. Con. Tab. 592. f. 1.	Ibid.	
——	——	obliquatus.	Ibid. Tab. 592. f. 2, 3. Hamites baculoides, Geol. Suss. Tab. 23. f. 6, 7.	Glynd.	Ibid.[y]
——	Nautilus	elegans.	Geol. Suss. Tab. 20. f. 1. Tab. 21. f. 5. Min. Con. Tab. 116.	Hamsey.	
——	——	expansus.	Min. Con. Tab. 458. f. 1. "N. elegans in a young state." Geol. Suss. Tab. 21. f. 1. 4.	Middleham.	

[s] This species has not been figured.

[t] Cast of the base of the shell.

[u] Quere if not a variety of Auricula incrassata?

[x] This shell is closely allied to, if not identical with, V. ambigua of Hordwell cliffs. It is attached to an Ammonite.

[y] Craie chloritée.

CLASS.	GENUS.	SPECIES.	REFERENCES AND SYNONYMS.	IN SUSSEX.	ELSEWHERE.
Mollusca.	Ammonites	Mantelli.	Geol. Suss. Tab. 22. f. 1. Min. Con. Tab. 55.	Middleham.	
———	———	Sussexiensis.	Ibid. Tab. 20. f. 2. Tab. 21. f. 10. A. Rhotomagensis, Desc. Geol. Env. de Paris, Tab. 6. f. 2. (Edit. 1822.) Min. Con. Tab. 515.	Hamsey.	Rouen.
———	———	varians.	Geol. Suss. Tab. 21. f. 2. 7. Min. Con. Tab. 176.	Ibid.	Ibid.
———	———	cinctus.	Ibid. 116. Min. Con. Tab. 564. f. 1. . .	Middleham.	
———	———	falcatus.	Ibid. Tab. 21. f. 6. 12. Min. Con. 579. f. 1.	Ibid.	
———	———	curvatus.	Ibid. Tab. 21. f. 18. Min. Con. Tab. 579. f. 2.	Hamsey.	
———	———	complanatus.	Ibid. 118. Min. Con. Tab. 569. f. 1. . .	Ibid.	
———	———	rostratus.	Min. Con. Tab. 173.	Southerham.	
———	———	tetrammata.	Ibid. Tab. 587. f. 2.	Hamsey.	
———	Scaphites	striatus.	Geol. Suss. Tab. 22. f. 3, 4. 9. 11. S. obliquus, Min. Con. Tab. 18. f. 4—7. . .	Ibid.	Ibid.
———	———	costatus.	Geol. Suss. Tab. 22. f. 8. 12. Parkin. Org. Rem. iii. Pl. 10. f. 10. . . .	Ibid.	
———	Hamites	armatus.	Geol. Suss. Tab. 23. f. 3, 4. Min. Con. Tab. 168.	Ibid.	
———	———	plicatilis.	Ibid. Tab. 23. f. 1, 2. Min. Con. Tab. 234. f. 1.	Ibid.	
———	———	alternatus.	Ibid. Tab. 23. f. 10, 11. . . .	Ringmer.	
———	———	ellipticus.	Ibid. Tab. 23. f. 9.	Ibid.	
———	———	attenuatus.	Ibid. Tab. 23. f. 8. 13. Min. Con. Tab. 61. f. 4, 5.	Hamsey.	Folkstone.
———	Turrilites	costatus.	Ibid. Tab. 23. f. 15. Tab. 24. f. 1. 4, 5. Min. Con. Tab. 36.	Ibid.	Rouen.
———	———	undulatus.	Geol. Suss. Tab. 24. f. 8. Tab. 23. f. 14. 16. Min. Con. Tab. 75. f. 1, 2, 3. . .	Ibid.	
———	———	tuberculatus.	Geol. Suss. Tab. 24. f. 2, 3. 6, 7. Min. Con. Tab. 74.	Middleham.	Ibid.

CLASS.	GENUS.	SPECIES.	REFERENCES AND SYNONYMS.	IN SUSSEX.	ELSEWHERE.
Pisces.	Squalus	Mustelus.[z]	Geol. Suss. 32. f. 2, 3. 5, 6. 9. 11. . .	Hamsey.	Wilts.
———	———	Galeus.	Ibid. Tab. 32. f. 12. 14, 15, 16. . .	Ibid.	
			Ibid. Tab. 34. f. 10.	Ibid.	
Scales, &c. and Coprolites.[a]			Geol. Suss. Pl. 9. f. 4. 5. 7. 8. 11.		

4. *Firestone or Upper Green Sand.* (Craie chloritée ou Glauconie crayeuse.)

This division contains so many of the fossils common to the marl, that in the following list those organic remains alone are enumerated which have been noticed exclusively in the Firestone. Among the fossils abundant in both deposits are, Pecten orbicularis, Plicatula inflata, Terebratulæ, Nautilus expansus, Ammonites varians, A. Mantelli, wood, scales of fishes, &c.

CLASS.	GENUS.	SPECIES.	REFERENCES AND SYNONYMS.	IN SUSSEX.	ELSEWHERE.
Agamia.	Fucoides	Targionii.[b]	Ad. Brong. Hist. Veget. Foss. Pl. 4. f. 2. 6. Geol. Suss. 98.	Bignor.	Near Florence.
Polypi.	Millepora	Gilberti.	Geol. Suss. 106. No. 8.	Southbourn.	
———	Siphonia	Websteri.	Parkin. Introd. Org. Rem. 50. Tulip alcyonium, Webster, Geol. Trans. 1st series, ii. Pl. 28.	Ibid.	Isle of Wight.

[z] Teeth resembling those of the recent species are occasionally found.

[a] These substances have been known since the time of Woodward by the name of "*Iuli of Cherry Hinton*," and were supposed to be the amenta or cones of a species of fir. Their animal origin was first suggested by Mr. König, see Geol. of Suss. p. 104. Dr. Buckland has lately investigated the subject with his usual acumen and success; and the analysis of Dr. Prout having proved their animal nature beyond all doubt, Dr. B. proposes to distinguish these fossils by the term *Coprolite:* he has shown them to be the fæcal remains of fishes. I have one of these bodies in an Amia? lying on the air-bladder.

[b] Occurs in vast quantities near Bignor. In the same locality was also found the culm or stem of an undetermined plant.

CLASS.	GENUS.	SPECIES.	REFERENCES AND SYNONYMS.	IN SUSSEX.	ELSEWHERE.
Polypi.[c]	Spongia.[d]		. . . .	Southbourn.	
Radiaria.	Spatangus	Murchisonianus.	König. Icon. Sect. Foss. Cent. 2. . .	Ibid.	
Conchifera.	Cardita ?[e]		. . . .	Ibid.	
———	Arca	carinata.	Min. Con. Tab. 44. (Lower figures) . .	Ibid.	Devizes.
———	Plagiostoma.[f]		. . . .	Ibid.	
———	Gryphæa	vesiculosa.	Ibid. Tab. 369.	Hamsey.	
———	Ostrea	carinata.	Ibid. Tab. 365. White, Nat. Hist. Selbourne. Desc. Geol. Env. de Paris, Pl. 3. f. 11. (Edit. 1822.)	Southbourn.	Le Hâvre.
———	Terebratula	biplicata.	Min. Con. Tab. 90.	Ibid.	Cambridge.
Mollusca.	Trochus	Rhodani.	Desc.Geol.Env.de Paris, Pl.9.f.8.(Edit.1822.)	Ibid.	Lignerolle au-[dessus d'Arbe.
———	———	bicarinatus ?	Min. Con. Tab. 221. f. 2.	Ibid.	
———	Ammonites	planulatus.	Ibid. Tab. 570. f. 5.	Ibid.	
———	———	Catillus.	Ibid. Tab. 564. f. 2.	Ibid.	

5. *Galt, or Folkstone Marl.*

CLASS.	GENUS.	SPECIES.	REFERENCES AND SYNONYMS.	IN SUSSEX.	ELSEWHERE.
Agamia.	Fucoides.[g]		Geol. Suss. 83.	Norlington.	Bletchingley, [Surrey.
Phanerogamia.					

[c] Several species undetermined ; common in the rocks near the sea-houses.

[d] The inferior bed of marl which is in contact with the firestone at Southbourn, is almost entirely composed of ramose zoophytes, probably Milleporites, Madreporites, &c. so as to form a reef of corals. In this bed was found a long cylindrical zoophyte, partly composed of chert, of the same kind as those which occur in the vale of Pewsey, in Wiltshire.

[e] Much compressed ; possibly a Pholadomya.

[f] A small species undescribed ; it occurs also in the malm at Amberly.

[g] In layers of indurated red marl.

CLASS.	GENUS.	SPECIES.	REFERENCES AND SYNONYMS.	IN SUSSEX	ELSEWHERE.
(Dicotyledonous)[h]	Coniferous Wood.		. . . .	Near Willingdon.	Folkstone.
Polypi	Turbinolia	Königi.[i]	Geol. Suss. Tab. 19. f. 22. 24. . .	Ringmer.	Bletchingley.
Radiaria.	Spatangus.[k]		. . . .	Ibid.	Ibid.
Crustacea	Arcania.[l]		Ibid. Tab. 29. f. 7, 8. 14. . . .	Ibid.	Near Cambridge.
———	———[m]		Ibid. Tab. 29. f. 9, 10. . . .	Ibid.	
———	Etyæa.		Ibid. Tab. 29. f. 11, 12. . . .	Ibid.	
———	Corystes.		Ibid. Tab. 29. f. 13. 15, 16. . . .	Ibid.	
———	Astacus.[n]		Ibid. 98.	Ibid.	
Annelides	Dentalium	striatum.	Ibid. Tab. 19. f. 4. Min. Con. Tab. 70. f. 4.	Ibid.	Folkstone.
———	———	ellipticum.	Ibid. Tab. 19. f. 21. 25. Min. Con. Tab. 70. f. 6, 7.	Ibid.	Ibid.
———	———	decussatum.	Min. Con. Tab. 70. f. 5. . . .	Newtimber.	
Conchifera.	Fistulana	pyriformis.[o]	Geol. Suss. 76.	Willingdon.	
———	Arca.[p]		. . . .	Ringmer.	
———	Nucula	pectinata.	Ibid. Tab. 19. f. 5, 6. 9. Min. Con. Tab. 192. f. 6, 7.	Ibid.	Folkstone.
———	———	ovata.	Geol. Suss. Tab. 19. f. 26, 27. . .	Ibid.	Ibid.
———	Pecten	orbicularis.[q]	Min. Con. Tab. 106.	Ibid.	
———	Inoceramus	concentricus.	Geol. Suss. Tab. 19. f. 19. Min. Con. Tab. 305. Desc. Geol. Env. de Paris, Pl. 6. f. 11. .	Ibid.	Blackdown, [Rouen.

[h] Rolled fragments, probably of a species of Fir or pine, and generally perforated by Fistulanæ.
[i] Hitherto observed in galt only. [k] A fragment only. [l] The thorax. [m] Unknown, but belonging to the family Corystidæ. [n] Remains of the abdominal covering of two unknown species.
[o] At the junction of the Galt and Shanklin sand, imbedded in wood. [p] A very imperfect cast. [q] One example only.

CLASS.	GENUS.	SPECIES.	REFERENCES AND SYNONYMS.	IN SUSSEX.	ELSEWHERE.
Conchifera.	Inoceramus	sulcatus.	Geol. Suss. Tab. 19. f. 16. Min. Con. Tab. 306. Desc. Geol. Env. de Paris, Pl. 6. f. 12.	Ringmer.	Perte du Rhone.
——	——	gryphæoides.	Min. Con. Tab. 584. f. 1.	Ibid.	
Mollusca.	Ampullaria	canaliculata.	Geol. Suss. Tab. 19. f. 13.	Ibid.	Bletchingley.
——	Natica.		Ibid. Tab. 19. f. 31, 32.	Ibid.	Folkstone.
——	Cirrus	plicatus.	Min. Con. Tab. 141. f. 3.	Norlington.	Ibid.
——	Rostellaria	carinata.	Geol. Suss. Tab. 19. f. 10—14.	Ringmer.	Bletchingley.
——	Belemnites	Listeri.	Ibid. Tab. 19. f. 18. B. minimus, Lister Hist.	Ibid.	Ibid.
——	——	attenuatus.	Ibid. Tab. 19. f. 17. 23. Min. Con. Tab. 589. f. 2.	Ibid.	Folkstone.
——	Nautilus	inæqualis.	Ibid. Tab. 21. f. 14, 15. Min. Con. Tab. 40. (lower figure)	Ibid.	Ibid.
——	Ammonites	splendens.	Ibid. Tab. 21. f. 13. 17. Min. Con. Tab. 103.	Norlington.	Ibid.
——	——	auritus.	Min. Con. Tab. 134.	Ringmer.	Ibid.
——	——	planus.	Geol. Suss. Tab. 21. f. 3. (var. of A. varians?)	Ibid.	Ibid.
——	——	lautus.	Ibid. Tab. 21. f. 11. Min. Con. Tab. 309.	Ibid.	Ibid.
——	——	biplicatus.	Ibid. Tab. 22. f. 6. A. Deluci? Geol. Min. Pl. 6. f. 4.	Ibid.	Le Hâvre.
——	——	tuberculatus.	Min. Con. Tab. 310. f. 1, 2, 3. Geol. Suss. p. 92.	Ibid.	Folkstone.
——	——	lævigatus.	Ibid. Tab. 549. f. 1.	Ibid.	Ibid.
——	Hamites	attenuatus.	Ibid. Tab. 61. f. 4, 5. Geol. Suss. Tab. 19. f. 29, 30.	Ibid.	Ibid.
——	——	maximus.	Ibid. Tab. 62. f. 1.	Ibid.	Ibid.
——	——	intermedius.	Ibid. Tab. 62. f. 2, 3. Geol. Suss. Tab. 23. f. 12.	Ibid.	Ibid.
——	——	tenuis.	Ibid. Tab. 61. f. 1.	Ibid.	Ibid.
——	——	rotundus.	Ibid. Tab. 61. f. 2, 3.	Ibid.	Ibid.
——	——	compressus.	Ibid. Tab. 61. f. 7, 8.	Ibid.	Ibid.
Pisces.	Squalus	Mustelus? [r]	. . . .	Norlington. Ringmer.[s]	

[r] Teeth.

[s] Scales and vertebræ; very rare.

6. *Shanklin Sand.* (Lower Green Sand.)

CLASS.	GENUS.	SPECIES.	REFERENCES AND SYNONYMS.	IN SUSSEX.	ELSEWHERE.
Phanerogamia. (Dicotyledonous).[t]	Coniferous Wood.		Geol. Suss. 76.	Willingdon near Folkstone.	
Radiaria.	Spatangus.[u]		. . . .	Parham.	
Crustacea.			*Martin Geol. Mem. West Suss. 32. . .	Bignor Common.[x]	
Annelides.	Dentalium	one or more.	Geol. Suss. 72.	Parham.	
——	*Vermicularia	concava.	Min. Con. Tab. 57. f. 1—5. . . .	Pulborough.	
Conchifera.	Mya	plicata var.?	Ibid. Tab. 419. f. 3. M. intermedia, Geol. Suss. 74.	Parham: near [Margate.	
——	——	*Mandibula.	Martin. Geol. Mem. West Suss. 33. Min. Con. Tab. 43.	Pulborough.	Devizes.
——	*Pholadomya.		. . . .	Ibid.	
——	Corbula	Striatula.	Min. Con. Tab. 572. f. 2, 3. . . .	Parham.	
——	Tellina	æqualis.	Not figured	Ibid.	
——	——	inæqualis.	Min. Con. Tab. 456. f. 2. . . .	Ibid.	Blackdown.
——	Venus	parva.	Ibid. Tab. 518. f. 4, 5, 6. . . .	Ibid.	Shanklin.
——	Venus	angulata.[y]	Min. Con. Tab. 65.	Parham.	Blackdown.
——	——	Faba.	Ibid. Tab. 567. f. 3.	Ibid.	Shanklin.

[t] Rolled fragments of wood at the junction of the sand with the Galt.

[u] Fragment of a species too imperfect to be determined.

[x] "Crustaceous fossil like a shrimp." — *Martin*.

[y] Casts four inches and a half wide sometimes occur.

CLASS.	GENUS.	SPECIES.	REFERENCES AND SYNONYMS.	IN SUSSEX.	ELSEWHERE.
Conchifera.	Venus	ovalis.	Min. Con. Tab. 567. f. 1, 2.	Parham.	Feversham.
——	Thetis	minor.	Ibid. Tab. 513. f. 5, 6.	Ibid.	Shanklin Chine.
——	Cucullæa	decussata.	Ibid. Tab. 206. f. 3, 4.	Ibid.	Feversham.
——	Nucula	impressa.	Ibid. Tab. 475. f. 3.	Ibid.	Blackdown.
——	*Modiola	æqualis.	Ibid. Tab. 210. f. 2.	Ibid.	
——	——	bipartita.	Ibid. Tab. 210. f. 3, 4.	Ibid.	Osmington.
——	Mytilus	lanceolatus.	Ibid. Tab. 439. f. 2.	Ibid.	
——	*Pinna.		Martin Geol. Mem. West Suss. 32.	Pulborough.	
——	*Trigonia	Dædalea.	Min. Con. Tab. 88. T. clavellata, Geol. Suss. 73.	Parham.	Blackdown.
——	——	alæformis.	Ibid. Tab. 215. f. 2.	Ibid.	Ashford, Kent.
——	——	*spinosa.	Ibid. Tab. 86. Martin Geol. Mem. West Suss. 33.	Pulborough.	
——	Gervillia	aviculoides.	Ibid. Tab. 511. Geol. Suss. 74.	Parham.	
——	——	solenoides.	Ibid. Tab. 510. f. 1—4. Geol. Suss. 74.	Ibid.	
——	——	acuta.[z]	Ibid. Tab. 510. f. 5.	Ibid.	
——	*Inoceramus.[a]		Martin Geol. Mem. West Suss. 33.	Pulborough.	
——	Pecten	quadricostatus.	Min. Con. Tab. 56. f. 1, 2. (3?)	Parham.	Exeter.
——	——	obliquus.	Ibid. Tab. 370. f. 2.	Ibid.	
——	——	orbicularis.[b]	Ibid. Tab. 186. Martin Geol. Mem. West Suss. 33.	Pulborough. Parham.[c]	
——	*Orbicula.[d]		Martin Geol. Mem. West Suss. 32.	Pulborough.	
——	Terebratula	ovata.	Min. Con. Tab. 15. f. 3.	Parham.	
——	——	lata.	Ibid. Tab. 502. f. 1.	Ibid.	Devizes.

[z] Avicula?

[a] Lower beds.

[b] "Lower beds of green-sand."

[c] Upper valve? A flat shell with numerous striæ.

[d] Unlike O. reflexa, Min. Con. Tab. 506. f. 1.

CLASS.	GENUS.	SPECIES.	REFERENCES AND SYNONYMS.	IN SUSSEX.	ELSEWHERE.
Conchifera.	* Lenia.		Martin Geol. Mem. West. Suss. 32. . .	Pulborough.	
Mollusca.	Patella.[e]		In the Author's collection. . . .	Parham.	
———	Pileopsis.[f]		. . . .	Ibid.	
———	*Auricula.		Martin Geol. Mem. West Suss. 31. . .	Pulborough.	
———	Natica	canrena.	Parkin. Org. Rem. iii. Pl. 6. f. 2. . .	Parham.	Devizes.
———	* Turbo.		Martin Geol. Mem. West Suss. 31. . .	Pulborough.	
———	Rostellaria	Parkinsoni.[g]	. . . .	Parham.	Blackdown.
———	———	calcarata.	Min. Con. Tab. 349. f. 6, 7. . . .	Ibid.	Ibid.
———	———	with 2 processes.[h]	. . . .	Ibid. [borough.	[Kent.
———	Nautilus.[i]		Martin Geol. Mem. West Suss. 31. . .	Near Pul-	Broughton,
———	Ammonites	Goodhalli.[k]	. . . .	Willingdon.	Blackdown.

Wealden Formation.

1. *Weald Clay.* (Upper Division.)

CLASS.	GENUS.	SPECIES.	REFERENCES AND SYNONYMS.	IN SUSSEX.	ELSEWHERE.
Crustacea.	Cypris	Faba.[l]	Min. Con. Tab. 485.	Cooksbridge.	Isle of Wight.
Conchifera.	Cyclas	membranacea.[m]	Ibid. Tab. 527. f. 3. "Tilgate Fossils," 26.	Shipley.	
———	———	media.[n]	Ibid. Tab. 527. f. 2.	Cooksbridge.	

[e] Oval, conical, depressed; longest diameter one inch and a half, transverse one inch.

[f] Of the size of Patella Unguis, Min. Con. Tab. 139. f. 7.

[g] Probably a variety of the chalk marl species, Geol. Suss. Tab. 18. f. 1.

[h] Resembles very closely R. Pes Pelicani.

[i] Species not particularised.

[k] Mr. Martin mentions three species of Ammonites; neither particularised.

[l] In limestone, septaria, and shale.

[m] In blue clay.

[n] In septaria and shale.

CLASS.	GENUS.	SPECIES.	REFERENCES AND SYNONYMS.	IN SUSSEX.	ELSEWHERE.
Mollusca.	Paludina	vivipara.[o]	Lamarck Hist. Nat. Anim. sans Vert. Vivipara fluviorum, Min. Con. Tab. 31. f. 1. .	Laughton, near [Lewes.	Near Tilvester. [Hill, Surrey.
———	———	elongata.[p]	Min. Con. Tab. 509. Tilgate Fossils, 26. .	Near Cooks- [bridge.	Compton-Grange, Isle of Wight.
———	———	carinifera.[q]	Ibid. Tab. 509. f. 3.	Resting-Oak- [Hill.	
———	Melanopsis or Melania.[r]		Tilgate Fossils, 25.	Shipley, near West Grinstead.	
Pisces.[s]			. . . .	Cooksbridge.	
Reptilia.[t]			Martin Geol. Mem. West Suss. 41. . .	Resting-Oak- [Hill.	

[o] The remains of this species, associated with those of Cypris Faba, form extensive beds of limestone, known by the name of Sussex marble.

[p] In septaria, clay, and shale.

[q] Associated with the other species. Some of the smaller specimens of the last two species closely resemble Paludina tentaculata.

[r] A small delicate species, always in a mutilated state.

[s] Scales, bones, &c.

[t] Bones of saurian animals (very rarely) with Paludinæ and Cyclades.

2. *Hastings Beds.* (Middle Division),

Including the Horsted sand; Tilgate sand, grits, and clays; and the Worth sandstone.

CLASS.	GENUS.	SPECIES.	REFERENCES AND SYNONYMS.	IN SUSSEX.	ELSEWHERE.
Cryptogamia (Vascularia).	Calamites.[u]		. . . .	Near Tunbridge [Wells.	
———	Sphenopteris	Mantelli.	Ad. Brongn. Prodr. Hist. Veget. Foss. 50. Hymenopteris psilotoides, Geol. Trans. 2d series, i. 424. Tilgate Foss. Pl. 1. f. 3. *a. b.* Pl. 3. f. 6, 7. Pl. 20. f. 1, 2.	Tilgate Forest.	
———	———	Sillimani.	*Vide* page 239. of this volume.	[Hastings.	
———	———	Phillipsii.	——— 240.———.	[Worth.	
———	Lonchopteris	Mantelli.	Ad. Brongn. Prodr. Hist. Veget. Foss. 60. Pecopteris reticulata, Tilgate Foss. Pl. 3. f. 5. Geol. Trans. 2d series, i. 423.	Hastings.	Env. de Beauvais. Eridge Park.
———	Lycopodites?[x]		. . . .	Chiddingly.	
Phanerogamia (Monocotyledonous).	Cycadites	Brongniarti.	*Vide* page 238. of this volume.	Riegate.	
	Clathraria	Lyellii.	Tilg. Foss. Pl. 1. f. 1, 2. 7. Pl. 2. f. 1, 2, 3. Geol. Trans. 2d series, i. 423.	Tilgate Forest.	

[u] A compressed culm; nearly an inch in circumference, five joints in the length of four inches. In blue shale.

[x] A small delicate plant, carbonised; related to the recent Lycopodia? leaves of several species of Ferns, and other remains too imperfect to be determined, occur in the same locality.

CLASS.	GENUS.	SPECIES.	REFERENCES AND SYNONYMS.	IN SUSSEX.	ELSEWHERE.
(Monocotyledonous). (*Families uncertain.*)	Carpolithus	Mantelli.[y]	Tilg. Foss. Pl. 3. f. 1, 2. Geol. Trans. 2d series, i. 423. Ad. Brong. Prodr. Hist. Veget. 127.	Tilgate Forest.	
	Endogenites	erosa.	Tilg. Foss. Pl. 3. f. 1, 2. Geol. Trans. 2d series, i. 423.	Ibid. Hastings.	
(Dicotyledonous : ? ? *Families not determined.*[z])			. . . .	Ibid. Hastings, &c.	
Crustacea.	Cypris	Faba.[a]	Min. Con. Tab. 485.	Ibid.	Isle of Wight.
Conchifera.	Cyclas	media.	Ibid. Tab. 527. f. 2.	Tilgate Forest. Hastings.	
——	——	cornea ?[b]	. . . .	Tilgate Forest.	
——	——	membranacea.[c]	Ibid. Tab. 527. f. 3.	Hastings.	
——	Unio	porrectus.[d]	Min. Con. Tab. 594. f. 1. . . .	Tilgate Forest.	
——	——	compressus.[d]	Ibid. Tab. 594. f. 2.	Ibid.	
——	——	antiquus.[d]	Ibid. Tab. 594. f. 3, 4, 5. . . .	Ibid. Hastings.	
——	——	aduncus.[d]	Ibid. Tab. 595. f. 2. Tilg. Foss. Pl. 10. f. 11.	Linfield. Bolney.	
——	——	cordiformis.[d]	Ibid. Tab. 595. f. 1.	Tilgate Forest.	

[y] This fossil M. Brongniart supposes to be the seed-vessel of Clathraria Lyellii.

[z] Carbonised wood in small masses; doubtful if dicotyledonous; and lignite disseminated in sand, clay, grit, &c.

[a] With Cyclades, &c.

[b] A small species somewhat resembling Cyclas cornea, abundant in the calciferous grit.

[c] In shale, &c.

[d] Casts of these species occur in abundance in the grits and sandstones of the Forest, in many instances constituting entire layers of considerable extent and thickness, like the muscle-band of the coal measures, formed by a species of the same fresh water bivalve. See Pet. Derb. Pl. 27, 28.

CLASS.	GENUS.	SPECIES.	REFERENCES AND SYNONYMS.	IN SUSSEX.	ELSEWHERE.
Mollusca.	Succinea ?[e]		. . . .	Tilgate Forest.	
———	Neretina	Fittoni.		Ibid.	
———	Paludina	vivipara.	Tilg. Foss. Pl. 10. f. 8, 9. . . .	Ibid.	Tunbridge Wells.
———	———	elongata.[f]	Ibid. Pl. 10. f. 7.	Tilgate Forest.	
Pisces.	Lepisosteus.[g]	Fittoni.	Tilg. Foss. Pl. 5. f. 4. 15. Martin Geol. Mem. West. Suss. 48.	[Hastings. Billinghurst, Tilgate Forest.	
———	Silurus.[h]		Specimen with pectoral fin, and teeth. .	Heathfield.	
			Tilg. Foss. Pl. 10. f. 4. 6. . . .	Tilgate Forest.	
			Ibid. Pl. 5. f. 14. Pl. 15. f. 2. 6. . .	Ibid.[i]	
				Ibid.[k]	
				Ibid.[l] [Wells.[m]	
			Ibid. Pl. 10. f. 2.	Ibid. Tunbridge Hastings.[n]	

[e] A small species related to S. amphibia, in limestone with a group of Paludina elongata.

[f] Abundant in the grit.

[g] A genus allied to Esox. A fragment of the fore part of the body with the gills; nine inches long, seven inches broad, five inches thick, covered with rhomboidal scales. Detached scales are common in every bed of the Hastings formation. In the Museum of the College of Surgeons in London, there is a portion of the skin of a fish covered with scales of a similar character, from the Brazils, which Mr. Clift supposed to belong to a fresh-water genus allied to Esox.

[h] Radii or fin bones; three or more species.

[i] Teeth tricuspid, striated. They resemble some from the Stonesfield slate.

[k] Teeth tricuspid, smooth. They differ from the tricuspid teeth of Squali.

[l] Palates or dentes tritores, resembling some from Stonesfield, Oxfordshire.

[m] Jaws with hemispherical teeth.

[n] Scales, vertebræ, &c., of a small species too mutilated to admit of determination. In the argillaceous partings of the strata.

CLASS.	GENUS.	SPECIES.	REFERENCES AND SYNONYMS.	IN SUSSEX.	ELSEWHERE.
Reptilia.	Trionyx.	Bakewelli.	Tilg. Foss. Pl. 6. f. 1. 3, 4, 5. 8. Pl. 7. f. 4. 7. p. 60.	Tilgate Forest.	Jura limestone. [of Soleure.
———	Emys.[o]		Ibid. Pl. 6. f. 6, 7. Pl. 7. f. 3. Oss. Foss. v. 232.	Ibid.	Mæstricht.
———	Chelonia.[p]		Ibid. Pl. 6. f. 2. Pl. 7. f. 1, 2. 5. 8. Oss. Foss. v. 239.	Ibid.	
———	Plesiosaurus.[q]		Ibid. Pl. 5. f. 11. Pl. 9. f. 4, 5. . .	Ibid.	
———	Crocodilus	priscus.	Ibid. Pl. 10. f. 5. Oss. Foss. v. Pl. 6. f. 1. .	Ibid.	
———	Leptorhynchus.[r]		Ibid. Pl. 7. f. 5, 6. 8. Cuv. Oss. Foss. v. 127.	Ibid.	Caen.
			Ibid. Pl. 5. f. 1, 2. 7. 10. 12. Cuv. Oss. Foss. v. 142.	Ibid.[s]	
			Ibid. Pl. 8. f. 8.	Ibid.[t u]	
———	Cylindricodon.[y]		Ibid. Pl. 15. fig. 3, 4.	Ibid.	
———	Megalosaurus.[x]		Ibid. Pl. 9. f. 2, 3. 6. Pl. 18. f. 2. Pl. 19. f. 1, 2. 8. 12. 14, 15, 16. Geol. Trans. 2d series, i. Pl. 40, 41.	Ibid.	Stonesfield, n. [Oxford.
———	Iguanodon.[z]		Tilg. Foss. Pl. 4. Pl. 10. f. 12. Pl. 12. f. 1, 2, 3, 4. Pl. 16. f. 1. Pl. 18. f. 1. Oss. Foss. v. 351. Phil. Trans. 1825. . . .	Ibid.	Swanage, Isle [of Purbeck.

o A remarkably flat species. p Related to the fossil turtle of Mæstricht. q Vertebræ, teeth, &c.

r The fossil species of Caen. s The fossil species of the Jura limestone.

t A very small species resembling that figured in the Oss. Foss. vol. iii. Pl. 76. f. 8.

u A small species. x Teeth, vertebræ, and other bones.

y Dr. Jäeger of Stutgard has discovered teeth of this kind in the neighbourhood of that city; together with the teeth and jaws of two other phytivorous saurians; *vide* p. 292.

z Horn, teeth, vertebræ, phalanges, femur, tibia, fibula, clavicles, coracoid bone, ribs, &c.

CLASS.	GENUS.	SPECIES.	REFERENCES AND SYNONYMS.	IN SUSSEX.	ELSEWHERE.
Reptilia.	Hylæosaurus.		*Vide* p. 289. *et seq.*	Tilgate Forest	
———	. . .		. . .	Ibid.[a]	
———	Pterodactylus?		Tilg. Foss. Pl. 8. f. 1, 2, 3. 10, 11. 18.	Ibid.	Hastings.
Aves.[b]	. . .		Ibid.	Ibid.	Ibid.

Sauro-copros of Dr. Buckland. Small obscurely spiral masses, supposed to be fæcal.

3. *Ashburnham Beds.* (Lower Division.)

Argillaceous limestone alternating with schistose marls.

CLASS.	GENUS.	SPECIES.	REFERENCES AND SYNONYMS.	IN SUSSEX.
Cryptogamia. (Vascularia)	Sphenopteris	Mantelli.[c]	. . .	Pounceford
Conchifera.	Cyclas	media.[d]	Min. Con. Tab. 527. f. 2.	Ibid.
———	———	membranacea.[e]	Ibid. Tab. 527. f. 3.	Ibid. Ashburnham.
———	———	cornea.[f]	. . .	Maresfield. West Hothly. Ashburnham. Hastings. Framfield
———	Psammobia.	———	*Vide* p. 248.	Pounceford.
———	Mytilus	———	. . .	Ibid.

[a] Bones of other undetermined saurian animals.

[b] Bones referable to Birds or to a flying reptile. Some are decidedly the tibiæ of a wading bird.

[c] Lignite and imperfect traces of carbonised vegetables.

[d] Forms beds of limestone.

[e] Constitutes the principal portion of the argillaceous beds in some localities.

[f] This species resembles C. corneus of Lamarck, vol. vi. Entire beds of limestone are formed of it, associated with shells of the genus Unio. It occurs also in vast quantities in the grit.

CLASS.	GENUS.	SPECIES.	REFERENCES AND SYNONYMS.	IN SUSSEX.
Conchifera.	Unio[g]	antiquus.	*Vide* p. 250.	West Hothly.
Mollusca.	Paludina	elongata.[h]	. . . .	Barnett's Wood, near Framfield.
——	——	vivipara.[i]	. . . .	Ibid.
Pisces.[k]			. . . .	Darvel's Wood, near Battel.
Reptilia.	Megalosaurus.[l]		. . . .	Pounceford.
——	Crocodilus.[m]		. . . .	Darvel's Wood.

[g] Several species in limestone with Cyclades. [h] In limestone. [i] In limestone and shale.
[k] Scales detached, small vertebræ, very imperfect remains in shale. [l] Vertebræ; uncertain if from grit or shale.
[m] Vertebræ from the clay between the limestone, on the authority of Dr. Fitton.

*** This Catalogue was began at the suggestion of Dr. Fitton, and intended as a supplement to his Memoir on the South-east of England, read before the Geological Society on the 15th of June, 1827, and which will appear in a subsequent volume of the Transactions.

Results.

There have been discovered in the strata of Sussex (exclusively of the organic contents of the comparatively modern alluvial deposits) the fossilised remains of nearly four hundred species of animals and vegetables, of which the following arrangement exhibits a condensed view.

Vertebral Animals.

Mammalia.	Pachydermata,	4 species	belonging to as many genera.
	Cetacea,	1 ——	—— 1 genus.
Aves.	Of the tribe Grallæ,	1 or more species	—— 1 ——

Reptilia.	Testudinata,	3 or more species belonging to	3 or more genera.
	Sauria,	9 ——	5 ——
	Pterodactylus?		
Pisces.	. . . . .	24 ——	18 ——

Invertebral Animals.

Mollusca.	Multilocular	(Nautilidæ)†,	58 species	belonging to 8 genera.
	Simple,	(5 species freshwater)	63 ——	—— 29 ——
Conchifera.	. . .	(12 species freshwater)	125 ——	—— 40 ——
Anelides.	. . . . . .		14 ——	—— 4 ——
Crustacea.	. . . . . .		12 or more species	—— 10 ——
Radiara.	Echinidæ, ...		24 species	—— 5 ——
	Asteriadæ, ...		2 or more species	—— 1 genus.
	Crinoidæ, ...		3 ——	—— 3 or more genera.
Polypi.	. . . . . .		27 species	—— 10 ——

Vegetables.

Acotyledonous,	10 or more species belonging to	6 or more genera.
Monocotyledonous,	4 ——	3 ——
Dicotyledonous,	2 ——	2 ——

Total—Mammalia 5 species; Aves 1 or 2; Reptilia 12; Pisces 24; Mollusca 121, of which 5 are freshwater; Conchifera 125, of which 12 are freshwater; Annelides 14; Crustacea 12; Radiara 29; Polypi 27; Plantæ 16.

The geological distribution of the species above enumerated is shown in the following Table, and the zoological characters of the respective formations are thus established, so far as the present imperfect state of our knowledge will permit.

† Under this term the ancient multilocular genera are included. See Fleming Brit. Anim. p. 226.

A Tabular View of the Geological Distribution of the Fossils of Sussex, exhibiting the Zoological Characters of the Strata.

The strata are grouped according to their zoological characters, the Shanklin Sand being included in the Chalk Formation. The Purbeck would of course rank with the Hastings Deposits.

Organic Remains.		Tertiary Form.[11]			Chalk Formation.						Wealden Formation.			
The contents of the alluvial beds, as belonging to the modern epoch, are not enumerated.	Diluvium.	London Clay.	Plastic Clay.	Total.	Chalk.	Chalk Marl.	Firestone.	Galt.	Shanklin Sand.	Total.	Weald Clay.	Tilgate Strata.	Ashburnham Bed.	Total.
Mammalia[1]	5	—	—	—	—	—	—	—	—	—	—	—	—	—
Aves[2]	—	—	—	—	—	—	—	—	—	—	—	2?	—	2?
Reptilia – Testudinata – Marine	—	—	—	—	—	—	—	—	—	—	—	1	—	1
Reptilia – Testudinata – Freshwater	—	—	—	—	—	—	—	—	—	—	—	2	—	2
Reptilia – Sauri[3]	—	—	—	—	1	—	—	—	—	1	1	7	2	10
Reptilia – Enalio-Sauri[4]	—	—	—	—	—	—	—	—	—	—	—	1	—	1
Reptilia – Pterodactylus?	—	—	—	—	—	—	—	—	—	—	—	?	—	—
Pisces[5]	—	3	1	4	14	3	1	2	—	20	2	7	1	10
Mollusca – Multilocular[6]	—	3	—	3	15	23	2	16	4	60	—	—	—	—
Mollusca – Simple – Freshwater[7]	—	—	3	3	—	—	—	—	—	—	4	3	2	9
Mollusca – Simple – Marine	1	34	—	34	5	11	2	5	8	31	—	—	—	—
Conchifera – Freshwater[8]	—	—	4	4	—	—	—	—	—	—	2	8	4	14
Conchifera – Marine	1	25	3	28	31	26	6	8	32	103	—	—	—	—
Annelides	—	5	—	5	3	3	—	3	2	11	—	—	—	—
Crustacea[9]	—	—	—	—	6	1	—	5	1	13	1	1	—	2
Radiaria – Echinidæ	—	—	—	—	17	4	1	1	1	24	—	—	—	—
Radiaria – Asteriadæ	—	—	—	—	2	—	—	—	—	2	—	—	—	—
Radiaria – Crinoidæ[10]	—	—	—	—	3	—	—	—	-	3	—	—	—	—
Polypi	—	1	—	1	18	4	3	1	0	26	—	—	—	—
Plantæ – Terrestrial[11]	—	1	3	4	1?	1?	2	1	1	6	—	8	1	9
Plantæ – Marine	—	—	—	—	4	1	1	1	—	7	—	—	—	—
Number of Species	7	72	14	86	120	77	18	43	49	307	10	40	10	60[12]
Character of the Formations		Marine and Fresh-water.			*Marine.* Formed in a vast Ocean.						*Fluviatile.* An ancient Delta.			

[1] Teeth, bones, &c. [2] Detached bones only. Some of those supposed to belong to birds may perhaps be referred to Pterodactylus. [3] Three of the genera extinct. [4] Genus extinct.
[5] The remains too imperfect, in many instances, to admit of positive conclusions as to their marine or freshwater habitats.
[6] Not a vestige in the Hastings beds; seven genera extinct. [7] Although the species are but few, these shells occur in vast numbers. [8] In immense quantities.
[9] Cypris Faba; very abundant in the upper beds of the Wealden. [10] Two genera extinct.
[11] The vegetables are probably much more numerous, their characters being in many instances too imperfectly displayed to admit of accurate determination. [12] As, in a few instances, the same species occur in more than one subdivision of the same formation, the total amount here given rather exceeds the number of distinct species.

Summary.

Diluvium.—Bones of Pachydermata and Cetacea.

London Clay.—Seventy-two species, of which sixty-two are marine shells; a large proportion of simple univalves.

Plastic Clay.—Fourteen species, of which ten are either terrestrial or freshwater.

Chalk.—Nearly three hundred species, which, with scarcely any exceptions, are marine. Fifty-eight species of multilocular Mollusca, and twenty-four of Echinidæ.

The Wealden.—About sixty species, which, with but few exceptions, are either terrestrial or fluviatile: Reptiles, Testacea, and Vegetables. Neither Echinidæ, Zoophyta, nor Marine Mollusca, occur in these deposits.

LIST OF THE WOOD ENGRAVINGS.

DESCRIPTION

OF

PLATE I.

Fig. 1. The internal axis of the stem of *Clathraria Lyellii*, from Tilgate Forest; the original is three feet in length; described p. 233.

2. The external surface of the stem of *Clathraria Lyellii*, showing the *cicatrices* formed by the bases of the leaves: the original is 15 inches long; described p. 234.

3. Magnified view of a portion of the *frond* of a species of *Lonchopteris*; p. 243.

4. The surface of *Endogenites erosa*, slightly magnified; p. 236.

5. Magnified view of a portion of a polished transverse section of *Endogenites erosa*; p. 236.

6. Portion of the internal part of the stem of *Clathraria Lyellii*: the tranverse direction assumed by the fibres on the upper part of the specimen indicates the situation of a floral axis; p. 235.

7. A small specimen of *Endogenites erosa*; p. 236.

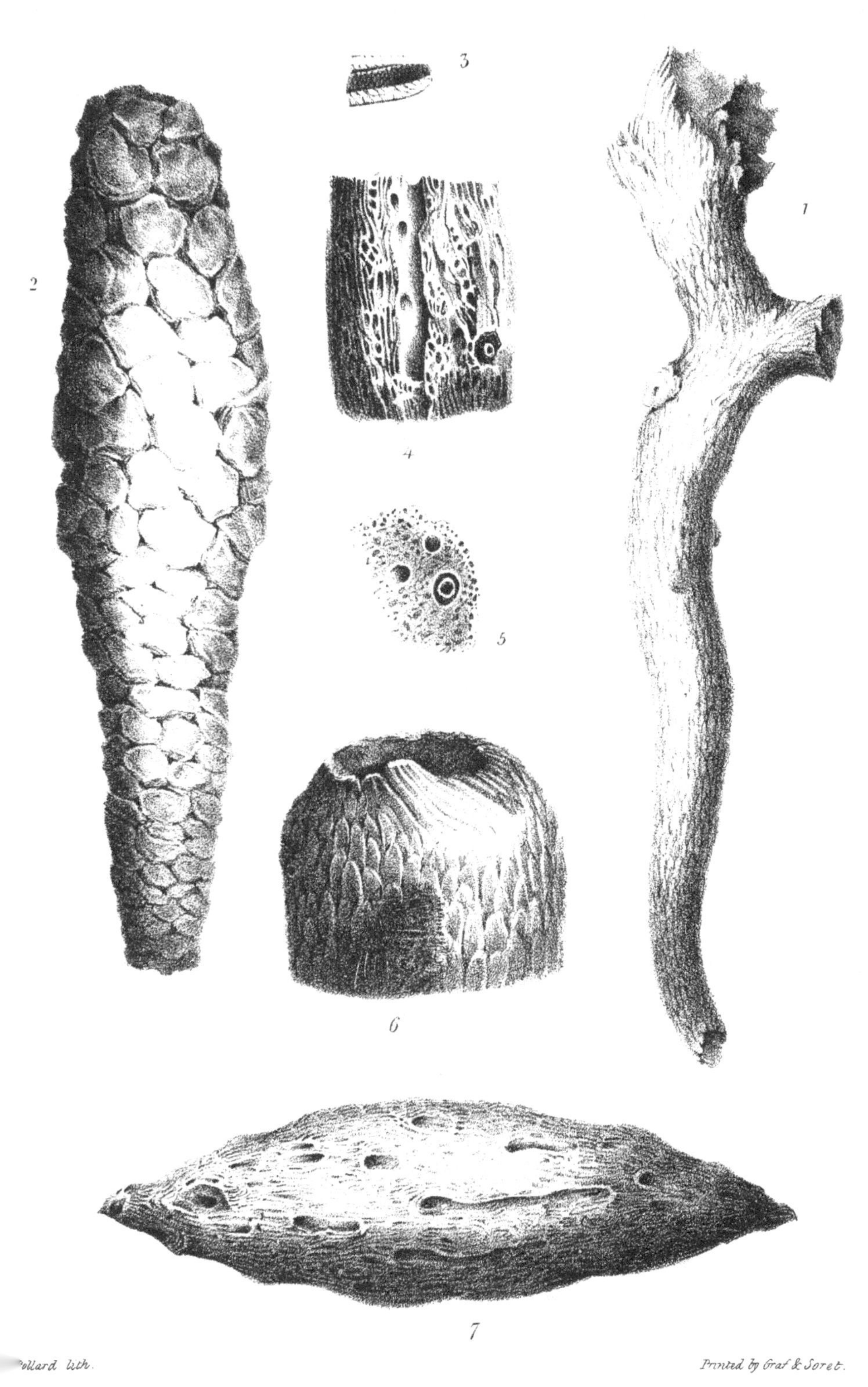

Pollard lith. Printed by Graf & Soret.

London, Published by Longman & C°.

Plate II.

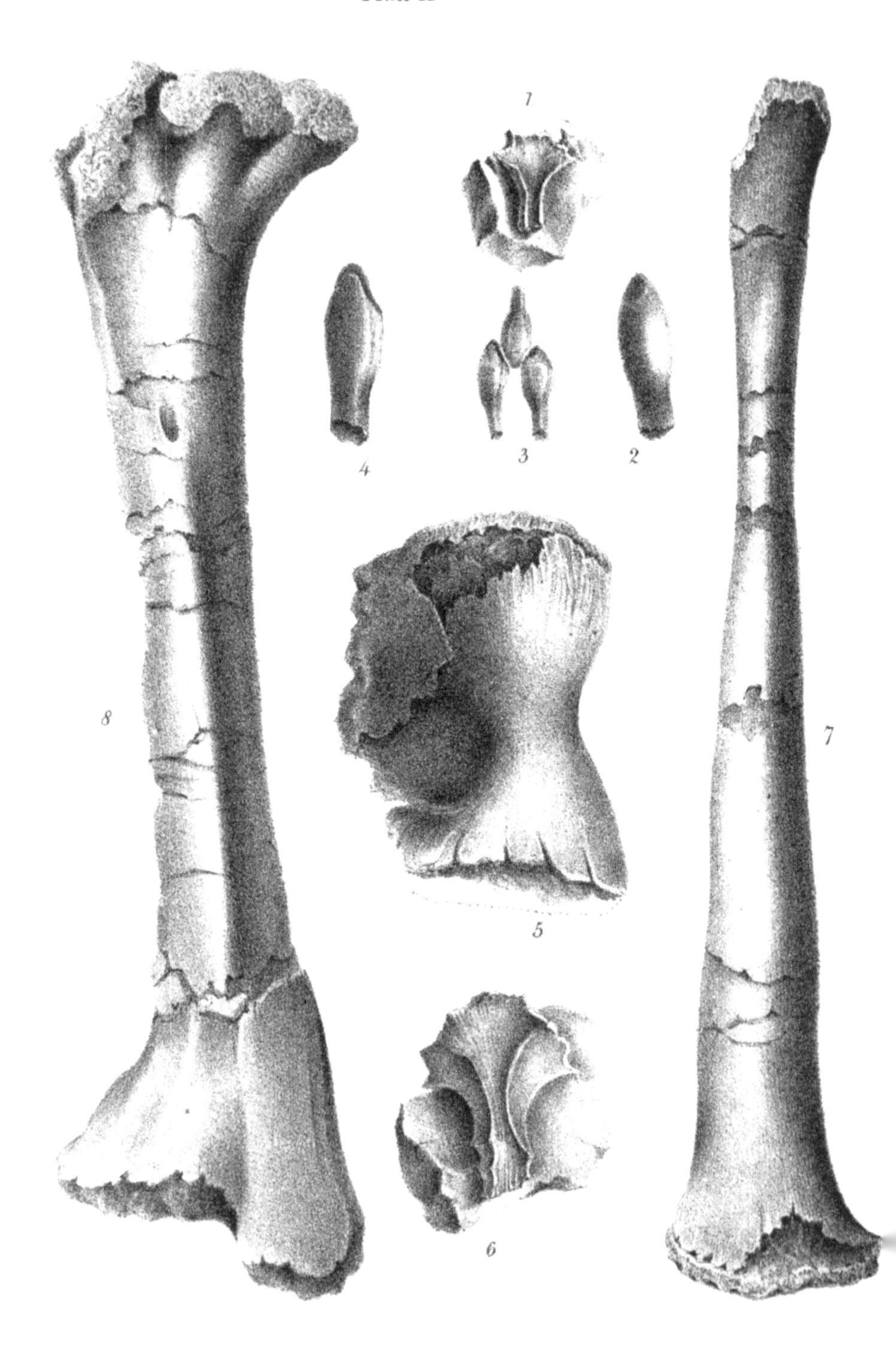

F.P. Ward lith.

Printed by Graf & Sore

London, Published by Longman & Co.

DESCRIPTION

OF

PLATE II.

Fig. 1. The *Os frontis* of a species of Crocodile, from Tilgate Forest; p. 264.

2. 4. Teeth of the *Cylindricodon;* p. 292.

3. Three teeth of the *Cylindricodon*, placed to illustrate the manner in which the crown of the tooth becomes worn away laterally; p. 293.

5. The *Os tympani* of a reptile, probably of the *Iguanodon;* p. 305.

6. The *Os frontis* of a Saurian animal, described p. 305.

7. The *Fibula*, or small bone of the leg of an Iguanodon: the original is 27½ inches long; p. 310.

8. The *Tibia*, or large bone of the leg of the same individual; the original is 30 inches in length, and 22 inches in circumference at the *femoral* or upper extremity, and 19 inches at the lower; described at p. 310.

DESCRIPTION

OF

PLATE III.

Fig. 1. The *Claw* or unguical bone of an Iguanodon, of the natural size; described p. 311.

2. The *Nail*, and fig. 3. the bone which it covered, of a recent *Iguana*, four feet long; also of the size of the originals, to compare with the fossil: p. 311.

4. The *Sternum*, or bone of the chest, of a young Iguanodon, from Tilgate Forest; p. 307.

5. The *Horn* of the *Iguanodon*, of the size of the original; p. 280. and 312.

Plate III.

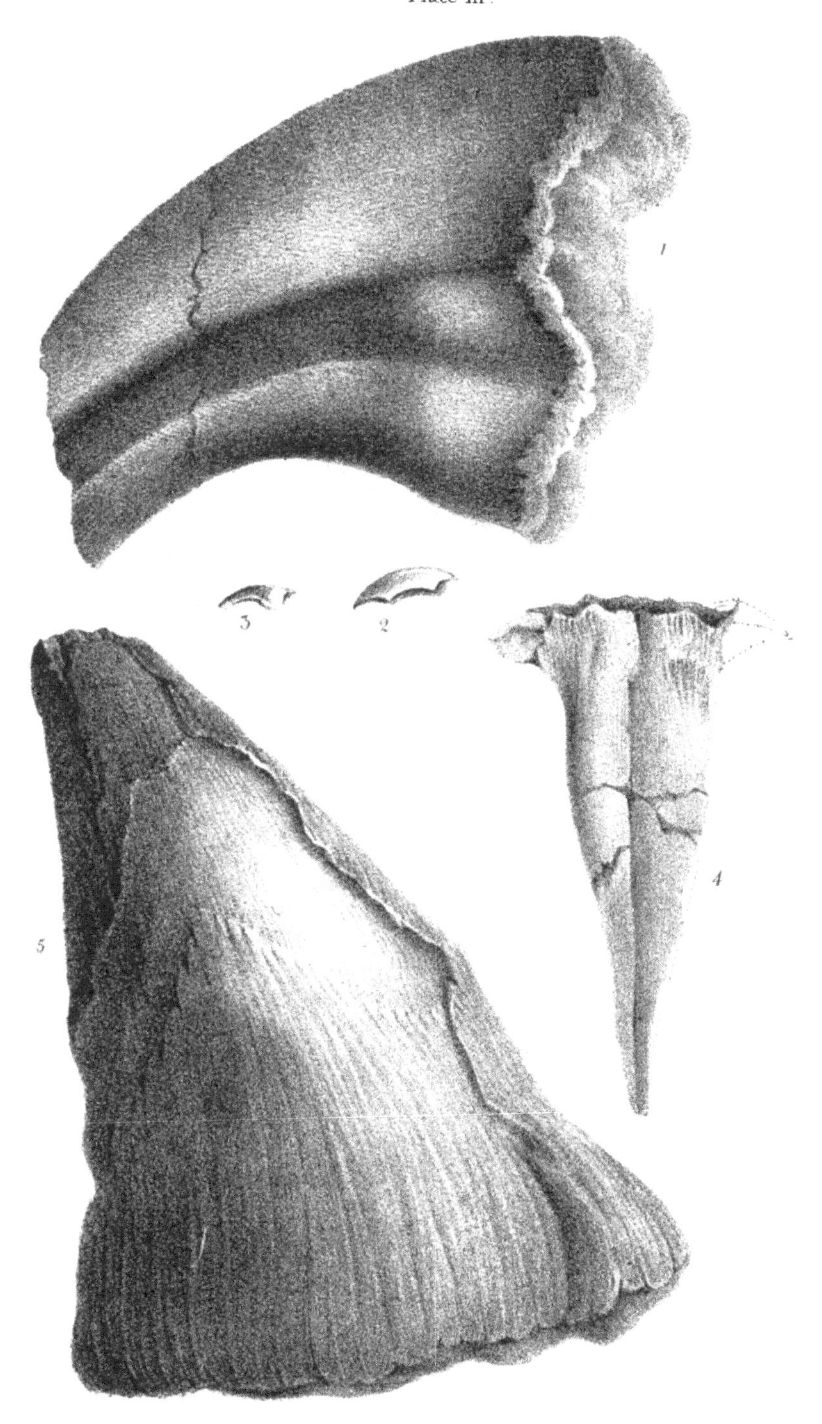

F. Pollard lith.

Printed by Graf & Soret

Longman & Co.

Plate IIII.

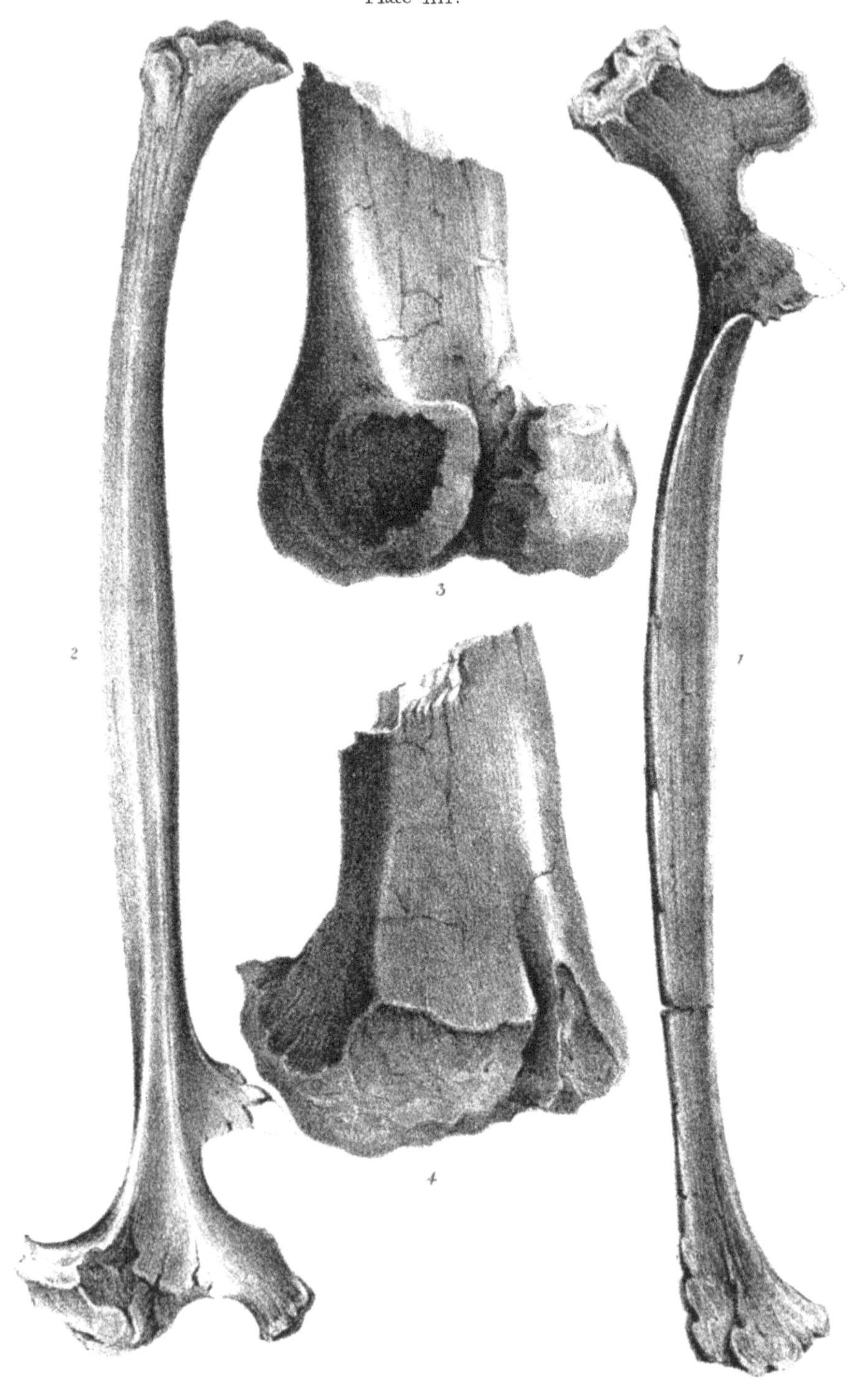

F. Pollard lith.

Printed by Graf & Soret

London. Published by Longman & Co.

DESCRIPTION

OF

PLATE IV.

Fig. 1. A fossil bone from Tilgate Forest, supposed to be the *Clavicle* of the Iguanodon: the original is 29 inches long; described at p. 308.

2. The representation of the opposite side of the same specimen.

3. The lower portion of the *Femur* or thigh bone of an Iguanodon, showing the *condyles;* the circumference of the original at this part is 27 inches; p. 310.

4. Anterior view of the same specimen. This bone was found imbedded in clay, in juxtaposition with the two leg bones figured in Plate II.; and must have belonged to a reptile nearly 70 feet long: described at p. 310.

DESCRIPTION

OF

PLATE V.

This plate represents the extraordinary and highly interesting fossil discovered in Tilgate Forest, by the Author, in the summer of 1832. The specimen exhibits the anterior part of the skeleton of an unknown extinct reptile, which has been named the HYLÆOSAURUS, or *Fossil Lizard of the Weald:* vide p. 289, *et seq.*

Letters of Reference.

a a; the *cervical vertebræ.*
b b; the *dorsal vertebræ.*
*b**; the *first dorsal* vertebra.
c; the *fifth* rib.
d; the transverse process of the *sixth* dorsal vertebra.
e; the *tenth* dorsal vertebra.
f f; the *Omoplates.*
g g; the *Coracoid* bones.
h h; the *dorsal spines*, or appendages.
h h* h**; the *three dorsal spines*, described at p. 323.
i; a displaced dorsal spine.
k; a *dermal* bone, near the first right rib.
l; a broken and displaced dorsal vertebra.
m m; the *glenoid* or *humeral* cavities; the sockets for the arm-bones.

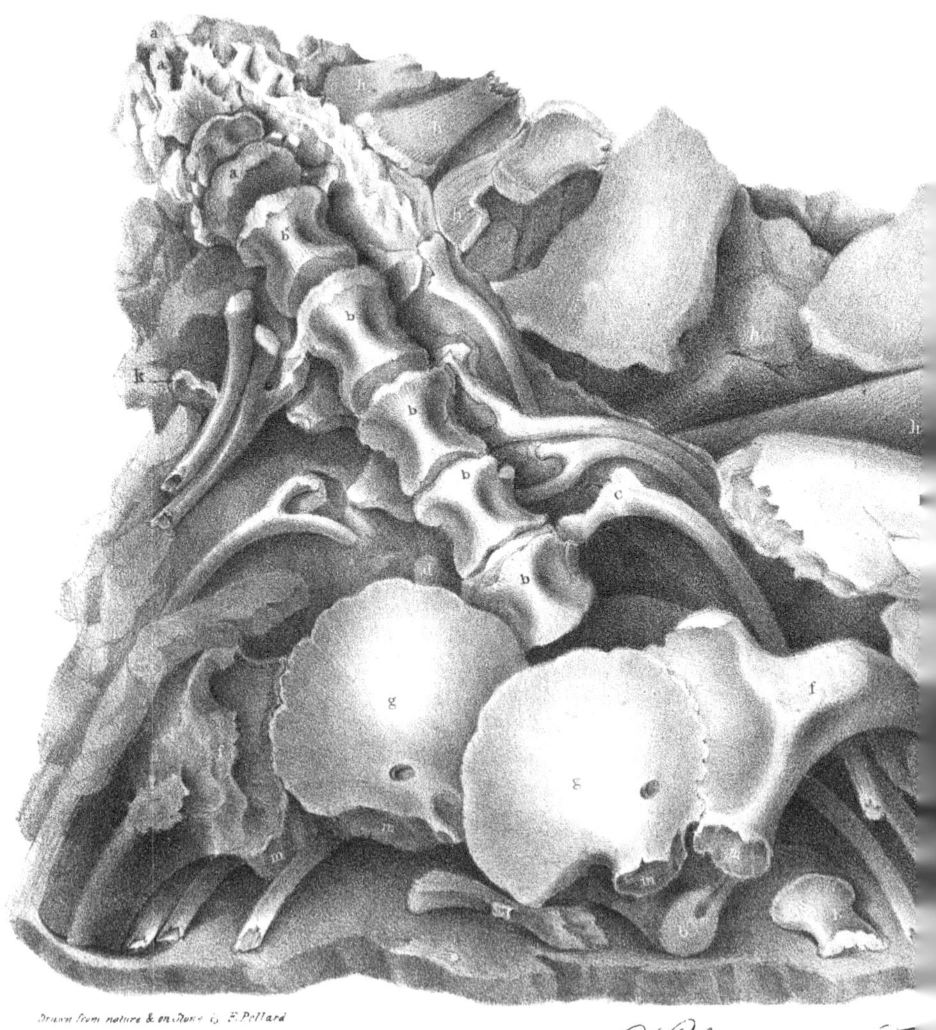

Drawn from nature & on Stone by F. Pollard

The remains of the Skeleton of the *Hylæosaurus* or Fo

Discovered by Mr. MANTEL

Plate V

Printed by Graf & Soret.

…zard of the Weald from Tilgate Forest, in Sussex.

…riginal is 4½ feet in length.

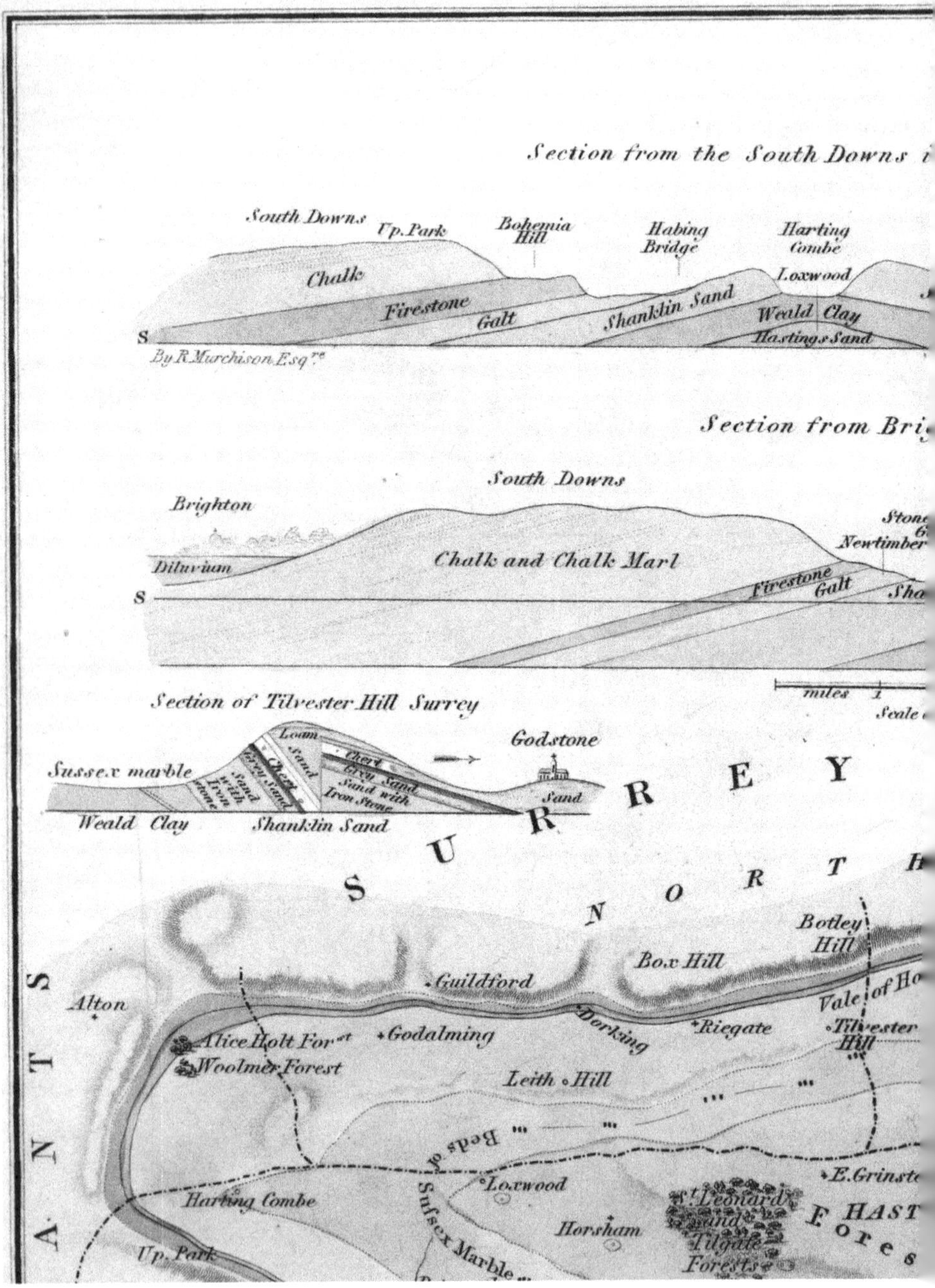
Section from the South Downs i
South Downs
Up. Park
Bohemia Hill
Habing Bridge
Harting Combe
Loxwood
Chalk
Firestone
Galt
Shanklin Sand
Weald Clay
Hastings Sand
S
By R. Murchison Esq.re
Section from Bri
South Downs
Brighton
Diluvium
Chalk and Chalk Marl
Newtimber
Firestone
Galt
S
miles 1
Scale
Section of Tilvester Hill Surrey
Loam
Sand
Chert
Grey Sand
Sand with Iron stone
Chert
Grey Sand
Sand with Iron Stone
Godstone
Sand
Sussex marble
Weald Clay
Shanklin Sand
SURREY
NORTH
HANTS
Botley Hill
Box Hill
Guildford
Alton
Vale of Ho
Dorking
Riegate
Tilvester Hill
Alice Holt For.st
Godalming
Woolmer Forest
Leith Hill
Beds of Sussex Marble
E. Grinste
Loxwood
Harting Combe
St. Leonards and Tilgate Forests
HAST
Fores
Horsham
Up. Park

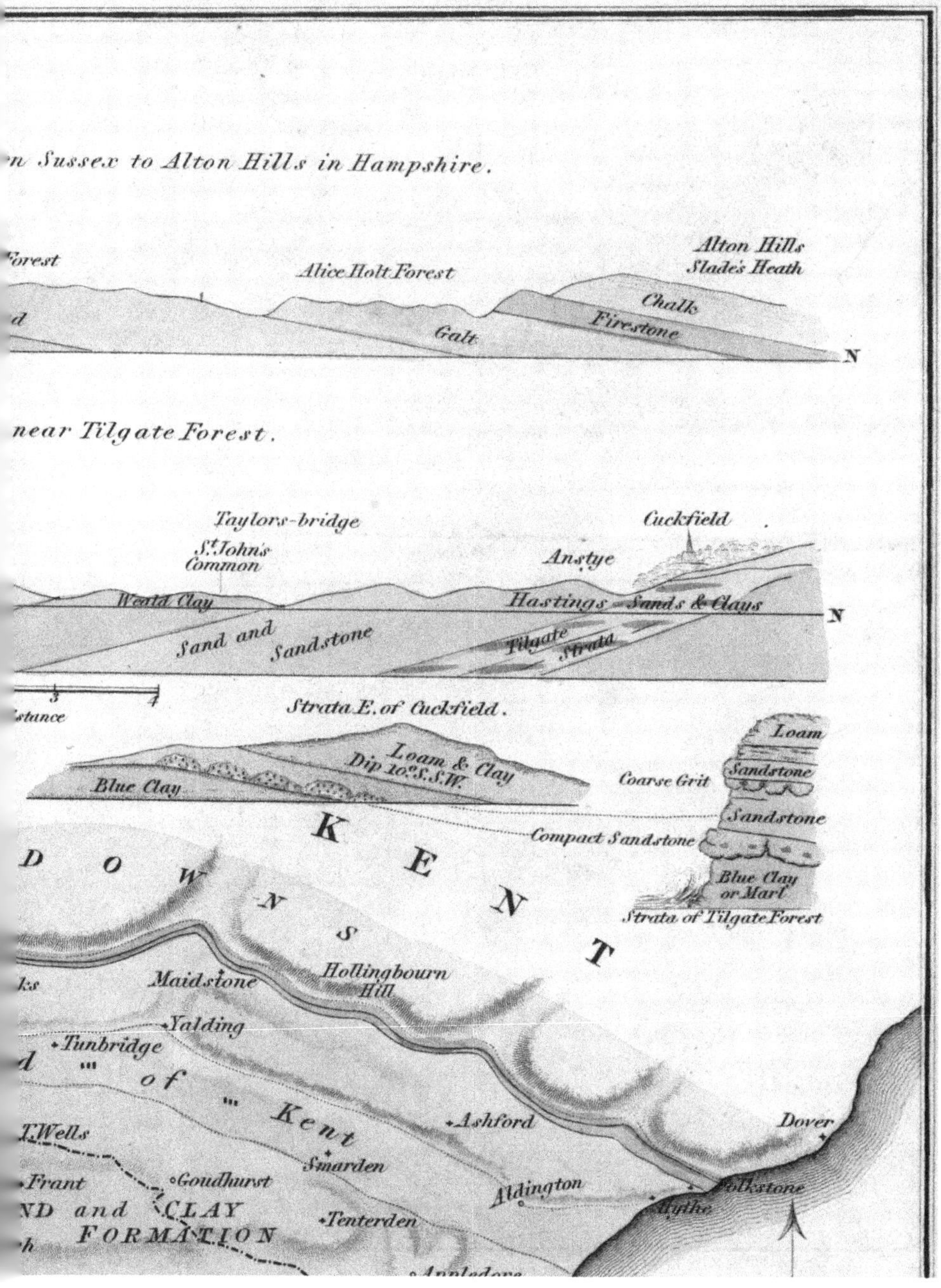

n Sussex to Alton Hills in Hampshire.

Forest
Alice Holt Forest
Alton Hills
Slade's Heath
Chalk
Firestone
Galt
d
N

near Tilgate Forest.

Taylors-bridge
St. John's Common
Cuckfield
Anstye
Weald Clay
Hastings Sands & Clays
Sand and Sandstone
Tilgate Strata
N
3
4
stance

Strata E. of Cuckfield.

Loam & Clay
Dip 10°. S.S.W.
Blue Clay

Loam
Sandstone
Coarse Grit
Sandstone
Compact Sandstone
Blue Clay or Marl
Strata of Tilgate Forest

K E N T
D O W N S
ks
Maidstone
Hollingbourn Hill
Yalding
Tunbridge
d of Kent
Ashford
Dover
T. Wells
Smarden
Frant
Goudhurst
Aldington
Folkstone
Hythe
ND and CLAY
FORMATION
h
Tenterden
Appledore

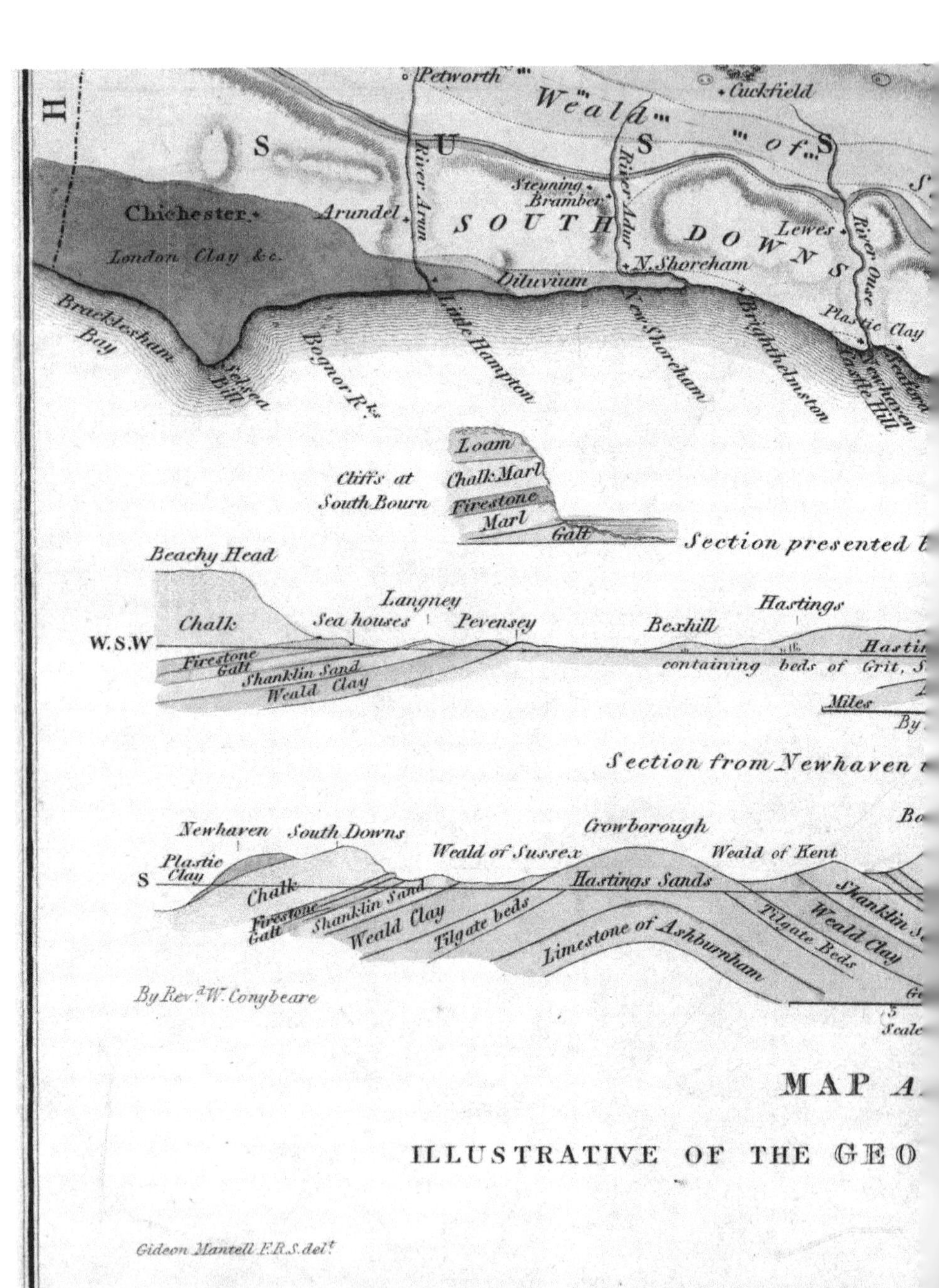
H
S
U
S
S
Petworth
Weald of
Cuckfield
Chichester
London Clay &c.
Arundel
River Arun
Steyning
Bramber
SOUTH
River Adur
DOWNS
Lewes
River Ouse
N. Shoreham
Diluvium
Brackleshan Bay
Selsea Bill
Bognor Rks
Little Hampton
New Shoreham
Brighthelmston
Plastic Clay
Newhaven
Castle Hill
Cliffs at South Bourn
Loam
Chalk Marl
Firestone
Marl
Galt
Section presented
Beachy Head
Chalk
Langney
Sea houses
Pevensey
Bexhill
Hastings
W.S.W
Firestone
Galt
Shanklin Sand
Weald Clay
containing beds of Grit,
Miles
Section from Newhaven
Newhaven
South Downs
Crowborough
Plastic Clay
Weald of Sussex
Weald of Kent
S
Chalk
Hastings Sands
Firestone
Galt
Shanklin Sand
Weald Clay
Tilgate beds
Limestone of Ashburnham
Tilgate Beds
Weald Clay
Shanklin
By Revd W. Conybeare
Scale
MAP A
ILLUSTRATIVE OF THE GEO
Gideon Mantell F.R.S. delt

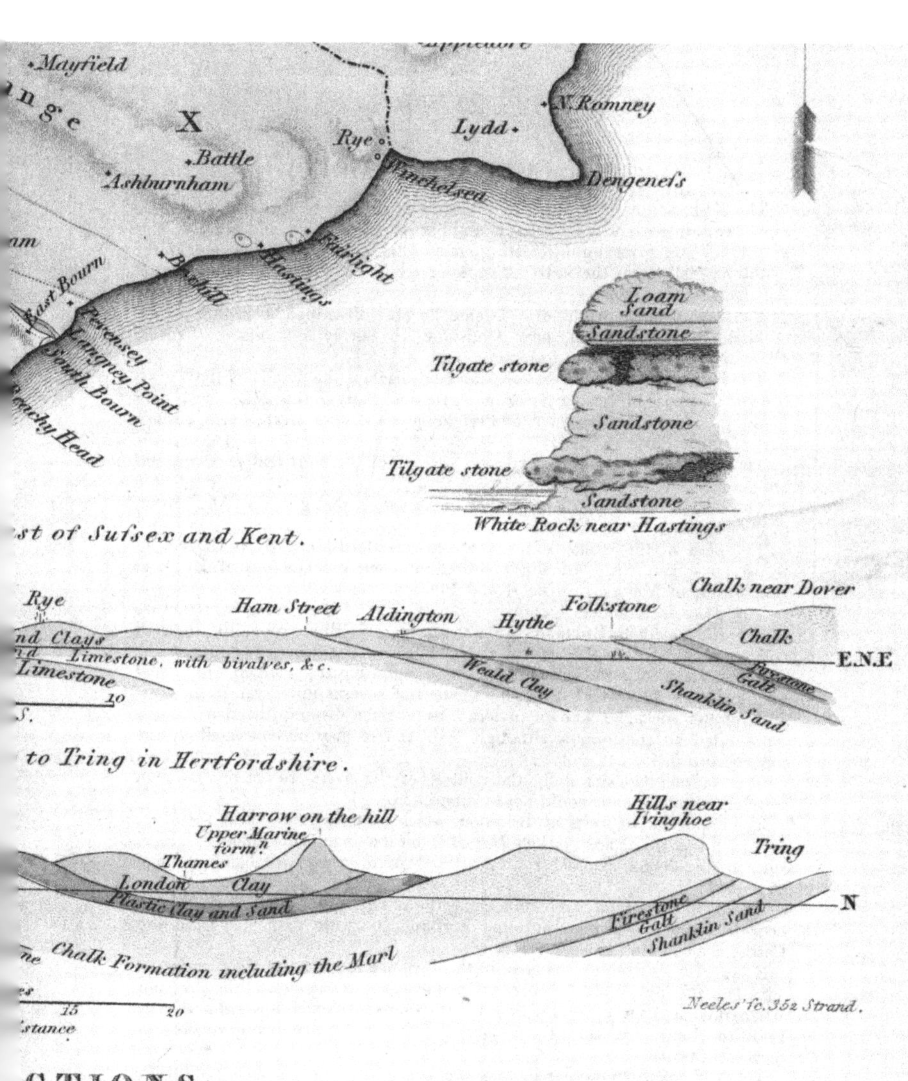

Mayfield
nge
X
Rye
Lydd
N. Romney
Battle
Ashburnham
Winchelsea
Dengeness
Fairlight
Hastings
Bexhill
East Bourn
Pevensey
Langney Point
South Bourn
Beachy Head
Loam
Sand
Sandstone
Tilgate stone
Sandstone
Tilgate stone
Sandstone
White Rock near Hastings
st of Sussex and Kent.
Rye
Ham Street
Aldington
Hythe
Folkstone
Chalk near Dover
nd Clays
Limestone, with bivalves, &c.
Limestone
Weald Clay
Chalk
Firestone
Galt
Shanklin Sand
E.N.E
10
to Tring in Hertfordshire.
Harrow on the hill
Upper Marine form.n
Thames
London Clay
Plastic Clay and Sand
Hills near Ivinghoe
Tring
N
Firestone
Galt
Shanklin Sand
Chalk Formation including the Marl
15
20
stance
Neeles' sc. 352 Strand.

CTIONS

OF THE S.E PART OF ENGLAND.

DESCRIPTION

OF

THE MAP AND SECTIONS.

The first section is by Mr. Murchison; it illustrates the geology of the north-western extremity of Sussex, and part of Hampshire, and is referred to at p. 183. It gives an interesting view of the structure of that part of the country which forms the extreme angle of the great southern denudation of the chalk of England.

The section from Brighton to Tilgate Forest is described at p. 208.

That of Tilvester Hill, near Godstone, in Surrey, exhibits the beds of *Chert* of the *Shanklin sand* formation, p. 178.

The sketch of the strata east of Cuckfield displays the details of the *Tilgate* beds in that part of Sussex; *vide* p. 211; and that of the *white rock*, near Hastings, shows the characters and relations of the same strata in the eastern part of the county: p. 194.

The cliffs at Southbourn expose a section of the *marl* and *firestone*, and are described at p. 162.

The section of the coasts of Sussex and Kent is that given by Dr. Fitton, in the "Annals of Philosophy."

The plan of the strata from Tring, in Hertfordshire, to Newhaven, in Sussex, is taken, with but slight alteration, from the Geology of England and Wales, of Messrs. Conybeare and Phillips.

The MAP is intended to convey a general idea of the geological structure of the South-east of England. For the purpose of illustration, the lines of colours which denote the several formations are continued throughout the whole district; for although in many places a stratum, or group of strata, may not appear on the surface through an extent of several miles, yet, if the situation of the rocks be known in localities not far distant, the deposits are represented as continuous, although their course may be concealed by the superficial soil.

The *Alluvial* deposits along the courses of the rivers are not marked; on so small a scale, details could not be attempted.

The important deposits at Brighton, which we have referred to the *Crag*, (see p. 30, *et seq.*), are marked *Diluvium* on the map, and coloured *green*, and are seen to extend from Rottingdean to near Littlehampton and Bognor.*

The *Tertiary* strata, stretching along the coast from Bracklesham Bay to near Worthing, and their outlying portions at Castle Hill and Chimting Castle, are coloured *brown*: see p. 45.

The *Chalk*, restricting the term in this instance to the *white chalk* and *chalk marl* (for the lowermost members of the formation, the *galt* and the *firestone*, are marked separately), is coloured *yellow*; and is seen extending from Beachy Head to Hampshire, from thence across to Surrey, and returning to the sea-coast, at Dover.

The *Firestone*, coloured *red*, and the *Galt*, *blue*, form a zone within the chalk hills.

The *Shanklin sands*, distinguished by a *pink* colour, constitute a broad belt between the *galt* and the *Wealden* beds.

The *Weald clay*, coloured *green*, lies between the boundary of the *Shanklin sand* and the *Hastings beds*, which are denoted by a *ferruginous* tint.

Note.—It may be necessary to observe, that in the above (as in most geological) sections, the altitudes of the strata are drawn on a much larger scale than their longitudinal dimensions: a method which is found indispensable to accommodate the plates to the size of the volume.

* Not *Bignor*, as is erroneously printed at p. 30. of this volume.

INDEX.

THE END.